Water Treatment Chemicals: Coagulants and Flocculants

Li Fengting

Water Treatment Chemicals: Coagulants and Flocculants

Chemical Industry Press Co., Ltd.

Springer

Li Fengting
College of Environmental Science
and Engineering
Tongji University
Shanghai, China

ISBN 978-981-96-8220-1 ISBN 978-981-96-8221-8 (eBook)
https://doi.org/10.1007/978-981-96-8221-8

Jointly published with Chemical Industry Press

The print edition is not for sale in China (Mainland). Customers from China (Mainland) please order the print book from: Chemical Industry Press.

The original submitted manuscript has been translated into English. The translation was done using artificial intelligence. A subsequent revision was performed by the author(s) to further refine the work and to ensure that the translation is appropriate concerning content and scientific correctness. It may, however, read stylistically different from a conventional translation.

This Springer imprint is published by the registered company Springer Nature Singapore Pte Ltd.
The registered company address is: 152 Beach Road, #21-01/04 Gateway East, Singapore 189721, Singapore

Preface

Water treatment agents, particularly coagulants and flocculants, play a vital role in various water treatment processes. Coagulation, specifically, is an essential step in these processes, whether it serves as the main clarification method, biochemical pretreatment, an enhanced treatment step, or a pretreatment for membranes. Importantly, there is no direct quantitative relationship between the dosage of water treatment agents used in coagulation and the levels of colloids or pollutants present. This lack of correlation makes the selection of appropriate water treatment agents a challenging task, especially when addressing unique water qualities. This book provides a detailed introduction to the principle of coagulation, production process, quality indicators, usage characteristics, application conditions, and process control of iron salts, aluminum salts, and other water treatment agents. And it comprehensively discusses the domestic and international market situation. I hope this work can benefit those engaged in water treatment engineering design, researchers of water treatment agents, and production engineers.

This book has served as a graduate textbook in environmental science, environmental engineering, and municipal engineering at Tongji University since 1999. It has been continuously used for many years and has achieved excellent results.

I would like to express my special gratitude to two of my predecessors for their guidance and support. They led me from the fields of polymer materials and chemical engineering to municipal engineering and environmental engineering. They are Prof. Chen Fujun of the former Shandong University of Architecture and Academician Tang Hongxiao of the Research Center for Eco-Environmental Sciences Chinese Academy of Sciences. Their earnest teachings are always in my heart, prompting me to continuously explore this field. During my research in the field of water treatment chemistry, I also received guidance from Mr. Li Runsheng of Zhongrun Water Industry Technology Development Co., LTD in Shenzhen City. With the support of the Ministry of Science and Technology of China, I have promoted China's advanced water treatment technology to Africa and countries along the Belt and Road. I am deeply grateful for the strong support from Academician Qu Jiuhui of the Research Center for Eco-Environmental Sciences Chinese Academy of Sciences and Tsinghua University.

With the rapid development of the national economy and the upgrading of the industrial structure in the past twenty years, the water treatment agent industry and coagulation technology have also developed rapidly. China has become the largest manufacturer of water treatment agents in the world. At the same time, with the improvement of technology and the revision and re-edition of standards, I have also tried to participate in the past 20 years. I would like to thank the strong support from Teng Houkai, Zhu Chuanjun, Yang Yuqi, Yang Yumei, Ming Yunfeng, Li Lin, Bai Ying, Zheng Shuzhong, Fang Yanping and other teachers of the CNOOC Tianjin Chemical Research and Design Institute Co., LTD. and Industrial Water Treatment Specialized Committee of The Chemical Industry and Engineering Society of China.

This book introduces the production and use of various water treatment agents currently in use. I hope that the publication of this monograph can contribute to the progress of the water treatment industry. This book could serve as a reference for those engaged in water treatment research and engineering, and is suitable for undergraduate and graduate students as a textbook.

I have updated most of the original chapters; Associate Professor Qiao Junlian and Professor Gu Yifan of Tongji University was responsible for updating Chap. 6 on Natural High Polymer Flocculants and Chap. 7 on Decolorizers; Associate Professor Yang Hong of China University of Mining and Technology contributed Chap. 5 on Microbial Flocculants, and Associate Professor Li Jie of Shanghai University contributed Chap. 8 on Adsorbents and participated in the update of some other chapters. Some of the achievements are the results of research by students who graduated from our team during their school years, including Lu Xuefei, Ouyang Qinghua, Zhao Yan, Wang Zheng, Wang Haifeng, Tao Xiaoping, Wang Xiaomin, Xu Meng, Yang Qian, Li Zhimei, Liu Yang, Li Jie, Hua Baolu, Li Zekun, Ma Xiaoliang, Lu Zhiheng, Qin Nianqiao and other students. These students and I have participated in the debugging of new technologies in many companies, and we have spent countless days and nights in the companies. Some of the results are still used in many water treatment agents and the water treatment industry, and some of the results have been extended to the new energy industry, such as the clean production of special raw materials in the battery industry and photovoltaic industry and water treatment engineering. Li Zekun, Li Zichen, Song Tingting, Yan Jiangyi, Li Chunling, Li Congwei, Geng Wuyue, Hao Wanle and other students from our team participated in text proofreading and illustration sorting. I would like to express my gratitude here.

Due to the tight schedule, there may be many inappropriate or imperfect places in the text, and I kindly ask readers to point them out for correction.

Shanghai, China
August 2024

Li Fengting

Contents

1 Water and Coagulation 1
1.1 Composition and Properties of Natural Water 1
1.1.1 Distribution of Water 1
1.1.2 Composition of Natural Water Bodies 1
1.1.3 Properties of Natural Water 10
1.1.4 Pollutants in Water 11
1.2 Chemical Treatment and Coagulation Process 13
1.2.1 Coagulation Process 13
1.2.2 Coagulation Principle 15
1.3 Hydrolysis Mechanism of Aluminum Salt and Iron Salt 27
1.3.1 Classification of Coagulants 27
1.3.2 Hydrolysis Mechanism of Aluminum and Iron Salts 29
1.4 Coagulation Kinetics 34
1.4.1 Coagulation Kinetics 34
1.4.2 Coagulation Parameters 42
1.5 Factors Affecting Coagulation 45
1.6 Methods of Measuring Zeta Potential 45
1.6.1 Zeta Potential and Precipitation 48
1.6.2 Zeta Potential and Flotation Process 49
1.6.3 Improvement of Zeta Potential Analysis Methods 49
1.6.4 Other Improvements in Water Quality Characteristics 50

2 Preparation and Quality Control of Inorganic Low Molecular Weight Coagulants 51
2.1 Aluminum Sulfate 51
2.1.1 Overview of Aluminum Sulfate 51
2.1.2 Production Materials of Aluminum Sulfate 54
2.1.3 Production Process and Equipment of Aluminum Sulfate 54

2.1.4 Preparation Methods of Aluminum Sulfate 55
2.1.5 Use of Aluminum Sulfate in the Water Treatment Industry 61
2.2 Ferrous Sulfate 62
2.2.1 Overview of Ferrous Sulfate Production 62
2.2.2 Synthesis Method of Ferrous Sulfate 62
2.2.3 Main Properties of Ferrous Sulfate 67
2.3 Ferrous Chloride 68
2.3.1 Synthesis Method of Ferrous Chloride 68
2.3.2 Main Properties of Ferrous Chloride 72
2.4 Ferric Chloride 74
2.4.1 Main Properties of Ferric Chloride 74
2.4.2 Synthesis Method of Iron Chloride 76
2.5 Ferric Chlorosulfuric Acid 78
2.5.1 Synthesis Method of Ferric Chlorosulfuric Acid 78
2.5.2 Main Properties of Ferric Chlorosulfuric Acid 79

3 Preparation and Quality Control of Inorganic Polymer Coagulants 81
3.1 Poly Aluminum Chloride 81
3.1.1 Distribution of Al(III) in Solution 81
3.1.2 Forms of Al_{13} 84
3.1.3 Environmental Effects of Polymeric Aluminum 86
3.1.4 Synthesis Method of Poly Aluminum Chloride 88
3.1.5 Resource Utilization of Hazardous Aluminum Ash and Production of Poly Aluminum Chloride 99
3.1.6 Drying Methods of Poly Aluminum Chloride 118
3.1.7 Main Properties of Polyaluminum Chloride 121
3.1.8 Preparation Method of Polyphosphoric Aluminum Chloride (PPAC) 122
3.1.9 Implementation of the New Standard and Industrial Impact 123
3.2 Poly Aluminum Sulfate 132
3.3 Polyferric Sulfate 137
3.3.1 Properties of Polyferric Sulfate 137
3.3.2 Synthesis Method of Polyferric Sulfate 137
3.4 Polyferric Chloride 145
3.4.1 Synthesis Principle 145
3.4.2 Synthesis Process 146
3.5 Polyferric Sulfate Chloride and Polyaluminum Ferric Sulfate 147

4 Polyacrylamide Polymers 149
4.1 Flocculation Theory and Application 149
4.1.1 The Necessity of Flocculation 149
4.2 Artificially Synthesized Bridging Flocculants 152
4.3 Natural Polymers 155
4.3.1 Factors Affecting Flocculation 156
4.3.2 Application of Flocculants 158
4.4 Application Fields and Selection of Polyacrylamide 163
4.4.1 Application Fields of Polyacrylamide 163
4.4.2 The Issue of Reasonable Selection of Polyacrylamide 164
4.4.3 The Molecular Weight Issue of Polyacrylamide 165
4.4.4 Quality and Salt Content Issues of Polyacrylamide 166
4.5 Polymerization Method of Acrylamide 167
4.6 Synthesis Methods of Acrylamide Cationic Polymers 169
4.6.1 Copolymer of Acrylamide and Diallyl Dimethylammonium Chloride 170
4.6.2 Copolymer of Acrylamide and Dimethyl Aminoethyl Methacrylate Chloride 171
4.6.3 Acrylamide-Dimethyl Aminoethyl Acrylate Chloride-Methoxyethyl Acrylate Copolymer 171
4.6.4 Preparation Method of N-Vinyl Formamide Homopolymer Dispersion Polymerization 172
4.6.5 Preparation Method of N-Vinyl Formamide-Ethyl Hexyl Acrylate Copolymer Dispersion 172
4.6.6 Acrylamide-Acrylic Acid N, N-Dimethyl Ethyl Ester Chloromethylamine Copolymer Acrylamide-N, N-dimethylaminoethyl Ester Ammonium Chloride Copolymer 172
4.7 Divinyl Propyl Dimethyl Ammonium Chloride Class Polymers 173
4.7.1 Raw Material Requirements 174
4.7.2 Brief Introduction to Synthesis Method 174
4.7.3 Polymer Production Process 175

5 Microbial Flocculants 179
5.1 Classification of Microbial Flocculants 180
5.2 Flocculation Mechanism and Influencing Factors of Microbial Flocculants 181
5.2.1 Flocculation Mechanism of MBF 181
5.2.2 Factors Influencing MBF Flocculation 185

5.3 Preparation and Production of Microbial Flocculants 189
5.3.1 Isolation and Screening of MBF-Producing Bacteria 190
5.3.2 Strain Identification 191
5.3.3 Optimization of the Fermentation Process 193
5.3.4 Extraction of MBF 197
5.3.5 Purification of MBF 200
5.4 Application of Microbial Flocculants in Water Environment 202
5.4.1 Raw Water Purification 202
5.4.2 Wastewater Treatment 202
5.4.3 Sludge Conditioning 202
5.4.4 Microalgae Harvesting 209
5.5 Development Trend of Microbial Flocculants 211
5.5.1 Developing Highly Applicable MBF Producing Bacteria 215
5.5.2 Developing Green and Low-Cost Alternative Culture Media 216
5.5.3 Intelligent Control and Method Improvement of MBF Production 217
5.5.4 Strengthening of MBF Flocculation Through Compounding, Combination, and Co-use 217
5.5.5 Strengthen MBF Safety Assessment and Application Research 219

6 Natural Polymer Flocculants 221
6.1 Starch 222
6.1.1 Overview 222
6.1.2 Structure and Properties of Starch 222
6.1.3 Applications 226
6.2 Galactomannan 234
6.2.1 Overview 234
6.2.2 Structure and Properties of Galactomannan 234
6.2.3 Applications 237
6.3 Cellulose 243
6.3.1 Overview 243
6.3.2 Structure and Properties of Cellulose 243
6.3.3 Applications 243
6.4 Chitosan 247
6.4.1 Overview 247
6.4.2 Structure and Properties of Chitin 248
6.4.3 Applications 251
6.5 Lignin 257
6.5.1 Introduction 257
6.5.2 Structure and Properties of Lignin 258
6.5.3 Applications 259

6.6 Tannin ... 263
6.7 Microbial Polysaccharides ... 267
6.8 Animal Glue and Gelatin ... 269
6.9 Moringa Seed Extracts ... 270

7 Decolorizing Agent ... 275
7.1 Coagulation Decolorization Treatment Technology ... 276
7.2 Types of Decolorizing Coagulants ... 277
7.2.1 Inorganic Coagulants ... 277
7.2.2 Organic Flocculants ... 279
7.3 Decolorizing Effect and Mechanism of Dicyandiamide Formaldehyde Condensate ... 281
7.3.1 Preparation Method ... 282
7.3.2 Production Process ... 283
7.3.3 Comparison with Other Coagulants ... 284
7.4 Factors Affecting the Coagulation Decolorization Effect of Melamine Formaldehyde Resin ... 286
7.4.1 Influence of Dye Molecular Structure ... 286
7.4.2 The Effect of the pH Value of the Dye Solution or Wastewater ... 286
7.4.3 The Impact of Stirring Speed ... 290
7.4.4 The Impact of Decolorizer Dosage ... 291
7.4.5 Preliminary Exploration of the Structure of the Decolorizer and the Flocculation Mechanism ... 291
7.5 New Developments in Coagulant Decolorizing Agent Research ... 292

8 Adsorbents ... 295
8.1 Principles of Adsorption ... 295
8.1.1 Phenomenon of Adsorption ... 295
8.1.2 Physical Adsorption and Chemical ... 296
8.1.3 Adsorption Isotherm and Adsorption Isotherm Equation ... 297
8.1.4 Adsorption Rate ... 301
8.1.5 Factors Affecting Adsorption ... 302
8.1.6 Static Adsorption and Dynamic Adsorption ... 304
8.2 Adsorbents and Activated Carbon ... 307
8.2.1 Adsorbents ... 307
8.2.2 Activated Carbon ... 313
8.3 Enhanced Coagulation and Adsorbents ... 332
8.3.1 Enhanced Coagulation ... 332
8.3.2 Adsorption Enhanced Coagulation ... 337

9 Selection and Configuration Methods of Coagulants or Flocculants 347
9.1 Purpose and Content of Chemical Treatment 347
9.1.1 Alkalinity 348
9.1.2 pH Value 350
9.1.3 Turbidity 352
9.1.4 Color 354
9.1.5 Temperature 354
9.1.6 Hardness 355
9.1.7 Odor and Smell 356
9.2 The Problem of Residual Aluminum in Water 356
9.3 Beaker Coagulation Experiment 358
9.3.1 Selection of Beaker Experiment Instruments 360
9.3.2 The Determination of Beaker Coagulation Test 363
9.3.3 Selection of the Final Treatment Process 369
9.3.4 Evaluation of Coagulants 369
9.4 Preparation, Measurement, and Addition of Medicinal Solution 370
9.4.1 Preparation, Measurement, and Addition of Medicinal Solution 370
9.4.2 Preparation of Medicinal Solution 371

References 373

Chapter 1
Water and Coagulation

1.1 Composition and Properties of Natural Water

1.1.1 Distribution of Water

Natural water is divided into surface water, groundwater, and precipitation.

Natural water bodies refer to the total of rivers, lakes, seas, etc. The total amount of water resources on earth is about 1.38 billion km^3, of which 97.5% is seawater (1.345 billion km^3). Freshwater only accounts 2.5%, most of which are polar ice and snow glaciers and groundwater. Only 0.01% is suitable for human use. Many countries in the world are facing a water crisis: 1.2 billion people are short of water, 3 billion people lack sanitary water facilities, and 3–4 million people die from water-related diseases each year. By 2025, the water crisis will spread to 48 countries, and 3.5 billion people will be trapped by water. By 2040, 33 countries will face extremely high water resource pressure. The deterioration of ecosystems and the destruction of biodiversity caused by the water crisis will also seriously threaten human survival. Like energy, water resources have become a strategic resource of a country.

1.1.2 Composition of Natural Water Bodies

Due to the different environments in which they exist or flow through, natural water bodies contain various substances in addition to water. According to their states of existence in water, these substances can be divided into three categories: suspended matter, colloidal matter, and dissolved matter.

L. Fengting, *Water Treatment Chemicals: Coagulants and Flocculants*,
https://doi.org/10.1007/978-981-96-8221-8_1

1.1.2.1 Main Ions

The main cations in natural water bodies are Ca^{2+}, Mg^{2+}, Na^+, K^+, etc. These ions come from their minerals, such as calcium feldspar ($CaA1_2Si_2O_8$), dolomite [$CaMg(CO_3)_2$], sodium feldspar ($NaAlSi_3O_8$), potassium feldspar ($KA1Si_3O_8$), etc. For example, the dissolution process of natural water on calcium feldspar and calcite ($CaCO_3$) is:

$$CaAl_2Si_2O_8(s) + H_2O + 2H^+ = Al_2Si_2O_5(OH)_4(s) + Ca^{2+}$$
$$CaCO_3(s) + CO_2 + H_2O = Ca^{2+} + 2HCO_3^-$$

The main anions in water bodies include Cl^-, $SO4^{2-}$, HCO^{3-}, $CO_3{}^{2-}$ ions, etc. The above four cations and four anions constitute the main components in water bodies, which we often refer to as the eight major ions. Generally, the eight major ions constitute 95–99% of the total ion content in natural water bodies.

In general river and lake water, the content of $HCO_3{}^-$ does not exceed 250 mg/L, and in a few cases, it can reach 800 mg/L. The content of Cl^- in various natural waters varies greatly, with ion content in river water being 1–35 mg/L, while in seawater it can be as high as 19.35 g/L.

1.1.2.2 Nutrients

Nutrients refer to elements related to biological growth, including nitrogen, phosphorus, silicon and other non-metallic elements, as well as some trace elements (manganese, iron, copper, etc.). The content of these elements generally ranges from μg/L to mg/L. Their existence is related to the acidity and redox properties of the water body.

(1) Nitrogen, Phosphorus

Nitrogen and phosphorus are essential nutrients for the growth and reproduction of aquatic organisms. However, for slow-flowing water bodies such as lakes, reservoirs, inland seas, and estuaries, an increase in nitrogen and phosphorus will lead to a large amount of algae reproduction, which will reduce or even deplete the dissolved oxygen in the water, endangering the survival of aquatic organisms. This phenomenon is called eutrophication. At this time, the water surface often appears blue, red, brown, milky white, etc., depending on the color of the dominant algae. This phenomenon is called water bloom in rivers and lakes, and red tide in the sea. After sewage discharge, a large amount of nitrogen and phosphorus pollutants are discharged into the receiving water body, and hundreds of water blooms and red tides occur each year in freshwater lakes and coastal areas. In 2007, large-scale water blooms broke out in Taihu Lake, Dianchi Lake, and Chaohu Lake in China, seriously affecting the drinking water safety of surrounding cities. Recently, many lakes that serve as water sources in China have mostly had eutrophication problems.

(2) Iron, Silicon

Iron in natural water is a common mineral element. In surface water, due to sufficient dissolved oxygen, iron often exists in the form of Fe(III), but the content is small. From this simplified calculation, it can be clearly seen that in most nearly neutral natural water bodies, the hydrated ions of Fe(III) can be ignored, and $Fe(OH)^{2+}$ is the main form of Fe(III).

$Si(OH)_4$ can also form the following polynuclear complex:

$$4Si(OH)_4 = Si_4O_6(OH)_6^{2-} + 2H^+ + 4H_2O \quad 1gK = -12.57$$

Under neutral or weakly acidic conditions, according to the above calculations, monomolecular orthosilicic acid is the main form of silicon. When the content of orthosilicic acid is high, monomolecular orthosilicic acid can polymerize into inorganic silicon polymers, forming colloidal particles.

1.1.2.3 Organic Matter

There are many types of organic matter in natural water bodies. Organic matter in water bodies is usually divided into two categories: non-humic substances and humic substances. Non-humic substances include carbohydrates, fats, proteins, vitamins, and other low-molecular-weight organic substances. Most of the organic matter in water bodies are brown or black amorphous humic substances. Their molecular weight ranges from several hundred to several thousand. The composition and structure of humic substances are not yet fully understood, and the classification and naming are not uniform.

Based on the solubility of humic substances in acid and alkali, they are divided into three main levels:

① Fulvic Acid (also known as Yellow Humic Acid): It can be represented by FA. The molecular weight of FA ranges from several hundred to several thousand, and it is soluble in both alkali and acid.
② Humic Acid (also known as Brown Humic Acid): It can be represented by HA. The molecular weight of HA ranges from several thousand to tens of thousands, and it is soluble in alkali but not in acid.
③ Humin (also known as Humic Black Substance): It is a component that is insoluble in both acid and alkali.

Existing data shows that these three types of humic substances are structurally similar. Their common characteristics include containing a large number of phenyl rings, as well as a large number of active groups such as hydroxyl (–OH), carboxyl (–COOH), and carbonyl (–C=O).

Due to the smaller molecular weight of fulvic acid, it has a higher oxygen functional group per unit mass. These active groups that give humic substances weak acidity, ion exchange properties, coordination distribution compounds, and redox chemical activities. Therefore, they have the ability to form stable water-soluble or water-insoluble compounds with metal ions in the water body, as well as the ability to interact with organic matter in the water body (including toxic substances).

1.1.2.4 Suspended Solids

Substances dispersed in a non-molecular or ionic state in the water body are collectively referred to as suspended solids. They include inorganic silt, hydraulic ash, coal washing, metallurgy, slaughter, fertilizer, chemical, construction, and other industrial wastewater and domestic sewage that contain suspended pollutants. After being discharged into the water body, in addition to making the water body turbid and affecting the photosynthesis of aquatic plants, they also adsorb organic toxins, heavy metals, pesticides, etc.

Due to the difference in particle size of suspended solids, their properties in the water body vary greatly. For larger particles, they can easily be separated from the water body by gravity, but it is difficult for small particles to settle. The relationship between particle size and sedimentation is shown in Table 1.1.

In colloid chemistry, people generally refer to particles with a size range of 10^{-9} to 10^{-7} m (1–100 nm) as colloids. It has been recently discovered that particles that have been present in the water environment for a long time actually have several micrometers to tens of micrometers of organic layers adsorbed on their surfaces, increasing their actual radius by a certain percentage, and giving their surfaces a certain flexibility, thereby increasing their hydrophilicity and dispersibility. During the coagulation process, the presence of this buffer layer often increases the consumption of the drug.

In colloid chemistry, people generally refer to particles with a size range of 10^{-9} to 10^{-7} m (1–100 nm) as colloids. It has been recently discovered that particles that have been present in the water environment for a long time actually have several micrometers to tens of micrometers of organic layers adsorbed on their surfaces, increasing their actual radius by a certain percentage, and giving their surfaces a certain flexibility, thereby increasing their hydrophilicity and dispersibility. During the coagulation process, the presence of this buffer layer often increases the consumption of the drug.

Table 1.1 Relationship between particle size and sedimentation

Particle size/cm	Approximate size	Natural sedimentation	
		Speed/(cm/s)	Time required to settle 1 m
0.05	Sand	10.4	10 s
0.01	Fine sand	0.42	4 min
0.001	Sludge (silt, clay)	0.0042	7 h
0.0001 (1 μm)	Bacteria	0.000042	28 d
0.00001	Colloid	0.00000042	8 y

The Yangtze River and the Yellow River are the largest sediment-carrying rivers in our country. The Yangtze River deposits 678 million tons of sediment into the East China Sea each year, with the smallest annual sediment discharge being 341 million tons, and the average annual sediment discharge being approximately 466 million tons. This trend has gradually decreased with the completion of the Three Gorges Dam. In 2023, the Yangtze River Hydrological Bureau monitored themajor hydrological control stations on the upper reaches of the Yangtze River: Qinghai Zhimenda, Yunnan Shigu, Sichuan Panzhihua, Xiangjiaba, Chongqing Zhutuo, Cuntan, Hubei Yichang, Shashi, Hankou in the middle reaches, and Anhui Datong Hydrological Station in the lower reaches. The average annual sediment content was found to be 1.03, 0.369, 0.032, 0.005, 0.056, 0.0.080, 0.006, 0.016, 0.065 and 0.066 kg/m^3 respectively. The relevant data from some domestic hydrological control stations are shown in Table 1.2.

It is relatively easy to remove particulates from natural water bodies, but many water bodies are both sources of water and recipients of wastewater, such as the Yangtze River, which has an annual runoff of about 970 billion cubic meters, and the annual discharge of wastewater in the Yangtze River basin is about 30 billion cubic meters. The discharge of wastewater brings great difficulty to the treatment of raw water. Tongji University conducted a survey in 2004 on the distribution of particulates in the raw water of the Yangtze River and the Huangpu River in several water plants in the downstream area of Shanghai. The distribution of particulates in the raw water of the Yangtze River in 2004 is shown in Table 1.3, and the distribution of particulates in a water plant in the raw water of the Huangpu River is shown in Table 1.4.

The table shows that 90% of the total amount of particulates in the raw water is less than 5 μm in diameter. Since the size of pathogens in water is also within this range, the removal of particulates at one time is of great significance. Especially in recent years, people's attention to Giardia (7–11 μm), Giardia cysts (1–5 μm), and Cryptosporidium (4–7 μm) has increased people's interest in the removal of particulates. Now we usually use the concept of turbidity to measure water quality, but since turbidity cannot fully reflect the number of particulates in the water after treatment, some countries have tried to measure water quality by the number of particulates.

1.1.2.5 Dissolved Gases

The gases generally present in natural water include oxygen, nitrogen, carbon dioxide, hydrogen sulfide, and methane, etc. These gases come from the dissolution of various gases in the atmosphere, the activities of aquatic plants and animals, and chemical reactions, etc. In the absence of chemical reactions in the water, at a certain temperature, the gas dissolution reaches equilibrium, and the solubility of the gas in the liquid is directly proportional to the partial pressure of the gas in the gas phase. It can be expressed by the mathematical formula:

Table 1.2 Relevant data from some domestic hydrological control stations

Hydrological control station		Zhimenda	Shigu	Panzhihua	Xiangjiaba	Zhutuo	Cuntan	Yichang	Shashi	Hankoui	Datong
Comparison of annual sediment transport	Multi-year averages (%)	135	− 37	− 96	− 100	− 95	− 94	− 99	− 98	− 89	− 87
	Comparison of 2023 with the average of the last 10 years (%)	74	− 45	− 40	− 54	− 67	− 66	− 85	− 78	− 47	− 58
	2023 versus 2012 (%)	201	146	125	65	52	− 25	− 29	− 16	− 6	− 33
Annual sediment transport/(100 million tons)	Multi-year averages	0.100	0.268	0.430	2.06	2.51	3.53	3.76	3.26	3.17	3.51
		1957–2020	1958–2020	1956–2020	1956–2020	1956–2020	1953–2020	1960–2020	1956–2020	1954–2020	1961–2020
	Average for the last 10 years	0.135	0.307	0.030	0.013	0.375	0.656	0.133	0.235	0.641	1.06
	2022	0.078	0.069	0.008	0.008	0.074	0.145	0.028	0.062	0.363	0.665
	2023	0.235	0.170	0.018	0.006	0.122	0.221	0.020	0.052	0.340	0.445
Average annual sediment content/(kg/m3)	Multi-year averages	0.745	0.631	0.754	1.440	0.946	1.030	0.869	0.831	0.448	0.392
		1957–2020	1958–2020	1966–2020	1956–2020	1956–2020	1956–2020	1953–2020	1950–2020	1954–2020	1951–2020
	2022	0.503	0.157	0.014	0.006	0.032	0.051	0.008	0.018	0.060	0.086

(continued)

Table 1.2 (continued)

Hydrological control station		Zhimenda	Shigu	Panzhihua	Xiangjiaba	Zhutuo	Cuntan	Yichang	Shashi	Hankoui	Datong
	2023	1.03	0.369	0.032	0.005	0.056	0.080	0.006	0.016	0.065	0.066
Mean annual mean particle size/(mm)	Multi-year averages		0.016	0.013	0.013	0.011	0.010	0.008	0.019	0.012	0.011
			1987–2020	1987–2020	1987–2020	1987–2020	1987–2020	1987–2020	1987–2020	1987–2020	1987–2020
	2022		0.012	0.008	0.018	0.012	0.013	0.012	0.035	0.012	0.021
	2023		0.009	0.011	0.017	0.012	0.012	0.011	0.022	0.013	0.012
Modulus of sediment transport/[t/(a km^2)]	Multi-year averages	72.6	125	166	449	361	407	374	/	213	166
		1957–2020	1958–2020	1966–2020	1956–2020	1966–2020	1950–2020	1950–2020	/	1954–2020	1961–2020
	2022	56.6	32.4	2.98	1.81	10.7	16.7	2.74	/	24.4	39.0
	2023	171	79.4	7.10	1.42	17.6	25.5	1.94	/	22.8	26.1

Table 1.3 Distribution of particulates in the raw water of the Yangtze River in 2004

Water plant of the Yangtze River raw water				
Particle size distribution/μm	Raw material		Disinfect water	
	Particle concentration/(pcs/mL)	Percentage/%	Particle concentration/(pcs/mL)	Percentage/%
1–2	24,115	27.58	1358	89.00
2–5	54,704	61.80	153	10.02
5–10	8717	9.85	10	0.66
10–15	593	0.67	2	0.15
15–20	73	0.08	1	0.09
20–25	8	0.01	1	0.09
25–30	1	0.00	0	0.00
> 30	0	0.00	0	0.00
Total	88,511	100.00	1526	100.00

Table 1.4 Distribution of particulates in a water plant in the raw water of the Huangpu River

Water plant of the Huangpu River raw water				
Particle size distribution/μm	Raw material		Disinfect water	
	Particle concentration/(pcs/mL)	Percentage/%	Particle concentration/(pcs/mL)	Percentage/%
1–2	143,229	52.90	4934	79.32
2–5	113,110	41.78	1090	17.53
5–10	13,039	4.82	172	2.77
10–15	1172	0.43	19	0.30
15–20	168	0.06	4	0.06
20–25	19	0.01	1	0.01
25–30	2	0.00	0	0.00
> 30	2	0.00	0	0.00
Total	270,741	100.00	6220	100.00

$$c = Kp \tag{1.1}$$

where c is the solubility of the gas in the liquid, generally referring to the mass of the dissolved gas in 1 kg of water, g;

p is the equilibrium partial pressure of the gas on the liquid surface, Pa;

K is Henry's constant, which is the characteristic constant of this gas–liquid system.

Table 1.5 Solubility of gases in water at different temperatures under a pressure of 101 kPa

Temperature/°C	O_2	N_2	CO_2
0	0.0489	0.0235	1.713
20	0.0310	0.0155	0.878
30	0.0261	0.0134	0.665
35	0.0244	0.0126	0.592

At a fixed temperature, the solubility of the gas is directly proportional to its gas phase partial pressure; when the gas phase pressure is fixed, as the temperature rises, the constant K decreases, and the solubility decreases. The solubility of a gas is usually expressed as the volume of gas that can be dissolved in 1 volume of liquid. The solubility of gases in water at different temperatures under a pressure of 101 kPa is shown in Table 1.5.

(1) Dissolved Oxygen

Oxygen dissolved in water is called dissolved oxygen. Dissolved oxygen exists in the water in a molecular state, mainly coming from the oxygen in the air and the oxygen produced by the photosynthesis of aquatic plants. The amount of dissolved oxygen in natural water bodies is related not only to the number of organisms and the amount of organic matter in the water body but also to the water depth. Gas–liquid exchange is relatively easy at the water surface, where the gas concentration is higher. As the water depth increases, the concentration of dissolved gases, especially oxygen, decreases, gradually transitioning from an oxygen-rich area to an oxygen-deficient area. Under normal conditions, the dissolved oxygen in surface water is 5–10 mg/L.

In the case of sufficient aeration of the water body, the water body can also be in a supersaturated state, such as the water body downstream of the Three Gorges Dam. In water bodies where algae bloom, the amount of oxygen released due to photosynthesis may also cause the oxygen in the water to reach a supersaturated state.

The dissolved oxygen in the water is mainly consumed by the respiration of organisms and the oxidation process of organic matter. When the water body is severely polluted by organic matter, the amount of dissolved oxygen in the water can even approach zero. At this time, organic matter decomposes under anaerobic conditions, causing fermentation and seriously deteriorating the water quality.

(2) Carbon Dioxide

Dissolved carbon dioxide is present in most natural bodies of water. Its main source is the decomposition products of the oxidation of organic matter in water bodies or soil. Carbon dioxide in the air can also dissolve in water. Under normal circumstances, carbon dioxide exists in the form of free and carbonate, especially when the water body is in a weakly alkaline state, it exists in the form of carbonate and bicarbonate.

(3) Hydrogen sulfide

Due to the organic pollution of the water body, under anaerobic conditions, sulfur-containing organic matter decomposes, and sulfates are reduced to hydrogen sulfide, causing the presence of hydrogen sulfide gas in the water body. Under different pH values, the proportions of various forms of hydrogen sulfide in water are different. When pH < 7, the form of hydrogen sulfide in water is mainly molecular; when pH < 5, there is actually no thiohydric ion ion in the water, only hydrogen sulfide; when pH > 7, the main form is thiohydric ion ion; when pH > 9, it exists in the form of sulfide salts in water.

1.1.3 Properties of Natural Water

1.1.3.1 PH of Natural Water

For most natural water bodies, their pH values range from 5 to 9. Due to the different regions the rivers flow through and the different bodies of water they receive, their pH values vary greatly, even at different time periods.

1.1.3.2 Alkalinity

Alkalinity refers to all substances in water that can neutralize with strong acids, that is, the total amount of substances that can accept protons. The substances that constitute the alkalinity of natural water can also be summarized into three categories: ① strong alkalis: such as sodium hydroxide, calcium hydroxide, etc.; ② weak alkalis: such as ammonia, aniline, etc.; ③ weak acid salts of strong alkalis: such as carbonates, bicarbonates, silicates, phosphates, sulfides, humic acid salts, etc. The hydrolysis of weak acid salts of strong alkalis can produce hydroxide ions or directly accept protons.

Alkalinity is composed of various components such as hydroxides, carbonates, and bicarbonates. According to the ionic state, it can be divided into three categories: ① hydroxide alkalinity, i.e., the content of hydroxide ions; ② carbonate alkalinity, i.e., the content of carbonate ions; ③ bicarbonate alkalinity, i.e., the content of bicarbonate ions. The alkalinity of natural water bodies is mainly the latter two.

The content of carbonate ions in the water body plays an important role in the migration and transformation of some metal ions, such as calcium, magnesium, strontium, barium, etc. The anions in the water body are: OH^-, HCO^{3-}, $CO_3{}^{2-}$; the cations are H^+ and various metal ions that constitute alkalinity. If the total equivalent number of these metal ions is equal to the total equivalent number of total alkalinity, according to the principle of electrical neutrality, the equivalent numbers of its cations and anions are equal, The equation of electrical neutrality can be listed:

$$[\text{Alkali}] + [\text{H}^+] = [\text{HCO}_3^-] + [\text{OH}^-] + 2[\text{CO}_3^{2-}]$$
$$\text{or } [\text{Alkali}] = [\text{HCO}_3^-] + [\text{OH}^-] + 2[\text{CO}_3^{2-}] - [\text{H}^+]$$
$$\text{Then: total alkalinity(CTC)} = \alpha([\text{Alkali}] + [\text{H}^+] - [\text{OH}^-])$$
$$\text{CTC} = \alpha[\text{Alkali}]$$

1.1.3.3 Acidity

The acidity of natural water bodies refers to the total amount of substances in the water that can neutralize with strong alkalis, or in other words, the total amount of substances that can release protons or produce hydrogen ions through hydrolysis. The substances that make up the acidity of water can be divided into three categories: ① Strong acids: such as hydrochloric acid, nitric acid, sulfuric acid, etc. The number of hydrogen ions in an ionic state before neutralization is called ionic acidity. For strong acids, the ion concentration equals the acidity; ② Weak acids: such as carbonic acid, hydrogen sulfide, various organic acids; ③ Strong acid weak base salts: such as ferric chloride, aluminum sulfate, etc.

1.1.4 Pollutants in Water

Non-toxic pollutants in water bodies include acids, alkalis, salts and other inorganic substances, proteins, oils, fats and various chemicals produced in industrial production and other organic substances. Although some chemicals have no biological toxicity, high concentrations can have adverse effects on aquatic ecosystems. For example, the discharge of domestic sewage and industrial cooling water into water bodies causes an increase in phosphorus content in the water, which can further lead to algal blooms, causing a sharp drop in dissolved oxygen in the water. In an oxygen-deficient state, the death of algae produces a large amount of algal toxins, threatening the safety of drinking water. The content of pollutants in water can also have adverse effects on industrial water use. For example, when there are a large amount of chloride ions and ammonium ions in the water, corrosion occurs in materials such as carbon steel and copper in the circulating cooling water system. If the ammonia nitrogen in drinking water exceeds the standard, it can cause a series of diseases such as gastritis. The classification of water pollutants is shown in Table 1.6.

1. **Toxic Pollutants**

Toxic pollutants refer to substances that, when they enter the body and accumulate to a certain amount, can cause biochemical and physiological changes in body fluids or tissues, causing temporary or permanent pathological conditions, and even life-threatening substances. At present, humans have synthesized more than ten million

Table 1.6 Classification of water pollutants

Category		Examples
Chemical substances	Organic	Halogenated hydrocarbons, phenols, carboxylic acids, sugars, oils, dyes, synthetic detergents, polycyclic aromatic hydrocarbons, heterocyclic substances, antibiotics
	Inorganic	Acids, alkalis, chlorine, heavy metal salts, nitrates, phosphates, cyanides, radioactive substances
Physical pollutants	Suspended matter	Foam, rotting plants, silt, sand, bacterial corpses
Biological pollutants	Microbial class	Bacteria, protozoa, fungi, algae, viruses

kinds of chemicals, and more than one hundred thousand kinds have been commercially produced, but less than 1% have undergone lifecycle and toxicological characteristic assessments. Therefore, the toxicity of many chemicals is unknown and requires in-depth research. Many chemicals and products were found to be toxic only after many years of use, such as the pesticide DDT, and the plastic micro-particle pollution that everyone is very concerned about now.

(1) Toxic Inorganic Pollutants

Mercury, lead, chromium and other heavy metals, arsenic, selenium, boron, fluorine and other non-metals, uranium, radium and other radioactive substances are all toxic inorganic pollutants. According to the nature, existing form and toxicity of pollutants, toxic inorganic pollutants can be divided into heavy metal pollutants, inorganic anions (such as NO_2^-, F^-, CN^-) and radioactive substances. They can cause direct damage or long-term cumulative damage to humans and ecosystems. The toxicity of mercury, cadmium, lead, chromium, and arsenic is significant, followed by beryllium, copper, selenium, zinc, nickel, tin, manganese, cobalt, silver, vanadium, etc. Although beryllium, arsenic, and selenium are not heavy metals, considering the toxicity of arsenic and selenium and some properties similar to heavy metals, and the close relationship between beryllium and human health, these three elements are often included in the category of heavy metals for discussion.

Heavy metal pollution has its own characteristics: First, although the content of heavy metals in the earth's crust is less than 0.1%, they are widely distributed, widely produced and used, and high concentration pollution may occur in local areas. Secondly, heavy metal pollutants have potential hazards. They are not only not decomposed by microorganisms, but can accumulate in organisms and convert some heavy metals into more toxic metal organic compounds in organisms, such as methyl mercury, phenyl mercury and other alkylated substances. Thirdly, the toxicity of heavy metal pollutants is not only related to the amount ingested into the body, but also closely related to their existing forms. The toxicity of the same form can vary greatly, such as the toxicity of alkyl mercury is significantly greater than that of Hg^{2+} salt, and the toxicity of Cr (VI) is much greater than that of Cr (III). In addition, the toxicity of some heavy metals is also affected by other co-existing substances, that is, the so-called synergistic effect, antagonistic effect and additive effect.

Synergistic effect refers to the effect that when two or more pollutants coexist, one pollutant can cause the toxicity of another pollutant to increase sharply. For example, when copper and zinc coexist, their toxicity is tens of times that of their separate existence. Antagonistic effect refers to the effect that when two or more pollutants coexist, their toxicity can cancel out part or most of each other. For example, zinc can inhibit the toxicity of cadmium; selenium can counteract the harm of harmful elements such as lead, mercury, steel, and arsenic to the human body. The additive effect is that when two or more heavy metal pollutants coexist, their toxicity is roughly the sum of the toxicity of each component.

(2) Toxic organic pollutants

Toxic organic substances mainly include acid compounds, organic pesticides, polychlorinated biphenyls, polycyclic aromatic hydrocarbons and other organic substances. The common feature of toxic organic pollutants is that they are chemically stable, have a long residual time and are difficult to return; because they are easily soluble in fat, they have strong accumulation and can accumulate in aquatic organisms up to hundreds of thousands of times the concentration in water, not only affecting the reproduction of aquatic organisms, but also harming human health through the food chain. These substances are also endocrine disruptors or environmental hormones that people are generally concerned about.

According to the surface water quality status, China divides water into five categories, and the main index basic project standard limit values are shown in Table 1.7. According to the requirements of drinking water sources in China, Class I, Class II and Class III water can be used as raw water for drinking water, but due to the general pollution of surface water in the eastern region, many areas use Class IV water as raw water.

1.2 Chemical Treatment and Coagulation Process

1.2.1 Coagulation Process

Given the characteristics of surface water, the coagulation process is to intensify the removal of colloidal substances with diameters in the range of 10^{-9} to 10^{-7} m (1–100 nm) and as many dissolved organic substances as possible. For particles with diameters less than 1 nm, adsorption or membrane filtration methods are now widely used (Fig. 1.1).

Colloidal surfaces generally carry negative charges, repel each other, exhibit characteristics of Brownian motion, and form stable suspensions. If colloidal or positively charged substances are added, they can neutralize the surface charge of the colloid,

Table 1.7 Surface water environmental quality standards (main indicators) basic project standard limit values (Unit: mg/L)

Serial number	Standard classification project	Class I	Class II	Class III	Class IV	Class V
1	Water temperature (°C)	The environmental water temperature change caused by human activities should be limited to a weekly average maximum temperature rise of ≤ 1, and a weekly average maximum temperature drop of ≤ 2				
2	pH value (dimensionless)	6–9				
3	Dissolved oxygen ≥	7.5	6	5	3	2
4	Permanganate index ≤	2	4	6	10	15
5	Chemical oxygen demand (COD) ≤	15	15	20	30	40
6	Five-day biochemical oxygen demand (BOD_5) ≤	3	3	4	6	10
7	Ammonia nitrogen (NH_3–N) ≤	0.15	0.5	1.0	1.5	2.0
8	Total phosphorus (as P) ≤	0.02	0.1	0.2	0.3	0.4
9	Total ammonia (in lakes and reservoirs as N) ≤	0.2	0.5	1.0	1.5	2.0
10	Copper ≤	0.01	1.0	1.0	1.0	1.0

Note The total phosphorus content in lakes and reservoirs is around 0–0.2 mg/L

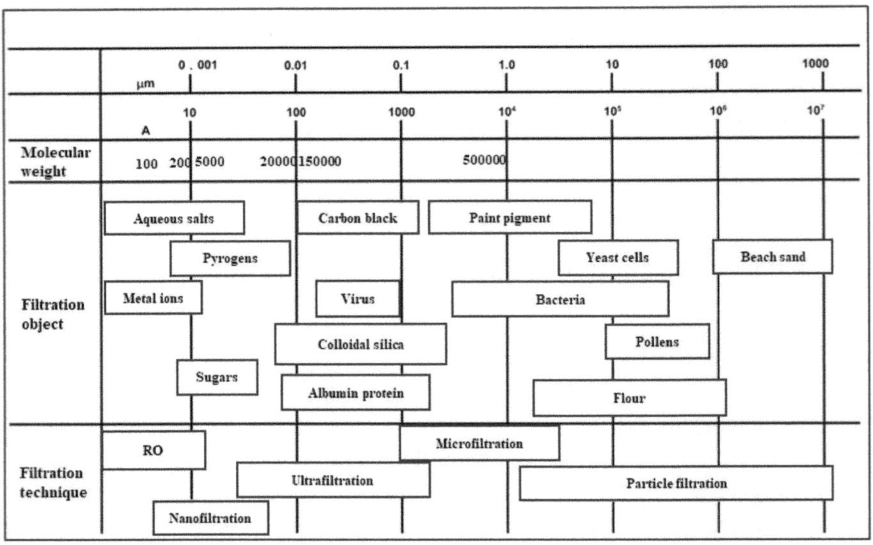

Fig. 1.1 Particle size and selection of filtration method

and the physical adsorption force (the Van der Waals force) exceeds the aforementioned repulsion, thereby causing the aggregation of colloidal substances. At the same time, the large specific surface area of alum flowers formed during the hydrolysis of inorganic salts adsorbs dissolved substances in the water body, achieving the purpose of purifying the water body.

The chemicals added during the coagulation process can generally be divided into inorganic coagulants and organic polymer flocculants. The history of using inorganic coagulants can be traced back to ancient Egypt. At that time, people used aluminum salts (complex salts prepared with aluminum sulfate, alkali metals, etc.) to treat drinking water, and today aluminum salts are still one of the most common coagulants. The main types of inorganic coagulants are still iron salts and aluminum salts. Organic flocculants are most representative of polyacrylamide.

In recent decades, with the advancement of technology, the boundaries between coagulants and flocculants have become increasingly blurred, especially inorganic–organic composite coagulants, such as polyaluminum chloride-polydiallyldimethylammonium chloride composites, polyamine-polyaluminum chloride composites, they not only have the properties of inorganic coagulants, but also have outstanding flocculation effects, avoiding the complexity of separate addition during use.

1.2.2 Coagulation Principle

A dispersed system is a system composed of two or more substances mixed together. The dispersed substance is called the dispersed phase. The substance that is continuous around the dispersed phase is called the dispersion medium. According to the size of the particles, the dispersed system composed of water and particles distributed in the water can be divided into three categories:

① Suspension: The particle diameter is larger than 10^{-7} m, also known as a coarse dispersed system.

② Colloidal solution: The particle diameter is between 10^{-9} and 10^{-7} m, also known as sol.

③ True solution: The particle diameter is less than 10^{-9} m.

Colloidal particles formed by the aggregation of colloidal molecules are called colloid cores. Their surfaces carry a charge due to the adsorption of certain ions, which can attract counter-ions in the solution. These counter-ions are influenced by two forces: one is the electrostatic attraction of the particle surface, which attracts counter-ions to the particles; the other is the diffusion force of the counter-ions' own thermal motion and the solvation force of the liquid on them. This forms an anion and cation layer on the particle surface, known as a double electric layer.

The double electric layer structure on the colloid surface (Fig. 1.2) is mainly composed of an inner layer close to the colloid surface and a diffusion layer. The former mainly depends on the adsorbed ions, and the latter depends on electrostatic

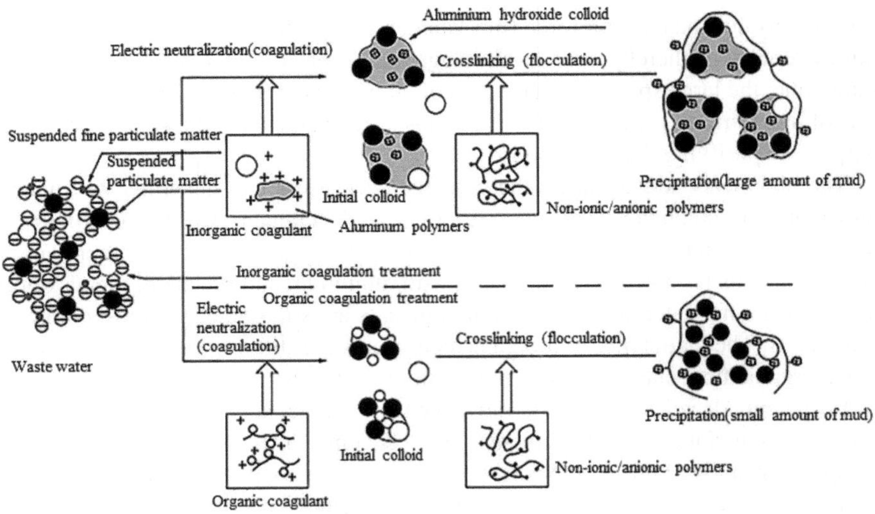

Fig. 1.2 Structure of the double electric layer on the colloid surface

forces and random thermal motion. The quantitative description of the double electric layer is very complex, and different assumptions often yield different results.

Our understanding of the charge distribution on and around the colloid surface has gradually deepened. So far, there are several models describing the distribution structure of charges on the colloid surface, among which Helmholtz, Gouy-Chapman, and Stern are three typical models.

(1) **Helmholtz model**

As early as 1879, Helmholtz pointed out that the internal structure of the double electric layer is similar to a parallel capacitor. However, this model does not distinguish between surface potential and electrokinetic potential; moreover, subsequent research has shown that according to the Helmholtz model, there will be no relative movement between the double electric layers, because the double electric layer as a whole is electrically neutral.

(2) **Gouy-Chapman model**

Gouy and Chapman pointed out that the counter-ions in the solution are subject to two opposing forces, one is electrostatic force; the other is thermodynamic force. When these two forces reach a balance, the counter-ions are not neatly bound to the colloid surface, but are diffusely distributed, and as the distance from the surface increases, the excess of counter-ions gradually weakens, until at a certain distance, the concentration of counter-ions equals the ion concentration, as shown in Fig. 1.3 for the distribution of ions in the diffusion layer.

The colloid core adsorbs ions and some counter-ions that are closely attached and move with it to form an adsorption layer. The other part of the counter-ions diffuses

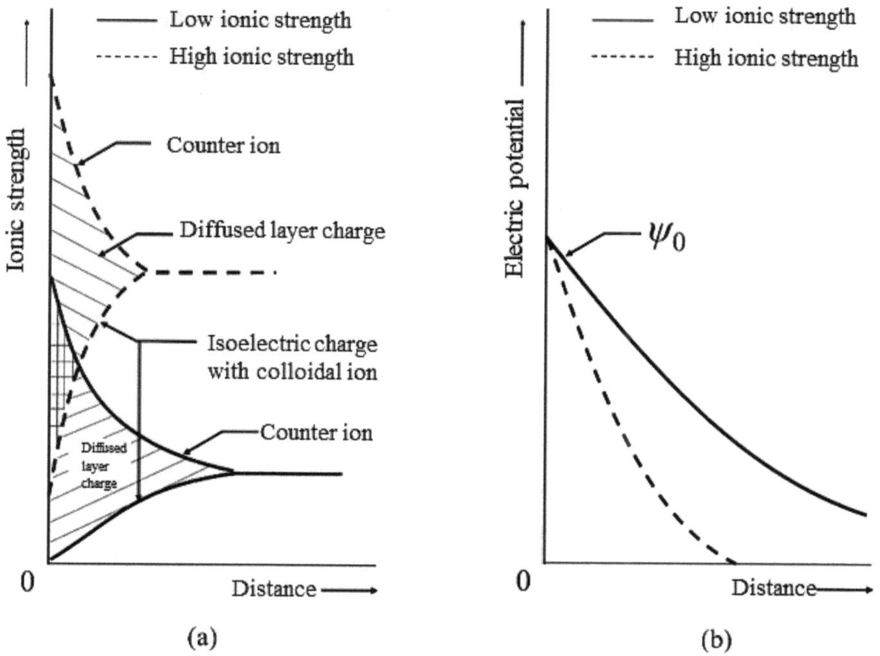

Fig. 1.3 Distribution of ions in the diffusion layer. **a** Distribution of ions in the diffusion layer; **b** Change of potential in the diffusion double electric layer

outward due to thermal motion and solvation, and can detach from the colloid core surface, thus forming a diffusion layer; the interface that can be detached is called the slip interface. The potential formed between the ions on the colloid core surface and their counter-ions is called the total potential, or Ψ potential. The potential difference between the slip interface and the solution interior is the ζ potential. The sign of the ζ potential value indicates the nature of the colloid particle's charge, its magnitude signifies the degree of charge on the colloid particle, and it can also reflect the thickness of the diffusion layer. If the ζ value increases, the diffusion layer becomes thicker; otherwise, the diffusion layer becomes thinner.

In order to represent the distribution of ions, Gouy and Chapman made the following assumptions:

① The solid surface is flat, infinitely large in the y and z directions, and the charge distribution on the surface is uniform;
② Ion diffusion only exists in the x direction, and the ion is considered as a point charge. It should follow the Boltzmann distribution in the double electric layer;
③ The number of positive and negative ion charges is equal, and the entire system is neutral;
④ The dielectric constant of the solvent is the same throughout the entire diffusion layer.

According to the Boltzmann distribution law, the charge distribution within the diffusion double layer can be represented by Eqs. (1.2) and (1.3):

The concentration of cations is:

$$n_{+} = n_{0} \exp[(zeV)/(kT)] \tag{1.2}$$

The concentration of anions is:

$$n_{-} = n_{0} \exp[(-zeV)/(kT)] \tag{1.3}$$

where n_0 is the ion concentration in the solution, mol/L;

n_+ is the concentration of cations in the solution, mol/L;

n_- is the concentration of anions in the solution, mol/L;

z is the valence of the anions and cations;

e is the amount of charge per unit, 1.602×10^{-19} C.

V is the potential at any point in the solution, V;

k is the Boltzmann constant, k = 1.380649×10^{-23} J/K;

T is the absolute temperature, K.

The charge density at any point in space can be represented by the following formula:

$$\rho = ze(n_{+} - n_{-}) \tag{1.4}$$

Substituting formulas (1.2) and (1.3) into the above, we get:

$$\rho = -2zen_{0} \sin h[(eV)/(kT)] \tag{1.5}$$

Since the hyperbolic sine can be represented as:

$$\sin h(x) = \left(e^{x} - e^{-x}\right)/2 \tag{1.6}$$

Using the Poisson equation, the relationship between charge density and potential in a spatial electric field can be represented as:

$$\nabla^{2} V = -\rho/\varepsilon \tag{1.7}$$

where ε is the dielectric constant of the medium, F/m;

∇^2 is the Laplacian operator.

For flat particles, formula (1.7) can be simplified to:

$$\nabla^{2} = \mathrm{d}^{2}/\mathrm{d}x^{2} \tag{1.8}$$

Combining formulas (1.5), (1.6), (1.7), and (1.8), the Poisson-Boltzmann equation can be represented as:

$$\nabla^2 V = [(2zen_0)/\varepsilon]\sinh[(eV)/(kT)] \quad (1.9)$$

When the surface potential is very low, i.e., eV << kT, formula (1.6) can be simplified as:

$$\sinh(x) \approx x \quad (1.10)$$

In this way, formula (1.9) is simplified into a linear equation:

$$\nabla^2 V = \kappa^2 V \quad (1.11)$$

where, $\kappa^2 = 2ze^2 n_0 / \varepsilon kT$ (Debye–Huckel theory)

Formula (1.11) can be used to calculate the potential of a single ion in a weak electrolyte solution. The potential of the solution in this case is:

$$V = V_0 \exp(-\kappa x) \quad (1.12)$$

where, $1/\kappa$ is the Debye length, i.e., the thickness of the diffusion layer.

Note that the potential corresponding to this diffusion layer thickness is only 1/e of the surface potential, which means that the diffusion layer thickness does not represent the distance from the surface potential V_C to the bulk potential zero, i.e., the entire thickness of the diffusion layer. The thickness of the diffusion layer is generally within a very narrow range of 1–1000 nm, and is independent of the size of the colloidal surface area.

The formula for calculating the thickness of the double layer is:

$$x_{DL} = \kappa^{-1} = [\varepsilon kT/(2n_0 z^2 e^2)]^{1/2} \quad (1.13)$$

where x_{DL} is the thickness of the double layer, nm;

ε is the dielectric constant of water;

n_0 is the number of electrolyte particles per unit volume;

When considering the sum of all ion concentrations in the solution, the thickness of the diffusion double layer (x_{DL}) can be expressed by the Debye length ($1/\kappa$):

$$1/x_{DL} = \kappa = \left[\left(e^2 \sum n_{i0} z_i^2\right)/(\varepsilon\varepsilon_o kT)\right]^{1/2} \quad (1.14)$$

where n_{i0} is the number of i ions in the solution;

z_i is the charge of the i ion;

ε is the dielectric constant, which is 80F/m for pure water and most dilute solutions;

ε_o is the permittivity of free space, 8.854×10^{-12} $C^2/(J\ m)$.

For a water solution at 25 °C, when the ion concentration is converted to molar concentration, the thickness of the diffusion layer is expressed as:

$$1/x_{DL} = \kappa = 2.32 \times 10^9 \left(c_i z_i^2\right)^{1/2} \quad (1.15)$$

where c_i is the ion concentration, mol/dm^3.

For a double layer composed of two types of ions with equal absolute charge |z| and opposite electricity, the Gouy-Chapman theory believes that its potential decreases with the increase of distance x, and its formula is:

$$d/dx = -[(2kT)/(ze)]\kappa \sinh[(ze)/(2kT)] \quad (1.16)$$

For a uniform electrolyte, its surface charge density (C m^{-2}) is:

$$\sigma_0 = [8n_0\varepsilon\varepsilon_o kT]^{1/2} \sin h(ze\Psi_0/(2kT)) \quad (1.17)$$

For a 1-1 type electrolyte, at a lower surface potential Ψ_0 (~ 25 mV), Eq. (1.16) can be simplified as:

$$\Psi(x) = \Psi_0 \exp(-\kappa x) \quad (1.18)$$

where $\Psi(x)$ is the potential at a distance x in the diffusion double layer, V;

Ψ_0 is the surface potential, V;

x is the distance (the distance between the ion and the colloid surface), m.

Equation (1.15) shows the effect of ion strength on the double layer: as the ion strength increases, κ also increases. Equation (1.18) indicates that when the surface potential (x) is 1/e of the initial surface value (~ 1/2.72), its double layer thickness is $1/\kappa$. The schematic diagram of the double layer model between the non-conductive surface and colloid particles in the electrolyte solution and the relationship curve between the potential drop and the distance from the colloid particles and the Debye length κ^{-1} are shown in Fig. 1.4.

Figure 1.4 shows that κ^{-1} is not the entire thickness of the diffusion layer, but only 1/e of the total thickness (~ 1/2.72). In Fig. 1.4, the flat surface is assumed to be a net negative charge, δ is the closest point of the compensating ion to the surface charge, the Ψ axis represents the absolute value of the potential, Ψ_0 is the surface potential, Ψ_δ is the stern potential, z is the Zeta potential. It is assumed that the shear plane is in the diffusion layer, and the potential drop curve from the surface ($\Psi = \Psi_0$) to the solution ($\Psi = 0$) is very obvious. The vertical dashed line indicates the position where the shear plane intersects with the potential curve, that is, the Zeta potential, ζ(mV). This potential is measured by electrokinetic analysis, and it can determine the electrophoretic mobility (μ_e, in units of $\mu m\ s^{-1}/(V\ cm^{-1})$) of particles and colloids in the Smoluchowski equation.

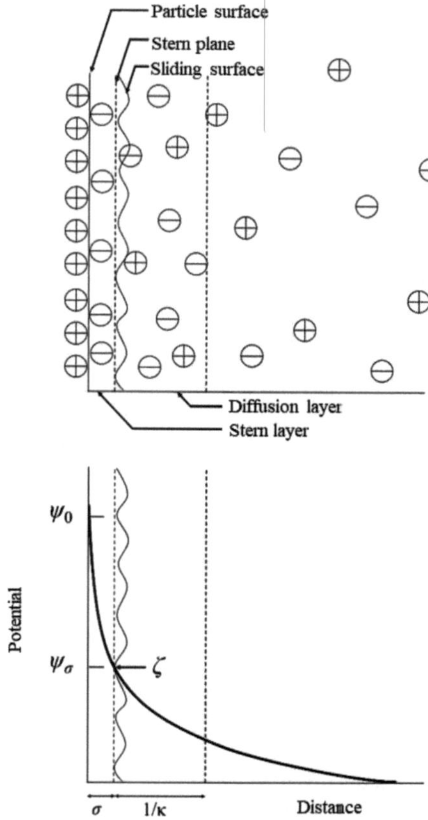

Fig. 1.4 Schematic diagram of the double layer model between the non-conductive surface and colloidal particles in the electrolyte solution and the relationship curve between the potential drop and the distance from the colloidal particles and the Debye length κ^{-1}

$$\mu_e = \varepsilon\varepsilon_0\zeta/\eta \tag{1.19}$$

where η is the viscosity, g cm^{-1} s^{-1}; μ_e is the electrophoretic mobility, $cm^2/(V\cdot s)$.

For convenience, the thickness of the diffusion layer corresponding to different types of electrolytes is plotted. According to the relationship shown in Eq. (1.15), the relationship curve between the molar concentration of electrolytes and the valence in a water medium with a dielectric constant of $\varepsilon = 80F/m$ and the thickness of the diffusion layer or Debye length ($1/\kappa$) is shown in Fig. 1.5. The types of electrolytes are: cationic charge: anionic charge is respectively 1:1, 1:2, 2:1, 1:3. The addition of these electrolytes will reduce the thickness of the double layer (this phenomenon is considered charge shielding), making the solution ions more easily approach the inner surface of the colloidal double layer, increasing the adhesion of the colloidal particles (and can reduce the electrophoresis or electroosmotic migration rate).

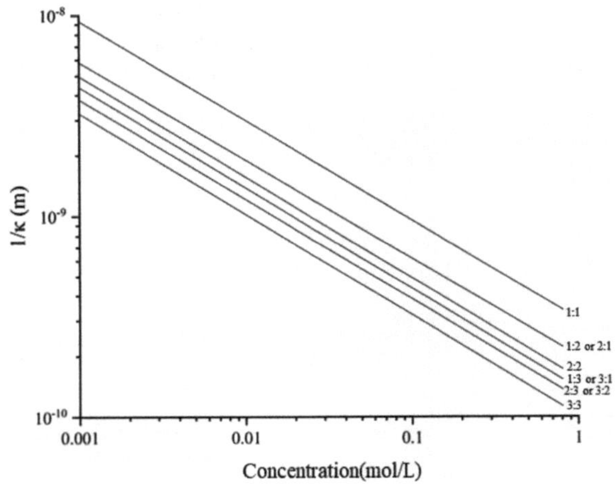

Fig. 1.5 The relationship curve between the molar concentration of electrolytes and the valence in a water medium with a dielectric constant of 80 at 27 °C and the thickness of the diffusion layer or Debye length (1/κ)

When the diffusion double layer is regarded as a flat plate, we can understand the relationship between the flat plate charge density σ and the distance between the flat plates.

Flat surface charge density:

$$\sigma = \varepsilon \frac{V_0}{\delta} \tag{1.20}$$

where V_0 is the surface potential, V;

δ represents the distance between the plates.

According to the principle of electrical neutrality, the number of charges on the solid surface is equal to the number of opposite charges in the diffusion double layer. The number of charges per unit area on the solid is equal to the number of excess counter ions in the volume of the liquid phase from the solid surface to infinity. Then the flat surface charge density is:

$$\sigma = -\int_{0}^{\infty} \rho \mathrm{d}x \tag{1.21}$$

Combining Eq. (1.7), the derivative can be obtained:

$$\sigma = \varepsilon \kappa V_0 \tag{1.22}$$

Comparing Eqs. (1.20) and (1.22), it can be seen that κ^{-1} is equivalent to the ε thickness in the flat plate model, so κ^{-1} is usually used as a measure of the diffusion double layer.

The potential in the diffusion layer decreases as the distance from the colloidal surface increases, as shown in Fig. 1.4. The rate of decrease is determined by the size of κ^{-1}. The increase in electrolyte concentration or valence will reduce the value of κ^{-1}, thin the double layer, and the potential decreases faster as the distance from the colloidal surface increases.

(3) **Stern model**

The Gouy-Chapman model distinguishes between surface potential and electrophoretic potential. The electrophoretic potential decreases as the ion concentration increases and is always the same sign as the surface potential, with a limit of zero. However, experiments have found that the electrophoretic potential will increase as the ion concentration increases, and sometimes it will be opposite to the surface potential.

Stern believed that ions in a solution should not be considered as point charges without volume. He proposed:

(1) Ions have size, and the distance between the ion center and the particle surface cannot be less than the radius of the ion;
(2) In addition to the electrostatic repulsion between ions and particles, there should also be Van der Waals attraction.

Stern divided the diffusion layer in the Gouy-Chapman model into two parts. In the region a few molecules thick near the colloid surface, counter ions are tightly bound to the colloid surface due to strong electrostatic attraction, forming the so-called "Stern layer". Outside the Stern layer is the diffusion layer where ions are diffusely distributed. The potential change within the Stern layer is similar to that in the Helmholtz plate model, dropping from the surface potential to the Stern layer. The potential of the Stern layer is called the Stern potential, and the potential in the diffusion layer conforms to the Gouy-Chapman model, dropping from the Stern potential to zero. The electrokinetic potential is the potential difference between the sliding surface and the solution body when the solid and liquid move relative to each other. Since the sliding surface is slightly outside the Stern layer, the electrokinetic potential is slightly lower than the Stern potential (Fig. 1.4). Under the same ζ potential, the surface potential is not necessarily the same. The structure of the double electric layer of colloidal particles in different solution systems also differs, as shown in Fig. 1.6 for the comparison of the structure and position of the double electric layer of colloidal particles in salt water and pure water solutions.

(4) **DLVO Theory**

In the 1930s, Deryagin and Landau, and Verwey and Overbeek independently completed the stability theory of lyophobic sols, known as the DLVO theory. This theory provides a quantitative description of the correlation between electrolytes and colloid stability. According to this theory, the interaction force V(d) between colloids includes electrostatic repulsion ψ_R (first term of Eq. 1.18) and Van der Waals attraction ψ_A (second term of Eq. 1.22).

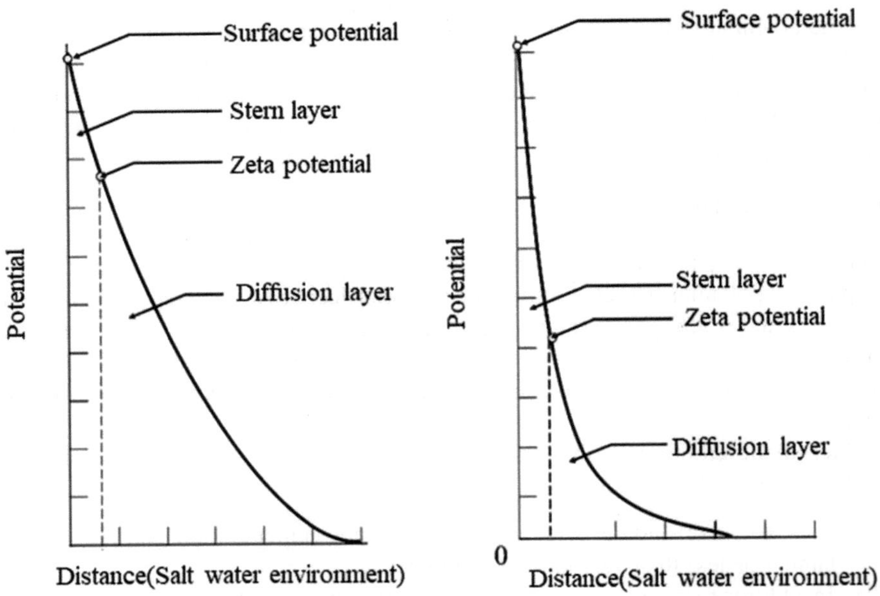

Fig. 1.6 Comparison of the structure and position of the double electric layer of colloidal particles in salt water and pure water solutions

$$\psi(d) = \psi_R - \psi_A = \exp\{-\kappa(d - 2R_s)\} - \frac{AR_s}{12d} \tag{1.23}$$

In the equation, A is the Hamaker constant, which generally ranges from 10^{-20} to 10^{-19} J. For oxides, A is generally 2×10^{-20} to 4×10^{-20}. The Hamaker constants of some materials are shown in Table 1.8.

The Hamaker constant is composed of the prefactor of the Van der Waals integral equation:

$$A = \left(\frac{\rho N_A}{M}\right)^2 \pi^2 \beta \quad \beta = \frac{3}{4} h\nu \left(\frac{\alpha}{4\pi\varepsilon_0}\right)^2$$

The order of magnitude of the Hamaker constant is mainly determined by the interaction difference "β", $\rho N_A/M$ is the reciprocal of the molecular volume, $\frac{\alpha}{4\pi\varepsilon_0}$ is about 10% of the atomic size, so:

$$A = \left(\frac{\rho N_A}{M}\right)^2 \pi^2 \beta = A \approx 3/4\pi^2 h\nu (0.1)^2 \approx 10^{-18}\,\text{J}$$

If the value of ψ_d is less than zero, the colloid becomes unstable and coagulates. Adjusting the electrolyte concentration in the solution can compress the double electric layer and reduce the effect of electrostatic repulsion. Gradually increasing the

Table 1.8 Hamaker constants of some materials (Unit: × 10^{-20} J)

Material	Vacuum	Water
Gold	40	30
Silver	50	40
Aluminum oxide	16.75	4.44
Copper	40	30
Pentane	3.8	0.34
Decane	4.8	0.46
Hexadecane	5.2	0.54
Water	3.7	–
Melted quartz	6.5	0.83
Crystalline quartz	8.8	1.70
Calcite	10.1	2.23
Calcium fluoride	7.2	1.04
Sapphire	15.6	5.32
Polymethyl methacrylate	7.1	1.05
Polyvinyl chloride	7.8	1.03
Polystyrene	7.9	1.3
Polystyrene	3.8	0.33

electrolyte concentration can effectively compress the thickness of the diffusion layer until the surface potential or ζ potential is reduced to a certain range to make the colloid completely unstable.

The Van der Waals force is more meaningful for colloids and gel particles than for atoms and molecules. In the adhesion and coagulation process of colloids and polymers, van der Waals forces play a significant role. The colloids and polymers mentioned here include surfactants (gel particles and liposomes) and large biological molecules, such as proteins and DNA.

Van der Waals attraction usually refers to the following three interactions:

① The interaction between two permanent dipoles;
② The interaction between a permanent dipole and an induced dipole;
③ The interaction of dispersion forces between molecules.

The potential energy of Van der Waals attraction can be calculated by the following formula:

$$V_A = \frac{AR_s}{12d} \tag{1.24}$$

where d is the distance between the gel particles, m;

Rs is the particle size, m.

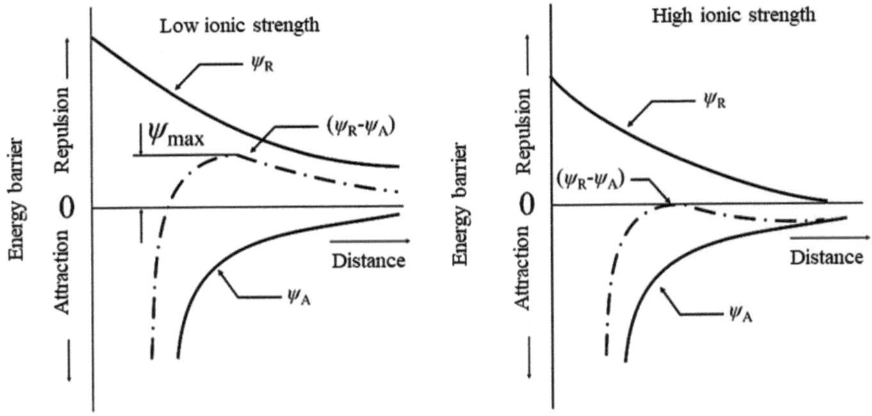

Fig. 1.7 Illustration of the total interaction energy between particles

Van der Waals attraction and the double electric layer repulsion of the solid–liquid interface coexist on the colloid. The total interaction energy between particles is shown in Fig. 1.7.

The sum of positive energy, i.e., the energy barrier, causes the particles to repel each other, while the sum of negative energy causes the particles to attract each other, leading to system instability. In water treatment, the goal is to overcome the energy barrier through different measures, destabilize the colloid, and thus achieve the purpose of removing these suspended particles. Adding electrolytes to the colloid solution can only change the repulsion between particles, but it has no effect on the Van der Waals molecular attraction.

The relationship between colloid surface charge and stability is shown in Table 1.9, which lists the control range of ζ potential for removing colloids in natural water bodies. The surface potential or charge density is influenced by the electrolyte in the solution and basically follows the Schulze-Hardy rule, critical coagulation concentration is very sensitive to the valence state of the anion, and high-valence anions have stronger coagulation ability. For the same colloidal solution, the minimum coagulant dosage (polymerization value) follows the following ratio relationship, Na^+:Ca^{2+}:Al^{3+} = 1:1/10:1/1000 (molar ratio). This is also the reason why we choose high-valence iron salts and aluminum salts in our water treatment process. In n the coagulation process, in order to achieve the purpose of destabilizing colloids, it is generally not necessary to completely reduce the colloidal surface potential to zero. The surface potential or ζ potential can achieve the effect of destabilization and polymerization within a certain range.

The colloidal surface potential depends on the surface charge density and the thickness of the double electric layer. The surface charge density depends on the concentration of ions in the solution (i.e., ions closely related to the surface) that affects the potential. For alumina colloids, the H^+ ion can determine the ζ potential, so the colloidal surface ζ potential is determined by the pH value. The relationship

Table 1.9 Relationship between colloidal surface potential and stability

Stability	Average colloidal surface potential/mv
Excellent stability	– 100 to – 60
Good stability	– 60 to – 40
Moderate stability	– 40 to – 30
Mild dispersion	– 30 to – 15
Coagulation	– 15 to – 10
Strong coagulation or precipitation	– 5 to + 5

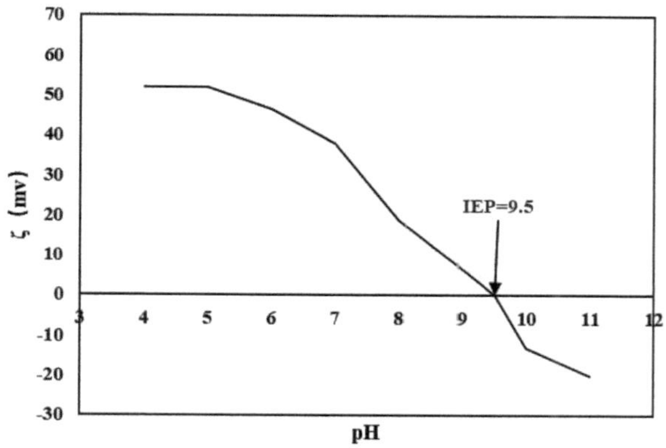

Fig. 1.8 Relationship between ζ potential of alumina colloid surface and pH value

between the ζ potential of the alumina colloid surface and the pH value is shown in Fig. 1.8.

The ζ potential decreases with an increase of the pH value and reaches the isoelectric point (IEP) when the pH value is 9.5. The ζ potential is affected by the surface potential and the concentration of electrolytes in the solution. Under the same ζ potential, the surface potentials may not be the same.

1.3 Hydrolysis Mechanism of Aluminum Salt and Iron Salt

1.3.1 Classification of Coagulants

There are many types of coagulants and flocculants currently, with inorganic coagulants mainly being iron salts and aluminum salts, and organic flocculants mainly being various polyacrylamides, polyamines, etc. Table 1.10 lists the common coagulants or flocculants on the market.

Table 1.10 Types of coagulants and flocculants

Category	Name	National or Industry Standard in China
Inorganic low molecular weight coagulant	Aluminum sulfate	GB/T 31060-2014, Water treatment agent Aluminum sulfate [S]
	Ferric sulfate chloride	/
	Ferric sulfate	GB/T 14591-2016, Water treatment agent Poly ferric sulfate [S]
	Ferrous sulfate	GB/T 10531-2016, Water treatment agent Ferrous sulfate [S]
	Ferric chloride	GB/T 4482–2018, Water treatment agent Ferric chloride [S]
Inorganic high molecular weight coagulant	Poly aluminum chloride	GB 15892-2020, Drinking water poly aluminum chloride [S] GB/T 22627-2022, Water treatment agent Poly aluminum chloride [S]
	Poly aluminum sulfate	HG/T 5006-2016, Water treatment agent Poly aluminum sulfate S]
	Poly aluminum ferric chloride	HG/T 5359-2018, water treatment agent Poly aluminum ferric chloride [S]
	Poly ferric sulfate	GB/T 14591-2016, water treatment agent Poly ferric sulfate [S]
	Poly ferric chloride	HG/T 4672-2022, water treatment agent Poly ferric chloride S]
	Poly aluminum ferric sulfate	HG/T 5565-2019, water treatment agent Poly aluminum ferric sulfate [S]
	Poly aluminum silicate chloride	GB/T 22627-2014, water treatment agent Poly aluminum silicate chloride [S]
	Poly aluminum silicate sulfate	HG/T 4537-2013, water treatment agent Poly aluminum silicate sulfate [S]
Inorganic polymer coagulant	Poly aluminum ferric silicate chloride	/
	Poly ferric silicate sulfate	/
	Poly aluminum ferric silicate sulfate	/
Synthetic polymer flocculant	Polyacrylamide	HG/T 5750-2020, water treatment agent Emulsion type anionic and nonionic polyacrylamide S] HG/T 31246-2019, water treatment agent Emulsion type cationic polyacrylamide [S] GB/T 31246-2014, water treatment agent Polyacrylamide [S]

(continued)

Table 1.10 (continued)

Category	Name	National or Industry Standard in China
	Dimethylamide-epi-chlorohydrin polyamine	GB/T 26369-2020, quaternary ammonium disinfectant hygiene requirements [S]
	Poly(dimethyldiallyl ammonium chloride)	GB/T 33085-2016, water treatment agent Poly(dimethyldiallyl ammonium chloride) [S]
	Acrylamide- dimethyldiallyl ammonium chloride copolymer	GB/T 31246-2014, water treatment agent Technical conditions and test methods for cationic polyacrylamide [S]
	Polyacrylic acid	GB/T 10-533-2014, water treatment agent Polyacrylic acid [S]
	Dicyandiamide-formaldehyde condense	HG/T 4817-2015, water treatment agent Dicyandiamide-formaldehyde condense [S]
Natural polymer flocculant	Modified cationic starch	/
	Chitosan	/
	Modified lignin	HG/T 3507-2008, sodium lignosulfonate dispersant [S]

1.3.2 Hydrolysis Mechanism of Aluminum and Iron Salts

All ions in the water will undergo different degrees of hydrolysis. Al^{3+} and Fe^{3+} can react with six water molecules to form an octahedral structure, such as $Al(H_2O)_6{}^{3+}$. This octahedral structure can further hydrolyze, that is, the hydroxyl group replaces the water molecule, or it is called the deprotonation process of the coordinated water molecule. For convenience, the hydration layer is often omitted (hydration shell), only represents metal ions and hydroxyl ions:

$$Al^{3+} \rightarrow Al(OH)^{2+} \rightarrow Al(OH)_2^+ \rightarrow Al(OH)_3 \rightarrow Al(OH)_4^-.$$

During the hydrolysis process, each step of the above reaction removes a proton, causing the equilibrium to shift to the right. Due to the low solubility of aluminum hydroxide, precipitation occurs within a moderate pH range. If the pH is further increased, soluble sodium aluminate will form. Similar reactions occur for iron salts, but the solubility of iron hydroxide is much lower.

Because insoluble hydroxides often are polynuclear polymer, determining the hydrolysis constant (hydrolysis constant) is very difficult, and the data currently published varies greatly. The hydrolysis constant can be determined through the following process:

$$M^{3+} + H_2O \rightleftharpoons M(OH)^{2+} + H^+ \quad K_1 \tag{1.25}$$

$$M(OH)^{2+} + H_2O \rightleftharpoons M(OH)_2^+ + H^+ \quad K_2 \tag{1.26}$$

$$M(OH)_2^+ + H_2O \rightleftharpoons M(OH)_3 + H^+ \quad K_3 \tag{1.27}$$

$$M(OH)_3 + H_2O \rightleftharpoons M(OH)_4^- + H^+ \quad K_4 \tag{1.28}$$

$$M(OH)_3 \rightleftharpoons M^{3+} + 3OH^- \quad K_5 \tag{1.29}$$

For the hydrolysis products of aluminum and iron, their most stable solid forms are gibbsite and goethite, respectively. However, the equilibrium time required to form these substances is very long (several months or even longer). The dissolution equilibrium of amorphous precipitates is often represented by the solubility constant K_{Sam}, which is faster and more convenient in the actual use of coagulants. However, the accurate value of the solubility constant of amorphous hydroxides has not yet been obtained, only estimates. Table 1.11 lists the solubility constants and hydrolysis constants (pK values).

It is generally believed that the range of aluminum hydrolysis constants is narrower than that of iron, which is mainly caused by the different number of water molecules. During the hydrolysis process, the coordination number of aluminum decreases from 6 to 4, while that of iron remains at 6. The equilibrium of dissolved hydrated iron and aluminum ions with amorphous hydroxide was showed in Fig. 1.9. Fe_T and Al_T represent the total dissolved iron and aluminum, respectively, and the dashed line represents the concentration corresponding to the Ksp of $Fe(OH)_3$ and $Al(OH)_3$.

As can be seen from Fig. 1.19, in aluminum salt solution, the main ion species changes from Al^{3+} to $Al(OH)_4^-$, and the pH range changes only by 1 unit. For hydrated iron ions, the change from $Fe(OH)^{2+}$ to $Fe(OH)_3$ spans 8 pH units.

High-charge metal ions can form simple monomers during hydrolysis, as well as a series of polymeric clusters. For example, between monomeric aluminum ions and aluminum hydroxide, there exists a series of cationic polymers, or polymers.

When using aluminum sulfate as a coagulant, electro-neutralization is the main coagulation mechanism, so effective rapid stirring is crucial for the success of the coagulation process. The pH value is also very critical to the coagulation process of aluminum sulfate, because the solubility of different forms of aluminum salts varies under different pH values. The relationship curve between the solubility of aluminum salts and pH value is shown in Fig. 1.10.

Table 1.11 Hydrolysis and solubility constants of Al^{3+} and Fe^{3+} (25 °C, ion strength is 0)

	pK_1	pK_2	pK_3	PK_4	pK_{Sam}
Al^{3+}	4.95	5.6	6.7	5.6	31.5
Fe^{3+}	2.2	3.5	6	10	38

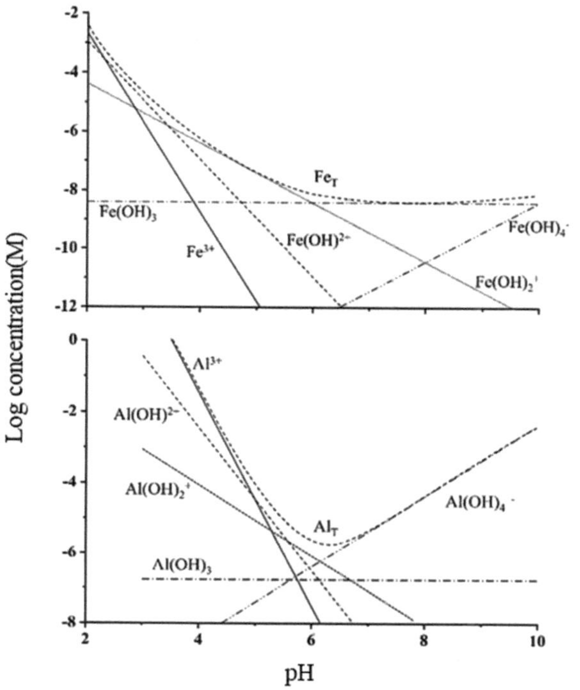

Fig. 1.9 Equilibrium of hydrated iron ions and hydrated aluminum ions with amorphous hydroxides

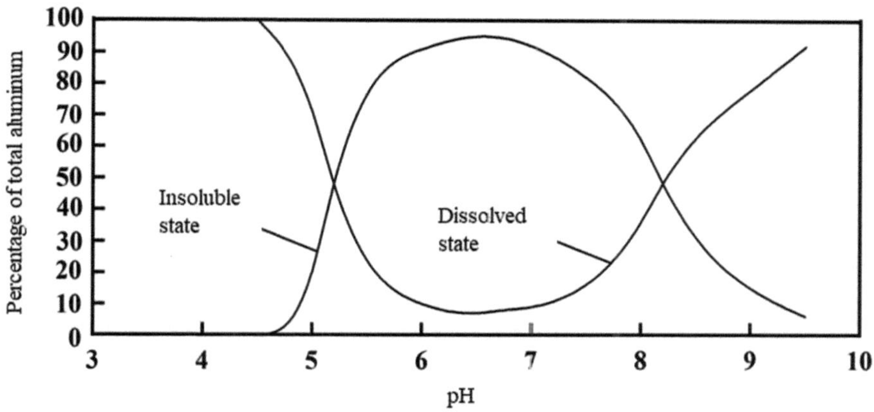

Fig. 1.10 Relationship curve between the solubility of aluminum salts and pH value

If the pH value of the mixed liquid is between 4 and 5, the aluminum salts mainly exist in the forms of $Al(OH)^{2+}$, $Al_8(OH)_4^{4+}$, and Al^{3+}. However, the best sweeping flocculation effect occurs when the aluminum salts in the water are mainly in a negative electric form, and at this time the pH value is between 6 and 8.

Amixthaxjah has conducted in-depth research on the coagulation process of aluminum sulfate. He believes that there are two main coagulation mechanisms in the coagulation process of aluminum salts: one is the adsorption of soluble polymers formed by the hydrolysis of aluminum salts on colloids, using electro-neutralization to destabilize the colloids, also known as adsorption; the other is the sweeping capture of colloid particles by aluminum hydroxide precipitates, also known as sweeping capture. Figure 1.11 shows the flocculation mechanism of aluminum sulfate and its relationship with Zeta potential.

Al^{3+} and its various hydrolysis forms have a dissolution–precipitation equilibrium relationship with precipitated $Al(OH)_3$. In a saturated solution, the saturation concentration of ions in various forms directly depends on the pH value of the solution. The straight lines in Fig. 1.11 represent the saturation concentrations of various soluble compounds in the saturated solution at different pH values. Beyond these concentrations, the soluble state becomes precipitation. That is, on the right side of the straight line, as the pH value increases, the solubility of $Al(OH)_3$ also increases. When it rises to a certain level, the lowest solubility will appear.

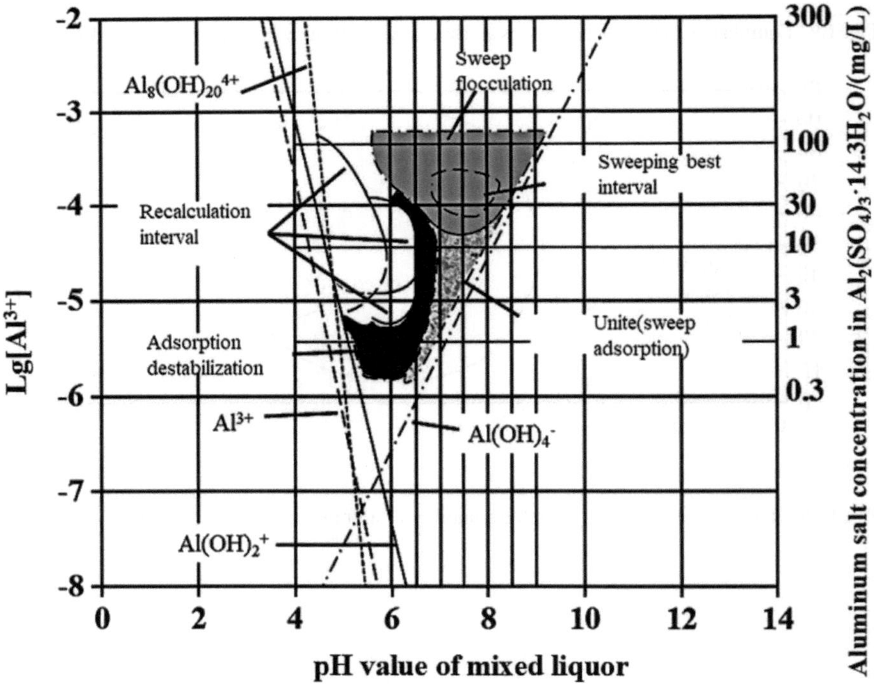

Fig. 1.11 Flocculation mechanism of aluminum sulfate and its relationship with Zeta potential

As can be seen from Fig. 1.11, when aluminum salts are used as coagulants, the pH value of the solution plays a very important role in its coagulation process. When the pH value is between 4 and 5, aluminum salts exist in the form of cations, such as Al^{3+}, $Al(OH)^{2+}$, $Al_8(OH)_{20}{}^{4+}$. When the pH value is between 6 and 8, aluminum appears in the form of anions, such as $Al(OH)_4{}^-$, and the best sweeping coagulation zone will appear. The range of the best sweeping coagulation zone is that the dosage of alum is 20–50 mg/L, and the pH value is 6.8–8.2. Within this range, the best coagulation sedimentation effect can be produced with the lowest dosage of coagulant. The above analysis is based on the hydrolysis characteristics of aluminum sulfate solution to obtain the hydrolysis region diagram. In fact, the hydrolysis process of Al^{3+} is largely affected by the anionic coagulant aid. The most obvious contrast is the difference in coagulation effect between polyluminum chloride and aluminum sulfate. With the same dosage of aluminum content, the former's coagulation effect is far better than the latter. Generally, for the same water, the pure aluminum dosage of poly aluminum chloride (PAC) can achieve the same coagulation effect when it is 1/3–1/2 of the dosage of aluminum sulfate. This huge difference is due to the presence of anions, but currently there is less research on the coagulation mechanism of anions.

Adding water-soluble chitosan, polyamine, etc., to aluminum salts or iron salts can also affect the hydrolysis diagram of aluminum salts. Furthermore, Fig. 1.11 cannot reflect the kinetic process of coagulation, and different coagulants have significant differences in coagulation areas and kinetics.

Figure 1.12 shows the coagulation process of polyaluminum chloride compounded with polyamines and other polymers (820 A–C). From Fig. 1.12, it can be seen that after adding polymers, the hydrolysis coagulation speed and floc average particle size are much higher than ordinary polyaluminum chloride, especially for low-temperature and low-turbidity water, the time for polyaluminum chloride with added polymers to form alum flowers can be 2–5 min earlier than regular polyaluminum chloride.

The wider pH value adaptation range makes iron salt an ideal coagulant. However, the choice of coagulant should not only consider the simple pH value range, but also consider the more complex cost-effectiveness and residual concentration. The influencing factors are as follows:

(1) Comparison of cost-effectiveness between aluminum salts and iron salts: ① Polyaluminum chloride and aluminum sulfate in aluminum salts are cheap and readily available; ② Aluminum salts have good treatment effects, and the final turbidity is low, which is difficult for iron salts to achieve; Due to the low viscosity and turbidity of winter water, the density of iron salts is somewhat larger, and the compounding effect of iron and aluminum salts is better; ③ Compared with aluminum salts, iron salts have a relatively higher removal rate for COD; ④ Iron salt coagulants perform poorly at low pH values; ⑤ When the raw water pH value is relatively high, it will increase the residue of aluminum silicate.

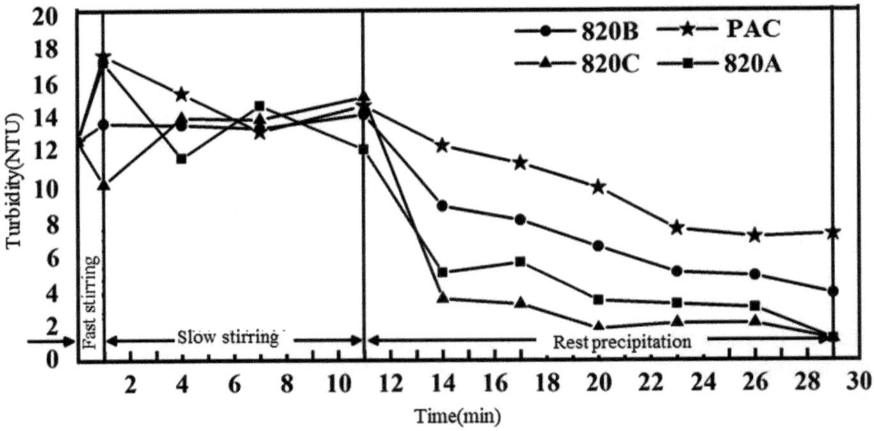

Fig. 1.12 The coagulation process of polyaluminum chloride compounded with polyamines and other polymers (820 A–C)

(2) Comparison of residues between aluminum salts and iron salts: Compared with aluminum salts, the residues of iron salts in the water body are more beneficial to human health.

Therefore, aluminum salts have been widely used in drinking water treatment, and more than 95% of enterprises in our country use polyaluminum chloride for drinking water treatment, with a small number of enterprises using aluminum sulfate.

1.4 Coagulation Kinetics

1.4.1 Coagulation Kinetics

1.4.1.1 Purpose of Coagulation

The purpose of coagulation is to neutralize the surface charge of particles to destabilize them, aggregate them to form large flocs, and then settle out. The initial coagulation flocs are generally insoluble particles of molecular size, and the flocculation process is to aggregate the neutralized colloidal substances and these insoluble substances to form large particle flocs, which can be removed by sedimentation and filtration methods. The flocs formed during the coagulation process also have a trapping effect on small particles in the environment and can separate them from the water body through co-precipitation. Therefore, the coagulation process can very effectively remove small particles in turbid water samples, including pathogens, bacteria, etc.

The movement of suspended colloidal particles is caused by Brownian motion, fluid flow, and external forces. The movement of colloidal particles causes a difference in relative speed between colloidal particles, resulting in collisions between colloidal particles. In a stable suspension system, the electrostatic repulsion between colloidal particles prevents collisions between particles. The aggregation of colloidal particles can be achieved by adding coagulants or flocculants. Under this condition, a connection will form between colliding particles, thereby achieving the purpose of particle aggregation.

Flocculation includes two stages: coagulation and flocculation. Coagulation refers to the destabilization stage of the colloid, while flocculation refers to the stage where the destabilized colloid forms large flocculent particles. The flocculation caused by Brownian motion is called peri kinetic flocculation, and the flocculation caused by hydraulic action is called orthokinetic flocculation. In addition, between two particles of different sizes, the larger particle can catch up with the smaller particle during its faster descent, resulting in a collision that causes flocculation, which is called differential settling flocculation.

1.4.1.2 Peri Kinetic Flocculation

Particles with a diameter less than 1 μm cannot grow into noticeable large flocs. At this time, peri kinetic flocculation occurs, which will produce a kind of fine floc.

The rate at which the particle concentration of a single dispersed phase decreases due to Brownian motion collisions can be expressed as:

$$-\frac{dn}{dt} = k_p n^2 \tag{1.30}$$

At the beginning of the twentieth century, Smoluchowski derived $k_p = a_B 8 D_b \pi a$, where $D_b = \frac{kT}{6\pi\mu a}$ (Einstein-Stokes formula).

Where, k_p is effective collision rate constant;

D_b is diffusion coefficient, m^2/s;

μ is viscosity of water, N s/m^2;

a is the radius of the sphere, m.

Substituting the above two formulas into formula (1.30) gives:

$$-\frac{dn}{dt} = \frac{4a_B kT}{3\mu} n^2 \tag{1.31}$$

where, n is particle concentration, particles/m^3;

a_B is inter-particle adhesion efficiency factor.

Integration gives the following formula:

$$\frac{1}{n} - \frac{1}{n_0} = \frac{4k\alpha_B T}{3\mu} t \quad (1.32)$$

Substituting $n = \frac{n_0}{2}$ into the above formula gives the half-life:

$$t_{\frac{1}{2}} = \frac{3\mu}{4a_B k T n_0} \quad (1.33)$$

The above formula shows that: ① The higher the temperature, the shorter the sedimentation time, and the better the flocculation effect, because peri kinetic flocculation is caused by Brownian motion, and temperature is closely related to Brownian motion. ② Assuming $a_B = 1$, it means that the particles stick together after collision, and substituting T = 293 K, $K = 1.38 \times 10^{-23}$ J/K and $\mu = 1 \times 10^{-3}$ N S/m^2 into the above formula gives:

$$t_{\frac{1}{2}} = \frac{2 \times 10^{11}}{N(0)} \quad (1.34)$$

Assuming $N(0) = 10^6$ per/cm^3, the calculated $t_{1/2} = 2 \times 10^5$ s, which is about 2.3d. Therefore, the flocculation effect produced by Brownian motion is limited.

1.4.1.3 Orthokinetic Flocculation

After the small particles coagulate through Brownian motion, the particle size gradually increases. But soon, Brownian motion will stop, and the chance of collision will drop significantly. Therefore, at this stage, orthokinetic flocculation, which occurs after the colloidal particles collide due to hydraulic or external stirring, plays a major role. There is still no unified understanding of the theory of orthokinetic flocculation. The initial theoretical formula was derived under laminar flow conditions, which obviously does not match the actual turbulent flocculation process. In recent years, many experts and scholars have explored the orthokinetic flocculation process directly from the theory of turbulence. Orthokinetic flocculation produces large and easily separable flocs. During the sedimentation process, the flocs are easily captured in the sand filtration process. Net capture and sweep flocculation also occur during the orthokinetic flocculation process.

The following briefly introduces the situations of laminar and turbulent flow respectively.

(1) Co-directional flocculation under laminar conditions

The particle collision diagram under laminar flow condition is shown in Fig. 1.13.

In Fig. 1.13a represents two adjacent particles, with radii r_i and r_j, which are moving in the same direction at a certain moment due to stirring. Perpendicular to the direction of motion, the distance between the two particles is Δz, and the

Fig. 1.13 Schematic diagram of particle collision under laminar conditions

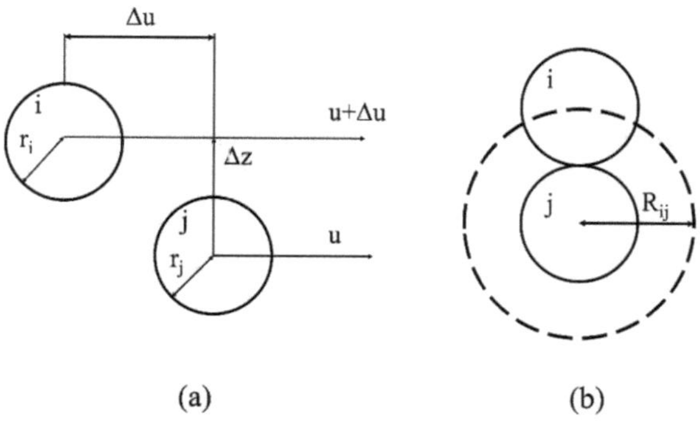

velocities are $u + \Delta u$ and u. That is, there is a velocity gradient $\frac{\mathrm{du}}{\mathrm{dz}} = \frac{\Delta u}{\Delta z}$ between the two particles.

In Fig. 1.13b represents a cross-section of a cylinder with a sphere of r_j as the center and R_{ij} as the radius, which is a cross-sectional view in the direction of motion. All particles within the range of radius R_{ij} will collide. Assuming that r_j is stationary, the flow rate q passing through this cylindrical cross-sectional area can be calculated as:

$$q = 2\int_0^{R_{rj}} x\frac{\mathrm{d}u}{\mathrm{d}z}\left[2(R_{ij}^2 - x^2)^{1/2}\right]\mathrm{d}x = \frac{4}{3}(R_{rj})^3\frac{\mathrm{d}u}{\mathrm{d}z} \tag{1.35}$$

Assume the concentrations of particles r_j and r_i in water are n_j particles/cm^3 and n_i particles/cm^3, and considering that the two particles are the same type, the number of collisions J between the two particles should be:

$$J = \frac{4}{3}n^2(R_{ij})^3\frac{\mathrm{d}u}{\mathrm{d}z} \tag{1.36}$$

Assuming that a portion of the particles will permanently stick together after collision, with a_0 representing the fraction of these sticking times in the total number of collisions, the rate of decrease in particle number due to collision dn/dt is:

$$-\frac{dn}{dt} = J = a_0\frac{4}{3}n^2(R_{ij})^3\frac{du}{dz} \tag{1.37}$$

Let φ be the total volume of n particles with a diameter of d at t = 0, then the above equation can be rewritten as:

$$-\frac{\mathrm{d}n}{\mathrm{d}t} = a_0\frac{8}{\pi}\phi\frac{\mathrm{d}u}{\mathrm{d}z}n \tag{1.38}$$

Integrating, we get:

$$n = n_0 \exp\left(-a_0 \frac{8}{\pi} \phi \frac{du}{dz} t\right) \tag{1.39}$$

where, n_0 is the number of particles at t = 0.

Substituting $n = n_0/2$ into the above equation, we can find its half-life:

$$t_{1/2} = \frac{0.693}{a_0 \frac{8}{\pi} \phi \frac{du}{dz}} \tag{1.40}$$

As can be seen from the half-life formula:

① Increasing the velocity gradient can shorten the settling time, but too large a velocity gradient will cause the floc to break, so in actual applications, the value of the velocity gradient is also limited.

② Compared with the same number of large particles and small particles, the order of magnitude difference in their settling half-life is $(d_{large}/d_{small})^3$, which can be derived from the factor ϕ.

In reality, the water flow is not laminar, but always in a turbulent state. For this reason, Camp and Stein calculated the G value by the power consumed by a unit volume of water flow that is instantaneously sheared and twisted to replace $G = \frac{\Delta u}{\Delta z}$, the derivation process is as follows:

In the stirred water flow, consider an isolated body "Δx·Δy·Δz" that is instantaneously sheared and twisted, the calculation of the velocity gradient is shown in Fig. 1.14.

In the process of shearing and twisting of the isolated body, the shear force does work. Assuming that in time Δt, the isolated body has twisted by an angle θ, so the angular velocity $\Delta\omega$ is:

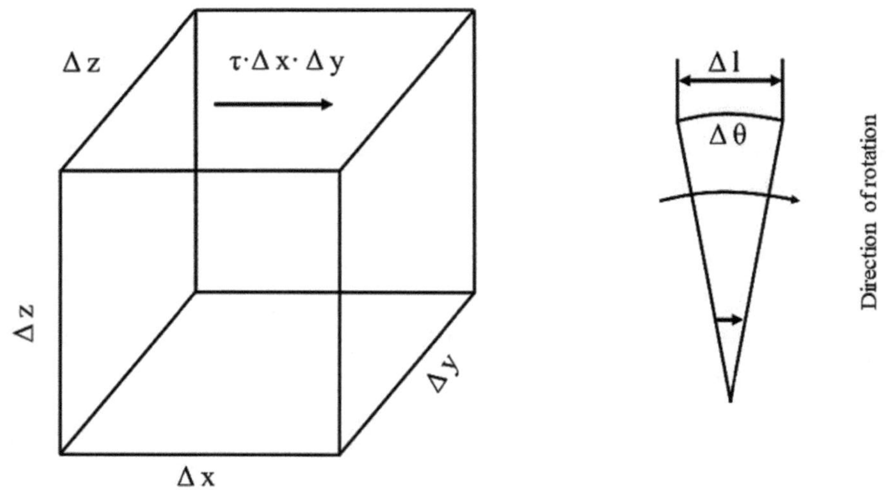

Fig. 1.14 Velocity gradient calculation diagram

$$\Delta\omega = \frac{\Delta\theta}{\Delta t} = \frac{\Delta l}{\Delta t} \times \frac{1}{\Delta z} = \frac{\Delta u}{\Delta z} = G \tag{1.41}$$

where, Δu is the twisting linear velocity, m/s,

G is the velocity gradient, s^{-1},

l is the length of side, m.

The torque *ΔM* is:

$$\Delta M = (\tau \Delta x \Delta y)\Delta z \tag{1.42}$$

where τ is the shear stress,

$\tau\Delta x\Delta y$ is the shear force acting on the isolator,

The power consumed by the isolator's torsion equals the product of torque and angular velocity, so the power "*p*" consumed per unit volume of water flow is:

$$p = \frac{\Delta M \cdot \Delta\omega}{\Delta x \cdot \Delta y \cdot \Delta z} = \frac{G \cdot \tau \cdot \Delta x \cdot \Delta y \cdot \Delta z}{\Delta x \cdot \Delta y \cdot \Delta z} = \tau G \tag{1.43}$$

According to Newton's law of internal friction, $\tau = \mu G$, substituting into Eq. (1.42):

$$G = \sqrt{\frac{p}{\mu}} \tag{1.44}$$

where μ is the dynamic viscosity of water, Pa s;

p is the power consumed per unit volume of fluid, W/m^3;

G is the velocity gradient, s^{-1}.

The *p* in the above equation depends on the actual situation. When mechanical stirring is used, it is provided by the mechanical stirrer. When a hydraulic flocculation tank is used, the *p* in the equation should be the energy consumption of the water flow itself: $pV = \rho gQh$, and $V = QT$. Substituting both into Eq. (1.43):

$$G = \sqrt{\frac{gh}{\nu \cdot T}} \tag{1.45}$$

where *h* is the water head loss in the coagulation equipment, m;

v is the kinematic viscosity of water, m^2/s;

T is the residence time of the water flow in the coagulation equipment, s.

(2) Co-directional flocculation under turbulent conditions

Although Camp and Stein replaced the G value with the power consumed by a unit volume of water flow that is instantaneously sheared and twisted, they still did not avoid the concept of laminar flow, that is, they did not clarify the particle collision rule from the law of turbulence. It should be pointed out that compared with co-directional and counter-directional flocculation, there is still very little research on flocculation under turbulent conditions.

According to Kolmogoroff's local isotropic turbulence theory, turbulence can be seen as composed of many vortices of different sizes. The energy applied externally causes large vortices, which transfer energy to medium-scale vortices, and finally, the energy is transferred to smaller vortices. Through this multi-level transfer process, the energy is finally converted into heat. He believes that in a fully developed turbulence, the anisotropy of the mainstream gradually disappears with the multi-level transfer process, so the statistical properties of smaller scale vortices can be determined without relying on the directional information of the mainstream. On this basis, Kolmogoroff introduced an important parameter—the micro-vortex scale "λ" to describe the size of an energy-dissipating vortex.

$$\lambda = \nu^{3/4}\varepsilon^{-1/4}. \quad (1.46)$$

where λ is the microvortex scale, m;

ε is the energy consumption rate per unit mass of fluid, W/m^3;

ν is the kinematic viscosity of the fluid, m^2/s.

If the scale of a certain area is smaller than λ, then the turbulence can be regarded as isotropic, and the size of λ is about 30–100 μm. When the micro-vortex scale is close to the alum particle scale, the flocculation reaction is most sufficient. Large vortices only carry colloidal particles and do not cause them to collide with each other, and scales that are too small are not enough to promote particle collisions. Therefore, only when the micro-vortex scale is close to or equal to the diameter of the colloidal particles, will it cause collisions between particles. Therefore, to improve the flocculation effect, it is necessary to effectively increase the proportion of the micro-vortex scale in turbulence.

The collision rate N_0 of particles under isotropic turbulence conditions can be expressed as:

$$N_0 = 8\pi dDn^2 \quad (1.47)$$

where, D is the sum of the turbulent diffusion and Brownian diffusion coefficients.

However, in turbulence, Brownian diffusion is much smaller than turbulent diffusion, so D can be approximated as the diffusion coefficient. The turbulent diffusion coefficient can be expressed as: "$D = \lambda\mu_\lambda$". Where λ is the vortex scale, and "μ_λ" corresponding to the "λ" scale. The fluctuating velocity can be expressed as:

$$\mu_\lambda = \frac{1}{\sqrt{15}}\sqrt{\frac{\varepsilon}{\nu}} \cdot \lambda \quad (1.48)$$

Assuming the vortex scale is equal to the particle diameter, i.e., "$\lambda = d$", substituting the above two formulas:

$$N_0 = \frac{8\pi}{\sqrt{15}}\sqrt{\frac{\varepsilon}{\nu}} \cdot d^3 \cdot n^2 \cdot a_0 \tag{1.49}$$

Although the above formula is derived under turbulent conditions and is theoretically more reasonable, there are many problems such as: ① In the actual process, the vortex scale is randomly changing, and the size of particles in the water varies and continues to increase during the coagulation process, which makes the premise of the above formula questionable. ② Equation (1.48) is only applicable to small vortices in the viscous region (all small vortex groups affected by the viscosity of water). These all greatly limit the application of the formula.

1.4.1.4 Differential Setting Flocculation

Analyzing with Fig. 1.13, assuming that $r_i > r_j$, their Stokes settling velocities are v_i and v_j, and the particle concentrations are n_i and n_j. Taking particle r_j as a reference, the particle flow rate J_i in the cylinder with a radius of R_{ij} can be expressed as:

$$J_i = \pi R^2 ij \left(v_i - v_j\right) n_i \tag{1.50}$$

Since there are n_j particles r_j, the total particle flow rate is:

$$J = \pi R_{ij}^2 (v_i - v_j) n_i n_j \tag{1.51}$$

Substituting the Stokes formula $v_p = \frac{1}{18}\frac{\rho_p - \rho_1}{\mu} g d_p^2$ into the above formula the rate of particle reduction due to differential sedimentation collision coagulation is:

$$-\frac{dn}{dt} = \frac{2\pi g}{9\mu}(\rho_p - \rho_1)(r_i + r_j)^3 (r_i - r_j) a_s n_i n_j. \tag{1.52}$$

where ρ_p is the particle density, g/cm^3;

ρ_1 is the water density, g/cm^3;

a_s is the adhesion efficiency factor;

d_p is the particle diameter.

1.4.2 *Coagulation Parameters*

(1) **Velocity Gradient**

In the early stage of the coagulation process, an appropriate velocity gradient is needed to make the particles collide with each other to form flocs. During the mixing process, the velocity gradient is generally maintained at 20–200 s^{-1}. Too high or too low velocity gradients are not conducive to the formation of flocs. If the turbidity load of the treated water is relatively high, the gradient can be kept higher; if the water turbidity is very low, the velocity gradient can be kept lower to avoid the flocs breaking under shear. In practical applications, the method of gradually reducing the gradient can be used, such as adopting a 2-stage or 3-stage flocculation method, gradually reducing the gradient, and maintaining the continuous growth of the flocs.

(2) **Detention Time**

From Fig. 1.15, which shows the relationship between floc particle size and time, it can be seen that it takes about 4 min for the floc to grow to 1 mm, and the floc reaches its maximum size at about 8 min. Increasing the reaction time often cannot further improve the size and density of the floc, and the general reaction time is controlled at 20 min. For low-temperature and low-turbidity water, a longer reaction time is often required, but for fixed facilities, the reaction efficiency can be improved by improving the performance of the coagulant.

(3) **G-T Value (Camp Number)**

In engineering design, we often use the G-T value or Camp number. In the 1930s, Professor Thomas R. Camp of MIT proposed this constant after summarizing data from more than 40 water treatment plants.

$$\text{G-T Value} = G \times t \tag{1.53}$$

where G is the velocity gradient, s^{-1};

t is time, s, so G-T value is dimensionless.

For example, in a good coagulation process:

Mixing tank: G = 150/s, t = 8 min = 480 s,

G-T Value = 150 × 480 = 72,000.

Flocculation tank:

Flocculation tank 1: G = 100/s, t = 10 min,

Flocculation tank 2: G = 30/s, time = 10 min,

Then, G-T Value = 100 × 600 + 30 × 600 = 78,000.

The general G-T value is within the range of 25,000–100,000.

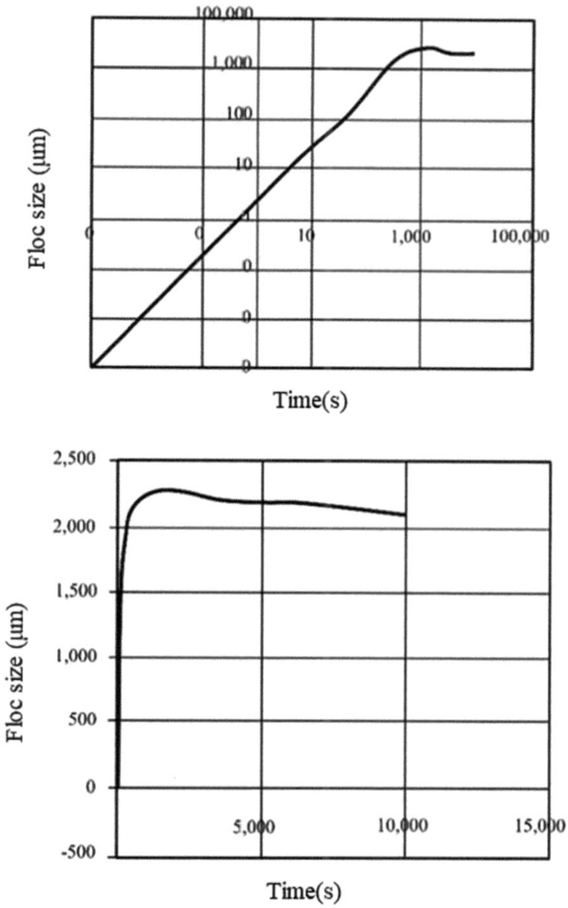

Fig. 1.15 Relationship between floc particle size and time

(4) **Flocculator Types**

In practical applications, there are two types of flocculators, namely hydraulic mixing and mechanical mixing.

Hydraulic flocculators, also known as mixing columns, have horizontal baffles installed on the sides. The water body continuously mixes during the descent and ascent process. The baffles are closer together at the inlet and farther apart at the outlet, ensuring a gradually decreasing mixing gradient (tapered mixing), promoting continuous floc growth and improving mixing efficiency.

(5) **Calculation Example**

Water plant inflow rate: 120 m^3/h.

Mixing well: 2 sets, diameter 1220 mm, height difference between inlet and outlet is 3.7 m.

Head loss after passing through the mixing well: 100 mm.

Time: 260 s.

Water flow rate = 120 m^3/h = 2 m^3/min.

Mixing well volume = 2 × (1.22 × 1.22 × π/4) × 3.7 = 8.65 (m^3).

Hydraulic detention time = 8.65 m^3/2 m^3/min = 4.33 min = 260 s.

Energy: 2 m^3/min = 2000 kg/min ≈ 33.33 kg/s.

The above water volume, under the action of gravity (9.81 m/s^2), drops 0.1 m to generate energy,

$$P = 33.33 \times 9.81 \times 0.1 = 32.7\,\mathrm{W}\ (\mathrm{kg/s} \times \mathrm{m^2/s} \times \mathrm{m} = \mathrm{N} \times \mathrm{m/s} = \mathrm{Watts})$$

Velocity gradient:

$$\begin{aligned} G = [\mathrm{Power}/(\mu \times V)]^{1/2} &= [32.7/(0.0013 \times 8.65)]^{1/2} \\ &= [2907.7]^{1/2} = 53.9/\mathrm{s} \\ G - T = G \times t &= 53.9 \times 260 = 14{,}020 \end{aligned}$$

The above calculation shows that the water body's residence time is short, the velocity gradient is small, and therefore the G-T value is very low, unable to form good flocs. It is suggested to increase the volume of the mixing well, shorten the spacing of the baffles, enhance the mixing strength, increase the head loss of the mixed water, that is, increase the energy input. If the head loss is doubled, then G is 76.3/s, and the G-T value reaches 20,000, which is still very low.

(6) **The effect of temperature on coagulation**

Temperature has a significant effect on the viscosity of the treatment, which means that to achieve the same mixing effect at low temperatures as at high temperatures, more energy must be input. Similarly, due to the increase in viscosity, the settling speed of the flocs and the head loss of the filtered water will increase at low temperatures. Table 1.12 lists the viscosity of water at different temperatures,

① When the particles are very small, peri kinetic flocculation plays a role, and the smaller the particle size, the more obvious this effect.

Table 1.12 The effect of temperature on coagulation

Temperature/°C	Viscosity/(mPa s)
0	1.79
20	1.00
40	0.65
100	0.28

② When the particles increase to a certain extent, Brownian motion is not enough to drive the particles to collide. At this time, the orthokinetic flocculation caused by hydraulic or external forces plays a leading role, and this stage also has differential settling flocculation at work, the larger the particle size, the more obvious these two effects.

③ In the actual process, most of them are the orthokinetic flocculation under turbulent conditions, but the current research in this area is far from enough. Although Kolmogoroff's micro-vortex theory appeared as early as the 1960s, it did not deeply explain the cause of coagulation kinetics, nor did it establish a relationship with the velocity gradient theory. It was not until the past decade that some researchers have done some research based on it. The orthokinetic flocculation under turbulent conditions still needs further conclusion.

1.5 Factors Affecting Coagulation

Cationic polymers can react with anions in the water body to form chemical bonds, thereby blocking the active groups on the polymer from adsorbing colloids in the water body. Therefore, for water bodies containing anions, to achieve the same coagulation effect, the amount of cationic polymer consumed will be higher, and this situation is more serious for wastewater treatment. For example, chloride ions, phenols, hydrogen sulfide, etc. in the water body will reduce the effect of cationic polymers, and the impact of pH value is also very significant, because each polymer can only perform well within a certain pH range. The effect of phenol on cationic polymers and the effect of pH value on cationic polymers can be shown in Figs. 1.16 and 1.17. In the treatment of industrial circulating cooling water turbidity, a large amount of anionic corrosion inhibitors, such as organic phosphates, acrylic polymers, etc., are often added, and their presence increases the use of coagulants.

The Zeta potential can also be used to evaluate the sensitivity of cationic polymers to changes in pH. Some polymers are very sensitive to changes in the pH of the solution. As we discussed earlier, pH affects both the surface properties of colloids and the properties of polymers.

1.6 Methods of Measuring Zeta Potential

According to the previous analysis, it has been understood that colloid carries charge in water, and particles move in the electric field due to the action of electric field. The application of electric field between particles in solution is shown in Fig. 1.18.

If an electric field is applied to a solution containing colloidal particles, the speed of the particles in the electric field is related to the strength of the electric field. Generally, the instruments on the market that measure this movement speed and its correlation with the electric field strength can be purchased, which is the Zeta

Fig. 1.16 The effect of phenol on cationic polymers

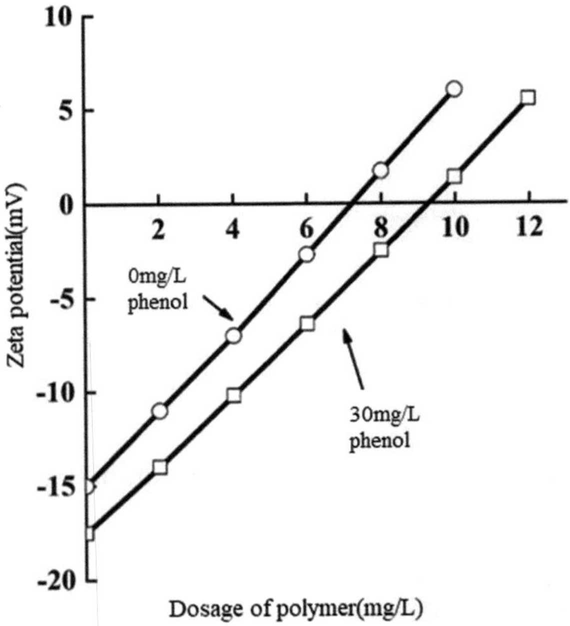

Fig. 1.17 The effect of pH value on cationic polymers

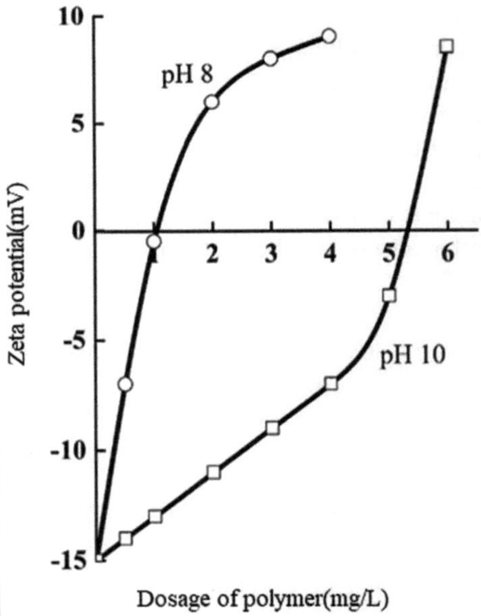

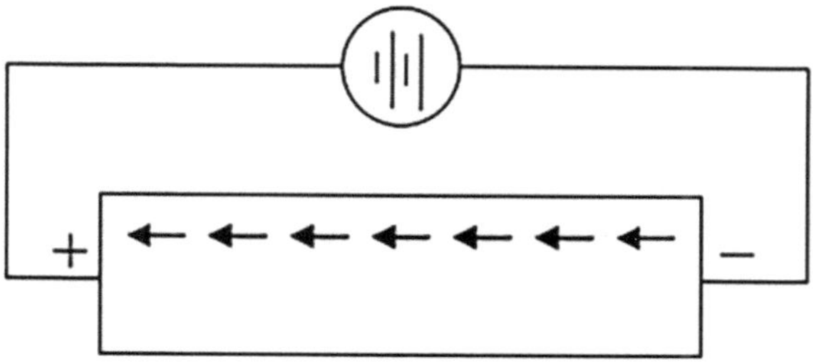

Fig. 1.18 Application of electric field between solution particles

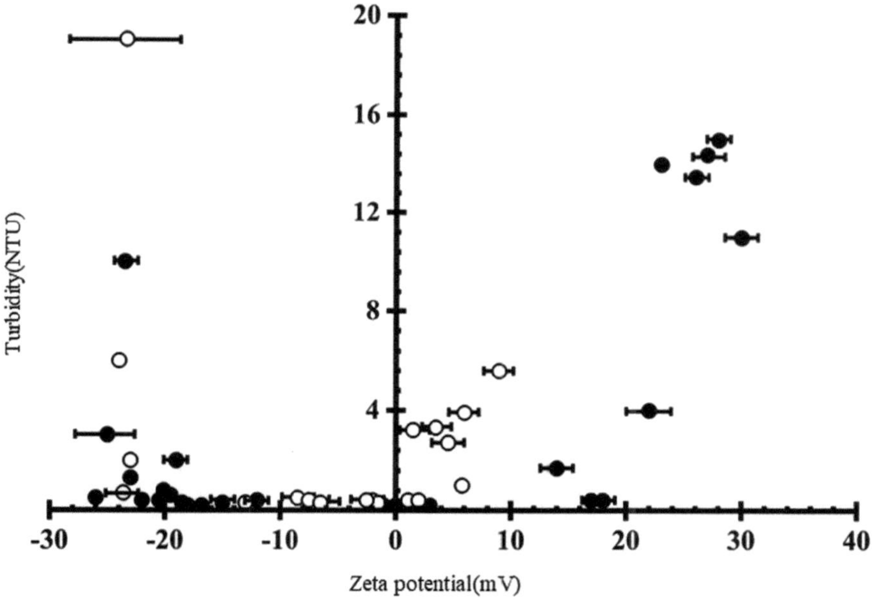

Fig. 1.19 Relationship between turbidity after coagulation and sedimentation of natural organic matter and Zeta potential (hollow circles represent the use of low charge density coagulants, solid circles represent the use of high charge density coagulants)

potential meter. The movement speed of the particles is observed and measured through a microscope. The observed speed V_E has the following relationship with the Zeta potential.

$$V_E = \frac{\varepsilon E \varsigma}{4\pi \mu} \tag{1.54}$$

where ε is the dielectric constant of the fluid, F/m;

E is the external electric field strength, V/m;

Table 1.13 Critical coagulation concentration of hydrophobic sols (Unit: $mmol/dm^3$)

As_2S_3		AgI		Al_2O_3	
LiCl	58	$LiNO_3$	165	NaCl	43.5
NaCl	51	$NaNO_3$	140	KCl	46
KCl	49.5	KNO_3	136	KNO_3	60
KNO_3	50	$RbNO_3$	126	K_2SO_4	0.3
CH_3COOK	110	$Ca(NO_3)_2$	2.40	$K_2Cr_2O_7$	0.63
$CaCl_2$	0.65	$Mg(NO_3)_2$	2.60	$K_2C_2O_4$	0.69
$MgCl_2$	0.72	$Pb(NO_3)_2$	2.43	$K_3[Fe(CN)_6]_6$	0.08
$MgSO_4$	0.81	$Al(NO_3)_3$	0.067		
$AlCl_3$	0.093	$La(NO_3)_3$	0.069		
$1/2Al_2(SO_4)_3$	0.096	$Ce(NO_3)_3$	0. 69		
$Al(NO_3)_3$	0.095				

ς is Zeta potential, mV;

μ is the viscosity of the medium, Pa s.

When electrolytes are added to the colloid solution, we can observe the trend of Zeta potential changes. The general dosage of coagulant is from 10 to 400 mg/L, and aluminum sulfate, polyaluminum chloride, etc. are typical coagulants. In the water treatment process, the Zeta potential is expected to zero in order to reduce the electrostatic repulsion between particles as much as possible.

Measurement of streaming potential: When forced through a capillary or porous plug, the electrolyte carries away the residual charge on the particle periphery or particle wall due to the pressure gradient, and the potential difference measured at both ends of the capillary is called the streaming potential. In the experiment, there is generally a critical concentration that can cause colloid coagulation, the minimum concentration C_f required for salt substances to cause coagulation. The critical coagulation concentration of hydrophobic sols is shown in Table 1.13.

1.6.1 Zeta Potential and Precipitation

The Zeta potential can affect the size and density of the flocs formed. An increase in density can cause particles to settle more quickly. A low Zeta potential reduces the electrostatic interaction between particles, causing them to come closer together and form denser flocs. When the Zeta potential is between + 3 mV and – 22 mV, the turbidity of the effluent is low and stable. When the Zeta potential is below – 22 mV, the repulsion effect between suspended particles is large, causing the particles to prefer to stabilize in the water, leading to a rapid increase in the turbidity of the effluent. By changing the coagulant, the range of Zeta potentials at which coagulation occurs can be expanded. The relationship between turbidity after coagulation and sedimentation of natural organic matter and Zeta potential is shown in Fig. 1.19. The higher the charge density of the coagulant, the wider the range of Zeta potentials that can be coagulated.

A low absolute value of Zeta does not guarantee good coagulation results. Since the Zeta potential corresponding to coagulation is within a region, good coagulation kinetics must be present after colloidal destabilization to ensure the growth and settling of flocs. For example, if the mixing conditions during coagulation are too intense, although flocs have formed, the particles are too small and contain a high amount of water, still not achieving good removal effects. Also, under the same raw water and kinetic conditions, when different coagulants are used the flocs formed, low absolute value of Zeta potential does not necessarily have good coagulation effect. Because if the coagulant contains a high polymer flocculant, although the ability of polymer flocculant to reduce Zeta is not as good as that of inorganic coagulant, the bridging effect of polymer is far more than the effect of reducing Zeta. That is, the flocs of Zeta potential – 10 mv are more aggregated and settled than the flocs of Zeta potential – 8 mv. The most typical case is the effect of polyaluminum chloride mixed with high polymer polymers on the treatment of low temperature and low turbidity water in the lower reaches of the Yangtze River.

1.6.2 Zeta Potential and Flotation Process

The flotation process is closely related to the density of bubbles and flocs. The Zeta potential during the flotation process affects the adhesion performance of bubbles and particles. Generally, measuring the Zeta potential of the solid phase is enough, but experiments have shown that the flotation process is affected by the Zeta potential of both bubbles and particles. Figure 1.20 shows the relationship between turbidity removal rate and Zeta potential when coagulating and floating high turbidity raw water.

Experimental data show that the lower the product of the Zeta potentials of bubbles and particles, the higher the turbidity removal rate. When the product of the Zeta potential of the effluent increases, the turbidity removal rate decreases. This indicates that whether the charge on the surface of particles or bubbles increases, the efficiency of coagulation or flotation will decrease. In the initial stage of coagulation flotation, because the particle size is very small and carries residual excess charge, the treatment effect at this moment is not good.

1.6.3 Improvement of Zeta Potential Analysis Methods

The effectiveness of modern experimental methods, the reliability and accuracy of measurement techniques have reconsolidated the application of Zeta potential

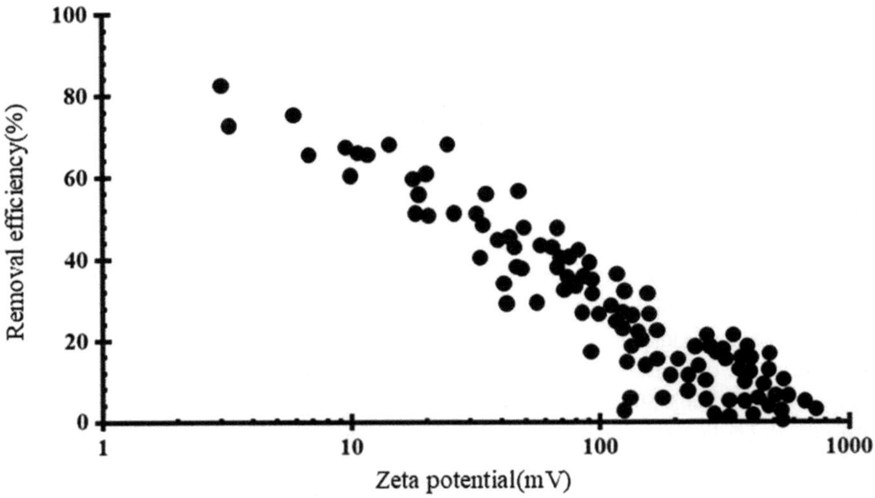

Fig. 1.20 Relationship between turbidity removal rate and Zeta potential when coagulating and floating high turbidity raw water

in judging and dealing with physical processes. The orderly use of Zeta potential parameters has become more feasible, especially with the improvement of physical properties in technology, making it possible to measure on-site.

1.6.4 Other Improvements in Water Quality Characteristics

Related fields of analysis have also made progress, which has given it a broader range of characteristics in terms of water. The current challenge is how to control the size of the Zeta potential operating system and how to operate it to improve its accuracy. This may require changing the chemical properties of the water, or even their physical actions themselves.

Chapter 2
Preparation and Quality Control of Inorganic Low Molecular Weight Coagulants

2.1 Aluminum Sulfate

2.1.1 Overview of Aluminum Sulfate

Since the invention of aluminum sulfate in the United States in 1884, it has a history of more than a hundred years. Different countries use different raw materials to produce aluminum sulfate. Japan uses all aluminum hydroxide as raw material, while the United States uses bauxite and clay (kaolin) as raw materials. In China, bauxite is the main raw material, followed by aluminum hydroxide, and some factories use clay and aluminum ash for production. Aluminum ash is the slag and scum produced in the process of smelting refined aluminum, and some non-ferrous metal processing enterprises in China produce about 4 million tons of aluminum ash each year.

Aluminum sulfate is divided into drinking water grade (Class I) and industrial water grade (Class II). The difference between Class I and Class II mainly lies in the different contents of heavy metals (see Table 2.1).

The heavy metal content of general drinking water products is about half of that of industrial water products. The appearance forms are divided into liquid and solid. Solid aluminum sulfate contains Al_2O_3 14 ~ 17.5% (in block and powder form), liquid aluminum sulfate contains Al_2O_3 6 ~ 8.5%, and they have products of various concentration specifications. The concentration of liquid aluminum sulfate in Japan is 8 ~ 8.2%, while in the United States it is 7.5 ~ 8.5%. When the concentration of Al_2O_3 is higher than 8.0%, crystallization often occurs as the temperature decreases.

Aluminum sulfate is widely used in papermaking, water treatment, leather making, oil clarification, petroleum decolorization, and the preparation of other aluminum salts. The raw materials used for the production of aluminum sulfate at home and abroad vary according to the grade of the product. Industrial grade aluminum sulfate mainly uses bauxite and clay mines, followed by aluminum hydroxide. The production process is generally the continuous leaching method of raw powder pressing.

L. Fengting, *Water Treatment Chemicals: Coagulants and Flocculants*,
https://doi.org/10.1007/978-981-96-8221-8_2

Table 2.1 National Standard in China "water treatment agent aluminum sulfate" (GB31060-2014)

Index item		Index			
		Class I		Class II	
		Solid	Liquid	Solid	Liquid
Mass fraction of aluminum oxide (Al_2O_3)/%	≥	15.60	7.80	15.60	6.50
Mass fraction of iron (Fe)/%	≤	0.20	0.05	1.00	0.50
Mass fraction of insoluble matter in water/%	≤	0.10	0.05	0.20	0.10
pH value (1% aqueous solution)	≥	3.0			
Mass fraction of arsenic (As)/%	≤	0.0002	0.0001	0.001	0.0005
Mass fraction of lead (Pb)/%	≤	0.0006	0.0003	0.005	0.002
Mass fraction of cadmium (Cd)/%	≤	0.0002	0.0001	0.003	0.001
Mass fraction of mercury (Hg)/%	≤	0.00002	0.00001	0.001	0.00005
Mass fraction of chromium (Cr)/%	≤	0.0005	0.0003	0.005	0.002

The distribution of aluminum sulfate production enterprises in China is mainly concentrated in the provinces of Shandong, Henan, Sichuan, Liaoning, Jiangsu, Hunan, and Yunnan. The liquid variety first appeared in Shanghai, where the paper-making and water purification industries are relatively developed. After the appearance of the liquid variety, it showed its great superiority: for the aluminum sulfate production enterprises, it eliminates the steps of solution evaporation concentration, condensation solidification, crushing packaging, etc., shortens the production process, saves a lot of energy consumption, improves labor productivity, and improves working conditions; especially in the transportation between users by railway, tank truck or ship, it is easy to realize mechanized operation, which is very convenient; for users, it can be directly pumped to the application process when used, and can be used after slight dilution according to the required concentration.

Aluminum sulfate is a coagulant widely used in our country from the 1950s to the 1970s. It can form flocs in a wide range of pH values. The Zeta potential of different concentrations of aluminum sulfate solution and the pH value of different concentrations of aluminum sulfate solution are shown in Figs. 2.1 and 2.2.

$$Al_2(SO_4)_3 + 6H_2O \rightleftharpoons 2Al(OH)_3 + 3H_2SO_4 \quad (2.1)$$

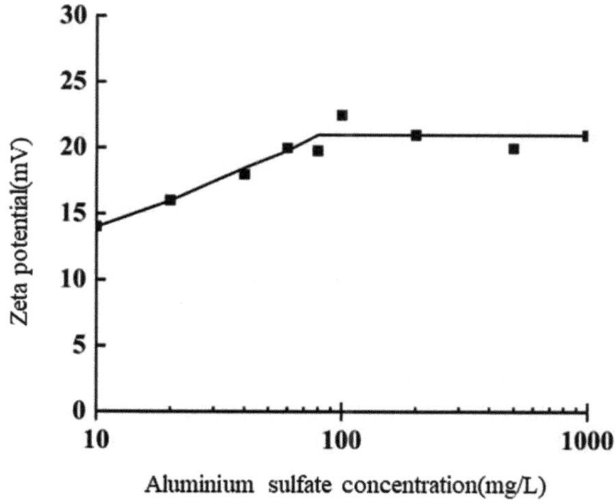

Fig. 2.1 Zeta potential of different concentrations of aluminum sulfate solution

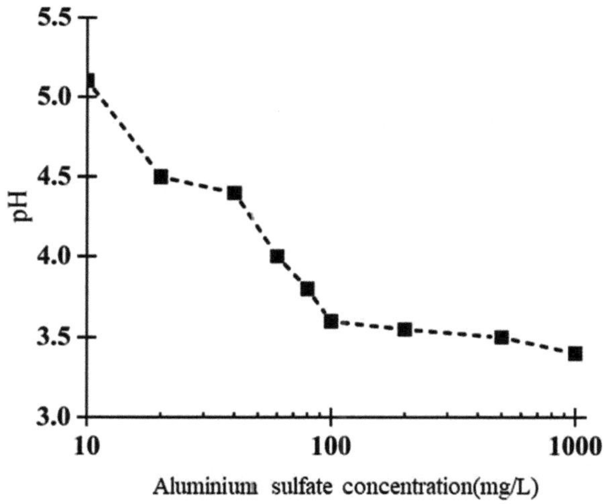

Fig. 2.2 pH value of different concentrations of aluminum sulfate solution

In the hydrolysis process of aluminum sulfate, one aluminum ion hydrolyzes to release three hydrogen ions, so it often consumes a large amount of alkalinity in the water body during use, causing the pH value to drop rapidly. Moreover, its water purification effect is generally 1/2 ~ 1/3 of the dosage of polyaluminum chloride (all are liquid commodities), so its market share is relatively small, but the use of aluminum sulfate in Europe, America and African countries is higher, which is also related to the structure of coagulant raw materials in our country. At present, the domestic use of aluminum sulfate is more in some water plants in Shanghai. They use a large amount of aluminum sulfate for coagulation in the treatment of Yangtze River water in summer, and use composite polyaluminum chloride and polyaluminum sulfate in winter.

2.1.2 Production Materials of Aluminum Sulfate

Although theoretically, there are many types of aluminum-containing raw materials available for the production of aluminum sulfate, in fact, the aluminum-containing raw materials with practical application value are only a few: alunite, alum stone, aluminum hydroxide, coal gangue, etc. It is understood that more than 90% of our country uses alunite (including hard and soft) as raw materials, and only a very small number of areas lacking alunite resources use coal gangue (containing Al_2O_3 generally less than 30%) as raw materials, such as the rich natural aluminum sulfate in the Naomao Lake area of Xinjiang Uygur Autonomous Region in China. The most ideal aluminum-containing raw material to produce aluminum sulfate is aluminum hydroxide, but in order to make rational use of resources and reduce production costs, it will not be used for general varieties except for the production of special aluminum sulfate.

Among the alunite, the most used in China at present is the hard alunite of one water and three water mixed birth. Some manufacturers in Shanxi Province and Henan Province in China are forced to use soft ore with low alumina content and difficult solid–liquid separation due to the lack of hard ore sources in the local area. As for other aluminum-containing raw materials, such as aluminum scraps, aluminum ash, etc., they can also be used to manufacture aluminum sulfate. The bauxite used for the production of aluminum sulfate in China is mainly the kaolin one-water soft alunite bauxite from Zibo of Shandong Province, followed by the low-iron one-water soft alunite from Sichuan Province, which is also one of the high-quality raw materials.

In China, coal gangue is also a rich resource for the production of aluminum sulfate. Coal gangue mainly contains kaolinite, with other minerals such as muscovite and quartz in smaller quantities.

2.1.3 Production Process and Equipment of Aluminum Sulfate

In terms of technological routes, 90% of the total production of aluminum sulfate in China is currently produced by the sulfuric acid-bauxite method. In the 1950s and 1960s, it was all produced by calcined bauxite-pressure reaction. Now, the raw powder pressure reaction is mostly used. The specific process is as follows: blocky bauxite is crushed to a fineness of 60 mesh or less after two-stage crushing and reacts with dilute sulfuric acid with a concentration of about 50% under a pressure of 0.3 MPa. After the reaction, the slurry is diluted, and the residue insoluble in sulfuric acid is repeatedly washed by gravity sedimentation and counter-current washing to increase its alumina recovery rate. The semi-finished product solution is evaporated and concentrated, and the concentrated solution that meets the alumina content requirements is condensed, solidified, crushed, and packaged. The process equipment of each factory is the same, only the specifications and materials of the key equipment are different.

2.1.4 *Preparation Methods of Aluminum Sulfate*

(1) **Production of Aluminum Sulfate by Bauxite Method**

Currently, the preparation of aluminum sulfate (AS) in China mainly uses two methods: the acidification of bauxite raw material method and the acidification of aluminum hydroxide raw material method. In the past, aluminum sulfate was often produced by directly reacting bauxite with sulfuric acid, obtaining aluminum sulfate solution through precipitation separation, and then further concentrating, cooling to obtain aluminum sulfate crystals. Because the ore contains more heavy metal elements and ferrous ions, the product produced is often yellow to light blue. This method is still used to produce iron-containing aluminum sulfate.

Bauxite can be used to produce aluminum sulfate by directly reacting with sulfuric acid in a reaction kettle, as follows:

$$Al_2O_3 + 3H_2SO_4 = Al_2(SO_4)_3 + 3H_2O \tag{2.2}$$

This process has the advantages of abundant raw materials and cheap prices, but the disadvantage is that bauxite contains more impurities such as iron elements, and these impurity ions are not easily separated. The common process of producing aluminum sulfate by the bauxite method is an intermittent reaction process, as shown in Fig. 2.3.

(2) **Production of Aluminum Sulfate by Coal Gangue Method**

The main useful mineral in coal gangue, kaolinite, accounts for about 86%, up to 95%, and the lowest is 82%. Other minerals with small amounts include: muscovite

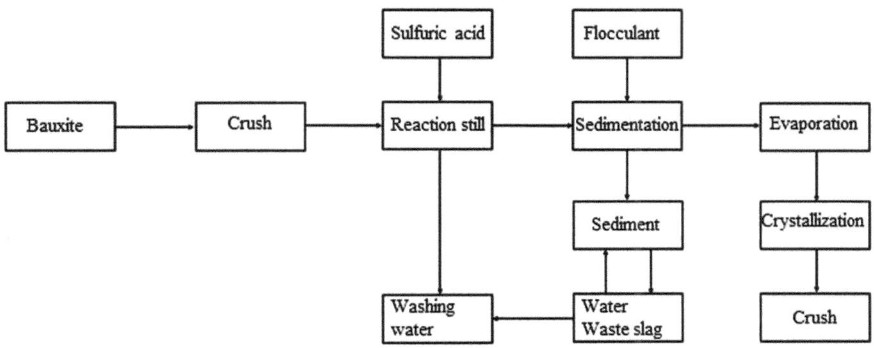

Fig. 2.3 Intermittent bauxite method aluminum sulfate production process

(1 ~ 2%), appearing as tiny flaky and sub-angular shapes interlayered with flaky kaolinite; quartz (3 ~ 5%), appearing as sub-angular and sub-rounded shapes. The chemical composition of aluminum sulfate coal gangue is shown in Table 2.2.

The production process of aluminum sulfate is mainly divided into three sections.

① Raw Material Section

This section mainly involves grinding the raw ore, preparing the slurry, and sending the slurry to the reaction kettle for use in the reaction section. That is, the raw ore (requiring Al_2O_3 content not less than 34.0%, Fe_2O_3 content not exceeding 2.0%, used for producing low-iron aluminum sulfate) is sent to the crusher to be crushed to a particle size less than 20 mm, then ground by a Raymond mill, with the grinding fineness required to be around 100 mesh. The ground coal gangue powder is mixed in a slurry tank, with the ore slurry concentration controlled at 50 ~ 55%, and the prepared ore slurry is sent to the reaction kettle. Alternatively, a wet grinding process can be used, which can reduce dust flying and significantly improve production efficiency. See Table 2.3 for a comparison of raw material consumption and energy consumption.

② Reaction Section

This section mainly involves reaction leaching, dilution, sedimentation separation, and sending the qualified aluminum sulfate solution to the high-level solution tank in the crystallization section, with the sediment entering the sedimentation tank. The reaction conditions are as follows: reaction temperature (140 ± 2) °C; pressure (0.3

Table 2.2 Chemical composition of aluminum sulfate coal gangue

Substance	Al_2O_3	TiO_2	Fe_2O_3	SiO_2
Raw material/%	33.68 ~ 36.73	0.83 ~ 1.32	1.10 ~ 1.85	47.12 ~ 48.13
Substance	MgO	Na_2O	K_2O	Loss on Ignition
Raw material/%	0.12 ~ 0.23	0.10 ~ 0.16	0.12 ~ 0.18	13.10 ~ 14.15

Table 2.3 Comparison of raw material consumption and energy consumption (per ton of solid product)

Material name	Main specifications	Unit	Dry Method	Wet method	Savings with wet method over dry method	Remarks
Aluminum ore	$Al_2O_3$38.7%	t/t	0.570	0.534	6.28	Comprehensive measurement
Sulfuric acid	98%	t/t	0.624	0.568	8.97	Comprehensive measurement
Electricity	–	kWh/t	49.56	26.54	46.45	Statistical data

± 0.10)MPa; time 4 ~ 5 h. If liquid products are being produced, the aluminum sulfate directly enters the finished product pool.

③ Concentration and Crystallization Section

This section mainly involves concentrating, crystallizing, and packaging the aluminum sulfate liquid sent from the reaction section. This section strictly controls the density of the concentrated liquid to reach 1.7 g/cm^3, and when the temperature is 110 °C, it enters the crystallization pool for crystallization. In addition, many manufacturers producing solid aluminum sulfate use drum dryers, steel belt dryers for crystallization processes, and spray drying processes.

(3) **Etching of aluminum foil products and comprehensive utilization of wastepaper containing aluminum foil**

In the electronics and mobile phone industries, it is often necessary to etch metal aluminum products. If it is pure etching without a coloring process, the aluminum sulfate obtained from the etched foil acid or hydrochloric acid and the precipitate from the subsequent neutralization have very high purity, which can meet the requirements for aluminum salts in drinking water products and even cosmetics.

$$2Al + 3H_2SO_4 \rightarrow A1_2(SO_4)_3 + 3H_2 \tag{2.3}$$

$$2Al + 6HCl \rightarrow 2AlCl_3 - 3H_2 \tag{2.4}$$

For etching acid with a relatively high concentration, an aluminum sulfate solution can be prepared. For hydrochloric acid etching solution, the reaction with aluminum foil can continue to produce aluminum chloride or high purity polyaluminum chloride (with an alumina content exceeding 20%, and a basicity greater than 80%). If the etching acid contains other impurities and cannot meet the requirements for producing high-purity polyaluminum chloride, it can further react with calcium aluminate to produce high-basicity polyaluminum chloride.

The comprehensive utilization of wastepaper containing aluminum foil is shown in Fig. 2.4. The wastepaper is crushed and washed, then heated in a high-temperature rotary kiln in two steps, respectively for drying (below 300 °C) and high-temperature decomposition (350 ~ 500 °C). The hot gas generated by the high-temperature furnace is dusted by a dust collector and then enters a steam boiler or heating furnace for heating. The solids containing aluminum produced by high-temperature decomposition are put into a reactor and mixed with sulfuric acid, hot water, and steam. The mixture is mechanically stirred by a stirrer. At the same time, the temperature is controlled at 80 ~ 100 °C. The reaction occurring in the reactor is as follows:

$$2Al + 3H_2SO_4 + 18H_2O \rightarrow A1_2(SO_4)_3 \cdot 18H_2O + 3H_2 \tag{2.5}$$

The aluminum sulfate solution (temperature 80 ~ 100 °C) is filtered by a filter press to separate a small amount of carbon-containing filter residue, and the liquid can be used to purify the solution. Activated carbon can be added to the filter press.

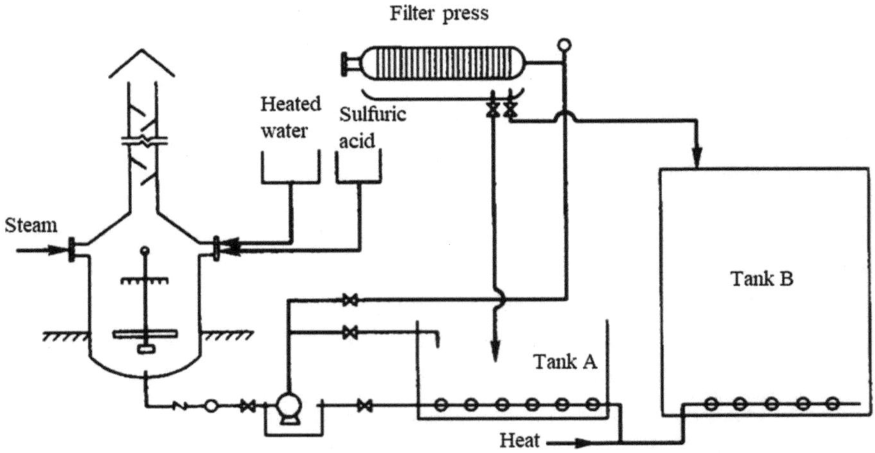

Fig. 2.4 Comprehensive utilization of wastepaper containing aluminum foil

The filtrate is added to storage tank A, which has a heating device or other heating evaporation device, to heat the filtrate to obtain a solution of suitable concentration (about 27%). Then the solution is added to storage tank B. If anhydrous aluminum sulfate is to be prepared, a heating device is required.

(4) **Production of aluminum sulfate by the aluminum hydroxide method**

Using aluminum hydroxide as a raw material and reacting with suitable concentration of sulfuric acid, the direct one-step method of producing iron-free aluminum sulfate crystals is widely used. The reaction chemical formula is as follows:

$$2Al(OH)_3 + 3H_2SO_4 = Al_2(SO_4)_3 + 6H_2O \tag{2.6}$$

The advantage of this method is that the raw material aluminum hydroxide is already iron-free when it leaves the factory, with low iron content and high product purity. Aluminum sulfate comes in two forms, solid and liquid. The Al_2O_3 content of solid aluminum sulfate is about 15 ~ 17%, and the liquid Al_2O_3 content is about 6 ~ 8%. The Al_2O_3 content in the raw material aluminum hydroxide is generally about 64 ~ 65%. Aluminum hydroxide reacts with a suitable concentration of sulfuric acid, releasing a large amount of heat. Through the self-heating process, aluminum hydroxide can completely dissolve, forming a high-temperature viscous aluminum sulfate liquid, which then flows on the moving stainless-steel belt, cools down to form a white, translucent sheet solid, and can be obtained by crushing it into white crystal products. The process of preparing acidified aluminum hydroxide raw materials is simple, the quality is easy to control and compared with the production process using bauxite as raw material, its production cost is slightly lower.

The domestic paper industry in China is the largest user of aluminum sulfate, followed by a large amount of aluminum sulfate exported to African countries through various channels as a coagulant for drinking water, used in conjunction with lime. The price of aluminum sulfate in African countries, especially in inland areas, is basically 3–4 times the domestic price in China.

Based on aluminum sulfate, some domestic scholars have done a lot of research on the preparation and properties of polyaluminum sulfate and polysilicon aluminum sulfate, including their preparation methods, molecular structure, and usage performance. The coagulation effect of these two types of polymers is better than that of aluminum sulfate, with a general salt base of 20 ~ 50%, and the alkalinity consumed is lower than that of aluminum sulfate.

Polyaluminum silicate sulfate is a new type of inorganic high polymer coagulant, which has both electro-neutralization and adsorption bridging functions, and its market share will gradually expand. Especially in recent times, the final disposal method of sludge in various regions is incineration. In order to avoid the corrosion of the furnace body caused by chloride ions brought by polyaluminum chloride and ferric chloride, aluminum sulfate coagulants are a good choice for sludge conditioning. Driven by the requirements of waste reduction in the industrial field, a large amount of iron salts, aluminum salts, or iron-aluminum composite salts are added as coagulants and sludge conditioners in the process of municipal sewage and sludge treatment. The formed sludge contains a large amount of iron ions and aluminum ions. After acid dissolution, iron and aluminum can be recovered and heavy metals can be separated. Iron salt and aluminum salt coagulants can also be prepared, including ferric sulfate, polyferric sulfate, aluminum sulfate, polyaluminum chloride, etc. Professor Li Fengting's team have conducted a comprehensive analysis of the sludge incineration ash from a certain place in Zhejiang Province, and its iron content is as high as 30%. After acid dissolution, iron salt coagulants can be prepared.

(5) **Production of Iron-free Aluminum Sulfate**

Drinking water treatment does not require the content of iron in aluminum sulfate purifiers. Under certain conditions, trivalent iron ions can promote the coagulation process, so the content of iron in aluminum sulfate for drinking water does not have a significant impact on the treatment effect. However, in the paper industry, because aluminum sulfate is used as a retention aid, the flocs formed directly enter the finished paper with the pulp, and trace amounts of iron will color the paper, so the content of iron ions in aluminum sulfate must be strictly limited.

In addition to large mines, there are a large number of small and medium-sized mines in China's bauxite, most of which are low-aluminum and high-iron ores (the ratio of aluminum to iron is less than 35). The products produced by conventional methods have a high iron content and are difficult to prepare low-iron products. At present, when producing aluminum sulfate at home and abroad, methods such as sulfide precipitation, oxidant oxidation, and ion exchange are usually used to remove iron. These methods are either too costly or too complex to control in production and are difficult to accept. The ferrous sulfate leached out by sulfuric acid is difficult to oxidize under acidic conditions, but under the condition of increasing the pH value of the solution, adding strengthening oxidants, such as potassium chlorate (sodium),

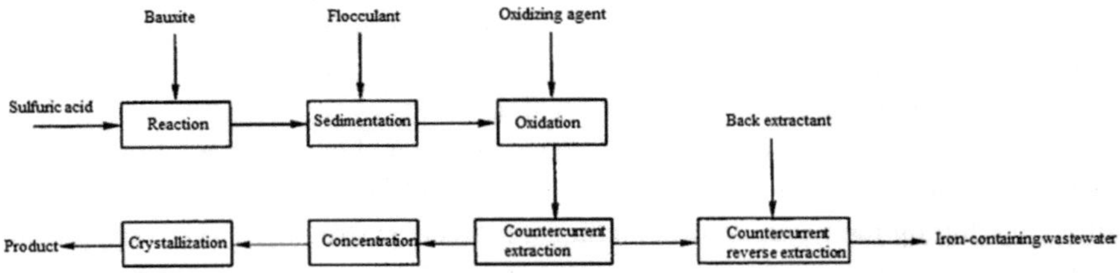

Fig. 2.5 Primary amine extraction of iron ions process

hydrogen peroxide, etc., can easily oxidize ferrous ions into trivalent iron ions, which can be hydrolyzed to form iron hydroxide colloids in the pH range of 3.5 ~ 4.5, and then the iron ions can be removed by precipitation filtration separation. In addition, the pH value of the final product of the reaction can be increased by adjusting the ratio of the material liquid during the production process, which is conducive to the oxidation of ferrous ions, but the oxidation speed of this method is relatively slow. In order to accelerate the separation of suspended particles and aluminum sulfate liquid, a certain amount of polyacrylamide can be added.

$$6FeSO_4 + NaClO_3 + 3H_2SO_4 = 3Fe_2(SO_4)_3 + NaCl + 3H_2O \quad (2.7)$$

There are also reports of using organic amines to extract iron ions, as shown in Fig. 2.5. The reaction mechanism is as follows: firstly, ferrous sulfate is oxidized to ferric sulfate, and then primary amine is added. The primary amine is alkaline. In a sulfuric acid medium, the primary amine first reacts with sulfuric acid to form an amine sulfate and then complexes with ferric sulfate in the leaching solution to form ammonium sulfate, which enters the organic phase.

$$3(RNH_3)_2SO_4 + Fe_2(SO_4)_3 \rightleftharpoons 2(RNH_3)_3Fe(SO_4)_3 \quad (2.8)$$

Add the extractant sodium chloride to the extract to precipitate ammonium sulfate.

$$2(RNH_3)_3Fe(SO_4)_3 + 6NaCl = 3(RNH_3)_2SO_4 + 2FeCl_3 + 3Na_2SO_4 \quad (2.9)$$

After extraction, the aluminum sulfate solution is concentrated and crystallized to obtain the final product, with an iron content of up to 0.0038%.

(6) **Production of aluminum sulfate by fly ash method**

Fly ash is the solid residue left after coal combustion in coal-fired boilers, mainly a by-product of coal-fired power plants. Its main components are silicon dioxide (about 20 ~ 40%), aluminum oxide (about 25 ~ 45%), and iron oxide (4 ~ 20%), as well as various other impurities, such as calcium oxide (1 ~ 12%), magnesium oxide (1 ~ 2%), titanium oxides, sodium oxide, and potassium oxide, as well as trace amounts of other metals, such as lithium, gallium, scandium, strontium, chromium, copper, etc. The aluminum oxide content in fly ash from Erdos and Wuhai regions in China generally exceeds 40%, and the leaching rate is relatively high, making it a high-quality raw material for the production of aluminum sulfate, aluminum chloride, and polyaluminum chloride. Many research institutions in China have conducted some research in this area, mainly aiming to prepare aluminum hydroxide and aluminum oxide to replace bauxite and provide raw materials for the electrolytic aluminum industry. The main problem faced by the hydrochloric acid dissolution method is that a large amount of iron and calcium ions enter the solution, and the cost of separating iron and calcium ions is relatively high. The technology of fly ash resource utilization also faces the competition of international bauxite prices. Although the cost of fly ash is relatively low, the separation of impurities is relatively high.

The process of producing aluminum sulfate from fly ash is as follows: The fly ash is thoroughly mixed with water after mechanical grinding, the unburned carbon black is removed by flotation, and then the magnetite powder is obtained by magnetic separation process. The fly ash residue after flotation to remove carbon black (which can also be not separated) and magnetic separation to remove iron is mixed with concentrated sulfuric acid to form a mixed solution, which is heated and pressurized in acid-resistant reaction equipment. The material after the reaction is subjected to solid–liquid separation, and aluminum sulfate crude liquid can be obtained by filtration. The aluminum sulfate crude liquid is evaporated at high temperature and concentrated, and after cooling, a concentrated solution of aluminum sulfate can be obtained.

In order to obtain high-purity aluminum sulfate products, organic alcohol can be used for acid washing and iron removal. The organic alcohol is mixed with the aluminum sulfate concentrate in a certain volume ratio, thoroughly stirred and mixed, and after filtration and drying, high-purity aluminum sulfate can be obtained.

2.1.5 *Use of Aluminum Sulfate in the Water Treatment Industry*

Currently, the production process of aluminum sulfate in China is intermittent, and the largest enterprise has reached an annual output of 200,000 tons of solid products. In China, the largest user of aluminum sulfate is domestic paper companies, and there is also a large number of exports. There are relatively few users of aluminum sulfate in the water treatment industry at present, and the largest user is the Shanghai Tap Water Company (using Al_2O_3 content 7.8% solution). The use of aluminum sulfate as a coagulant requires a larger alkalinity of the treated water. If the alkalinity is not enough, it needs to be supplemented. Aluminum sulfate is rarely used in sewage treatment at home and abroad, mainly because the amount of sludge produced by aluminum sulfate is relatively large. Many water plants in African countries use aluminum sulfate as a coagulant. When the pH value is below 6.5, lime is added to adjust the pH value. When using polyaluminum chloride with a relatively high salt base, not only is the dosage small, but also the secondary dosing is avoided. In the past nearly twenty years, Professor Li Fengting's team has been promoting the use of polyaluminum chloride and composite polyaluminum chloride technology in African countries.

2.2 Ferrous Sulfate

2.2.1 Overview of Ferrous Sulfate Production

At present, there are two main sources of ferrous sulfate in China, one is the by-product of titanium dioxide production, and the other is the mixture of ferrous sulfate and sulfuric acid produced on the surface of steel parts in the pickling process of the steel and machinery industries. In the 1950s, a two-step coagulation process mainly based on ferrous sulfate was used. Firstly, ferrous sulfate was added to the treated water, and then chlorine gas was added to oxidize the ferrous ions into ferric ions and play a coagulating role. Because the two-step treatment process is relatively complicated, this process has gradually been replaced by a one-step process. Ferrous sulfate is now rarely used as a coagulant in water treatment. However, in wastewater treatment, especially in the treatment of cyanide-containing wastewater and dyeing wastewater, ferrous sulfate is still widely used as a coagulant and decolorizer.

2.2.2 Synthesis Method of Ferrous Sulfate

(1) **Ferrous Sulfate By-product of Titanium Dioxide**

China has abundant titanium resources, accounting for about 50% of the world's reserves (340 million tons). Titanium ore mainly exists in the form of ilmenite, which is the most widely distributed, largest reserves, and most valuable ore. China's ilmenite is mainly produced in Hainan Province, Guangxi Province, and Guangdong Province, and it is weathered sand ore. The chemical structure of ilmenite is relatively complex, and its molecular formula is $FeO{\cdot}TiO_2$ (or $FeTiO_3$). The molecular formula of weathered ilmenite is $Fe_2O_3{\cdot}TiO_2$. Ilmenite is a black or steel-gray sand ore with metallic luster. The selected ore generally contains 46 ~ 60% TiO_2, 35 ~ 45% total iron, and also contains a small amount of manganese, vanadium, niobium, scandium and other trace elements. The indicators of Hainan ore are shown in Table 2.4.

Table 2.4 Hainan ore indicators/%

Substance	TiO_2	TFe	Fe_2O_3	SiO_2	Al_2O_3	MgO	CaO	Cr_2O_3	MnO	V_2O_5	P_2O_5
Indicator	51.15	34.7	–	0.86	0.17	0.31	0.23	0.009	1.9	0.36	0.10

In the process of producing rutile titanium dioxide from ilmenite using the sulfuric acid method, the sources of ferrous sulfate (green vitriol) are the trivalent iron and divalent iron contained in the ilmenite itself, and the iron scraps added in the process of reducing trivalent iron. Generally, the production of 1 ton of titanium dioxide will produce 3.5–4 tons of ferrous sulfate. According to the statistics compiled by the Secretariat of the Titanium Dioxide Industry Technology Innovation Strategy Alliance and the Titanium Dioxide Center of the National Chemical Productivity Promotion Center, in 2022, the titanium dioxide production of 41 full-process companies in China's titanium dioxide industry reached 3.861 million tons, the total output of rutile and anatase titanium dioxide and other related products produced by the whole industry increased by 71,000 tons year-on-year. In recent years, due to the huge demand for new energy batteries, especially lithium iron phosphate batteries, the production of ferrous sulfate has shifted towards the production of lithium iron phosphate, causing significant fluctuations in the market and prices of ferrous sulfate. According to the statistics of titanium dioxide, the production of ferrous sulfate in China in 2022 was around 14 million tons.

$$FeO + H_2SO_4 = FeSO_4 + H_2O \tag{2.10}$$

$$TiO_2 + H_2SO_4 = TiOSO_4 + H_2O \tag{2.11}$$

$$Fe_2O_3 + 3H_2SO_4 = Fe_2(SO_4)_3 + 3H_2O \tag{2.12}$$

$$Fe_2(SO_4)_3 + Fe = 3FeSO_4 \tag{2.13}$$

Ferrous sulfate is mainly produced when frozen ferrous sulfate is mixed with titanium sulfate solution. The main impurity in the ferrous sulfate crystals formed by crystallization is titanium sulfate, so it must be washed with water to recover the titanium liquid on the surface of the crystals. The degree of recovery of the titanium liquid reflects the level of titanium dioxide production technology. If the loss of titanium liquid is too large, the production cost of titanium dioxide will greatly increase. Generally, after freezing and crystallization, the content of ferrous sulfate is about 90%. In water treatment, ferrous sulfate is mainly used for decolorization of printing and dyeing wastewater and as a raw material for the production of ferric sulfate and polyferric sulfate. The quality of ferrous sulfate produced as a by-product of titanium dioxide production can generally meet the requirements for the production of polyferric sulfate water purifier, and the content of heavy metals is low.

(2) **Production of Ferrous Sulfate from Steel Pickling Waste Liquid**

Pickling is a very important treatment process in material protection and processing. Conventional steel parts are usually pickled with sulfuric acid and hydrochloric acid, and the pickling process produces a large amount of solution containing sulfuric acid and ferrous sulfate or hydrochloric acid and ferrous chloride. In the past, many companies treated this part of the waste liquid by neutralizing and precipitating with lime milk, which not only increased the treatment cost, but also did not make this part of valuable material resources. Some companies arbitrarily discharge it, causing not only a lot of environmental pollution, but also a great waste. These waste liquids are good resources for the production of ferrous sulfate, ferric chloride, polyferric sulfate, and polyferric chloride. For example, the sulfuric acid pickling waste liquid generally contains $FeSO_4 \cdot 7H_2O$ 15 ~ 20%, free acid 7 ~ 10%. By fully utilizing the pickling waste liquid, polyferric sulfate can be directly produced, thereby achieving the purpose of pollution control and waste utilization, and at the same time providing a good water treatment agent for wastewater treatment.

The iron filings generally used are collected from different machining units, so the sources of iron filings are very complex, often containing a large amount of alloy steel. After acid dissolution, the solution contains a large amount of heavy metal ions, such as lead, manganese, nickel, copper, chromium, cadmium, etc., as well as substances such as nitrogen and phosphorus.

The production process is as follows: the pickling waste liquid is added to the reactor, the liquid is heated to 90 ~ 100 °C, and then the iron filings are added, or the iron filings can be added before heating. Generally, the dosage of iron filings is not measured in the production process. When there are few bubbles in the liquid during the reaction process, a saturated solution of ferrous sulfate can be obtained. The liquid is filtered and placed in a crystallizing tank. With the decrease of temperature, a large number of crystals are precipitated in the solution, and when the temperature is reduced to room temperature, the crystals are no longer precipitated. The mother liquor is pumped out to obtain crude crystalline ferrous sulfate. After drying this crystal in a centrifuge, the product ferrous sulfate can be obtained.

Currently, there is also a method of putting the solution after the reaction into an open-air crystallization pool and letting it naturally evaporate and dry. This method occupies more land and has a longer production cycle. For hot mother liquor, a method of rapid crystallization by freezing can also be used.

If the mother liquor is left at a low temperature for a long time, small crystals will redissolve and precipitate on the surface of large crystals, growing into larger crystals. The purity of the formed large-grain crystals is relatively high. If the filtration process is not used before crystallization, a large amount of black solid will be produced at the bottom of the crystallization pool. Before crystallization, non-ionic polyacrylamide can be added to flocculate the ferrous sulfate solution, and then filtered and crystallized, so these carbon powders can be separated. High-purity ferrous sulfate crystals can be obtained by the method of multiple crystallizations. The proportion of metal ions in the pickling solution is basically consistent with the composition of the washed steel, especially carbon steel. When the obtained ferrous sulfate is used as an industrial water treatment agent, most of the heavy metal indicators can meet the requirements of the industry standard, but after the pickling process, if there is a hot dip galvanized process, due to the problem of pickling zinc removal of unqualified products, the mixed pickling waste liquid zinc exceeds the standard, and zinc ions in some areas are as high as 4×10^5 mg/L. This problem also exists for the hydrochloric acid pickling process.

When producing ferrous sulfate, attention must be paid to environmental issues and safety issues. A large amount of hydrogen gas is produced during the reaction of sulfuric acid and iron filings, so good ventilation must be ensured to quickly discharge the hydrogen gas to prevent aggregation and explosion. In addition, a large amount of acid mist, sulfur-containing, and phosphorus-containing gases are produced during the heating and large-scale evaporation of hydrogen gas. These gases have a very unpleasant smell and must be absorbed and treated.

(3) **Production of Food-Grade Ferrous Sulfate ($FeSO_4 \cdot 7H_2O$ and $FeSO_4 \cdot H_2O$)**

Ferrous sulfate is a widely used food additive. The ferrous sulfate in the by-product of titanium dioxide contains a certain amount of heavy metal ions, which must be refined to obtain food-grade and pharmaceutical-grade products. Generally, the method of reduction by active iron powder can be used, so that metal ions with lower activity than iron can be reduced, such as chromium, lead, copper, nickel, etc. Moreover, this method can also be applied to the removal process of heavy metals in ferrous chloride, pickling waste liquid ferrous sulfate solution.

The titanium oxide in ferrous sulfate, a by-product of titanium dioxide, can also be removed by hydrolysis reduction. The refining process of ferrous sulfate as a by-product of titanium dioxide is carried out in the following steps:: the by-product of titanium dioxide and ferrous sulfate are added to the container in a ratio of 1:1.25 (L) in sequence, heated and stirred to dissolve; concentrated sulfuric acid is added, the amount of which depends on the pH value of the solution, to produce a product with a purity of over 99%, and the titanium content is only 3.9×10^{-6} mg/L (calculated as 20% Fe). Combined with the treatment of reduced iron powder, the content of other heavy metals is also significantly reduced, and a high-purity ferrous sulfate solution can be obtained. The purity of the product after crystallization is higher and can be directly used for the manufacture of magnetic powder, high-purity iron oxide, various iron oxide pigments, pharmaceuticals, auxiliary agents and catalysts in the chemical field, as well as lithium iron phosphate which is currently in high demand.

(4) **Removal of heavy metal ions**

The most critical factor affecting ferrous sulfate is heavy metal ions. Table 2.5 shows the production indicators of ferrous sulfate produced by a unit using iron filings and steel pickling waste liquid.

As can be seen from Table 2.5, the heavy metal content in ferrous sulfate produced using acid washing waste liquid is far higher than the national standard. Therefore, the ferrous sulfate produced by the above process and the polyferric sulfate produced by it cannot meet the national standard of drinking water purification agent.

In order to reduce the heavy metal content in ferrous sulfate, recrystallization or chemical reduction precipitation can be used.

In addition, reduced iron powder can be added to the ferrous sulfate solution to precipitate out less active ions such as nickel, copper, chromium, cadmium, and mercury. The operation method is as follows: heating 30 ~ 40% ferrous sulfate solution to 80 ~ 100 °C, adding 1 ~ 2% reduced iron powder under stirring conditions, and the reaction lasted for 1 ~ 4 h. After treatment, the total heavy metal ions can

Table 2.5 Production Indicators of ferrous sulfate produced by a unit using iron filings and steel acid washing waste liquid/%

Item	$FeSO_4{\cdot}7H_2O$	$MgSO_4{\cdot}7H_2O$	$MnSO_4{\cdot}5H_2O$	$Al_2(SO_4)_3$	$CaSO_4{\cdot}H_2O$	$TiOSO_4$	Water insoluble	Others
Jiangxi	88.12	6.24	0.32	0.29	0.18	0.52	4.03	0.30
Nanjing	93.17	3.32	1.26	0.41	0.19	0.15	1.14	0.36
Shanghai	95.24	1.51	1.30	0.31	0.17	0.16	1.03	0.28

be reduced to below 20×10^{-6} mg/L. The same method can be used for ferrous chloride. The titanium content can also be reduced to 10×10^{-6} mg/L.

Ferrous sulfate is very easy to oxidize, especially when the pH value is high. During the crystallization and centrifugal filtration process, the crystals can be washed with low-concentration sulfuric acid to obtain blue-green crystals.

2.2.3 *Main Properties of Ferrous Sulfate*

Ferrous Sulfate has good coagulation effect, excellent decolorization ability, and functions of removing heavy metal ions, degreasing, dephosphorization, sterilization, etc., especially the decolorization of printing and dyeing wastewater and the removal of COD, the co-precipitation of iron oxide in electroplating wastewater has an obvious performance, and its price is cheap, especially suitable for wastewater treatment in printing and dyeing, electroplating, etc.

Technical requirements for ferrous sulfate: The appearance should be light green or light yellow-green crystals; the water treatment agent ferrous sulfate index should meet the requirements of Table 2.6.

Table 2.6 National Standard in China "water treatment agent ferrous sulfate" (GB10531-2016) index item requirements

Item		Index	
		Class I	Class II
Mass fraction of ferrous sulfate ($FeSO_4$ $7H_2O$) w_1/%	≥	90	87.0
Mass fraction of titanium dioxide (TiO_2) w_2/%	≤	0.75	1.00
Mass fraction of insoluble matter w_3/%	≤	0.5	0.50
Mass fraction of free acid (calculated as H_2SO_4) w_4/%	≤	1.00	2.00
Mass fraction of arsenic (As) w_5/%	≤	0.0002	0.001
Mass fraction of lead (Pb) w_6/%	≤	0.0004	0.002
Mass fraction of cadmium (Cd) w_7/%	≤	0.0001	0.0005
Mass fraction of mercury (Hg) w_8/%	≤	0.00002	0.0001
Mass fraction of chromium (Cr) w_9/%	≤	0.001	0.005

When this product's Class I is used as a raw material for producing drinking water treatment agents, it should comply with relevant laws and regulations

2.3 Ferrous Chloride

2.3.1 *Synthesis Method of Ferrous Chloride*

Currently, there are a large amount of steel hydrochloric acid pickling waste liquid and circuit board etching waste liquid in China, which are good raw materials for the production of ferric chloride and polyaluminum chloride.

In the electronics industry, ferric chloride solution is often used for etching electronic circuit boards. When its content reaches a concentration of 2%, the corrosiveness of ferric chloride decreases and is generally no longer used, and is discharged in the form of waste liquid, which contains a large amount of heavy metal ions such as nickel and copper. However, some manufacturers sell this harmful waste liquid containing heavy metals in the form of ferric chloride water treatment agents, which will inevitably cause great harm to the water environment in the long run. For this kind of ferric chloride solution containing heavy metals, if the heavy metal ions such as nickel and copper are removed through processing, it can meet the requirements of etchants or water treatment agents, providing a broad prospect for the application of this kind of waste liquid.

Nickel and copper metals can be separated from ferric chloride waste liquid by stepwise precipitation and iron powder reduction. The disadvantage of the neutralization reaction method is that it consumes a large amount of hydrochloric acid and alkali, and the generated iron hydroxide is dissolved again with acid, the treatment process is complex, the amount of labor is large, and the treatment cost is high. For manufacturers with an annual treatment volume of several thousand cubic meters of solution, it is somewhat difficult to analyze from an economic and operational reliability perspective. The reduced iron powder is mixed with ferric chloride etching or pickling waste liquid containing 0.005 ~ 2% nickel, and reacted at 40 ~ 95 °C for 1 ~ 4 h, causing the nickel in the waste liquid to deposit on the surface of the reduced iron powder, forming nickel-containing iron powder. The nickel content in the waste liquid is greatly reduced to below 10 mg/L, and at the same time, ferric chloride is reduced to ferrous chloride. The waste liquid from which nickel and other heavy metal ions have been removed is oxidized to obtain ferric chloride solution, which is used for drinking water or wastewater treatment, or can be used again as an etchant.

The main physical characteristics of the ferric chloride waste liquid provided by a manufacturer in the East China region are shown in Table 2.7.

For the waste ferric chloride liquid, the following two methods were adopted to remove nickel ions and prepare coagulants.

Table 2.7 Main physical characteristics of ferric chloride waste liquid

Item	Nickel content/%	Iron content/%	Boiling point/°C	Density/(g/cm^3)
Ferric chloride waste liquid	1.9	45	105	1.495

2.3.1.1 Concentration Crystallization Method

(1) Experimental Principle and Method

Hydrochloric acid vapor is introduced into the ferric chloride solution for the experiment, and it is heated to maintain the temperature at 35 °C. During the reaction, light yellow crystals precipitate. The crystals and mother liquor are separated by filtration. The mechanism is that as the common chloride ions increase, the solubility of nickel chloride becomes very small and precipitates.

After testing, the filtrate contains about 60% ferric chloride and 0.3 ~ 0.5% nickel chloride; the crystals contain 55 ~ 60% ferric chloride and 3 ~ 4% nickel. The crystallization method can achieve the purpose of reusing the ferric chloride solution for etching, and the separated mother liquor reacts with iron filings to reduce and deposit nickel, which can further reduce the nickel content, thereby meeting the requirements for heavy metal content in drinking water purifiers.

(2) Factors Affecting the Concentration Crystallization Process

The effect of the concentration of untreated ferric chloride solution on the crystallization effect: the untreated ferric chloride solution with an iron content of 45% is distilled under reduced pressure, concentrated to an iron content of 60 ~ 70%, cooled and maintained at an average temperature of 30 ~ 35 °C, and hydrochloric acid vapor is introduced, and a large amount of light-yellow crystals precipitate. The experiment shows that the concentrated solution can save the amount of hydrogen chloride used for absorption compared with the unconcentrated solution and is especially suitable for waste liquid containing ferrous chloride.

The effect of crystallization temperature on the crystallization effect: According to the experiment, it is appropriate to keep the average temperature of the ferric chloride waste liquid in contact with hydrogen chloride gas at 15 ~ 35 °C. According to the literature, when the temperature is below 15 °C, the cooling power cost increases, and when the temperature exceeds 35 °C, the solubility of nickel nitride increases, which is not suitable for precipitation.

2.3.1.2 Iron Powder Reduction Method

(1) Reaction Principle

This method uses the reductive nature of iron powder to displace and separate heavy metal ions in ferric chloride waste liquid. The main reaction equations are:

$$Fe + NiCl_2 = FeCl_2 + Ni \tag{2.14}$$

$$Fe + 2FeCl_3 = 3FeCl_2 \tag{2.15}$$

In order to prepare a nickel-free ferric chloride solution, chlorine can be used as an oxidant to oxidize ferrous ions. The reaction equation is:

$$2FeCl_2 + Cl_2 = 2FeCl_3 \tag{2.16}$$

(2) Experimental Method

Using the iron powder reduction method, iron powder is added to the waste liquid and heated to 40 ~ 60 °C to precipitate some metal ions, but at the same time, trivalent iron ions are also reduced. After filtration and separation, metal precipitates containing nickel and mother liquor are obtained. According to the analysis, the metal precipitate contains 10% ~ 20% nickel, and the mother liquor contains less than 40 mg/L nickel, meeting the national requirements for drinking water purifiers.

(3) Analysis of Influencing Factors

① Effect of Temperature: Literature suggests that as the temperature increases, the removal efficiency of nickel improves. Take 100 g of untreated ferric chloride waste liquid, add 20 g of iron powder, stir, react for 4 h, and change the reaction temperature. Maintain a constant temperature at 40 °C, 60 °C, and 80 °C respectively. After the reaction is complete, cool, filter, and separate. The experiment shows that after the ferric chloride solution reacts at 80 °C for 4 h, there is a large amount of residue after filtration, and the pH value of the mother liquor is 5.5–6.0. Possibly due to the high temperature, the hydrolysis reaction is intense, resulting in the precipitation of ferrous hydroxide and the mother liquor is turbid. Therefore, in subsequent experiments, the reaction temperature is set to 40–60 °C.

② Effect of Stirring Speed: Research has found that intense stirring can increase the solid–liquid contact area, accelerate the reaction rate, and allow the iron powder to fully react with the ferric chloride solution.

③ Effect of Reaction Time: Take 200 g of untreated ferric chloride solution, add 30 g of iron powder, stir at a speed of 1000 r/min, control the temperature at 40 ~ 60 °C, change the reaction time, and react for 0.5 h, 1 h, 2 h, 4 h respectively. Take 200 g of untreated ferric chloride solution, add 40 g of iron powder, control the temperature at 40–60 °C, under stirring conditions, react for 0.5 h, 1 h, 2 h, 4 h respectively.

The reactions of both iron powder dosages show: As the reaction time increases, the reaction between iron powder and nickel ions in the ferric chloride solution is more complete, and the concentration of nickel ions in the mother liquor after filtration and separation is lower. After 4 h of reaction, the concentration of nickel ions in the mother liquor is very low, and an overly long reaction time will inevitably increase the volume of the reactor. This experiment believes that the selected reaction time of 2–4 h is appropriate.

④ Effect of Iron Powder Dosage: in the experiment, take 200 g of untreated ferric chloride solution, maintain the temperature at 40–60 °C, react for 4 h, and change

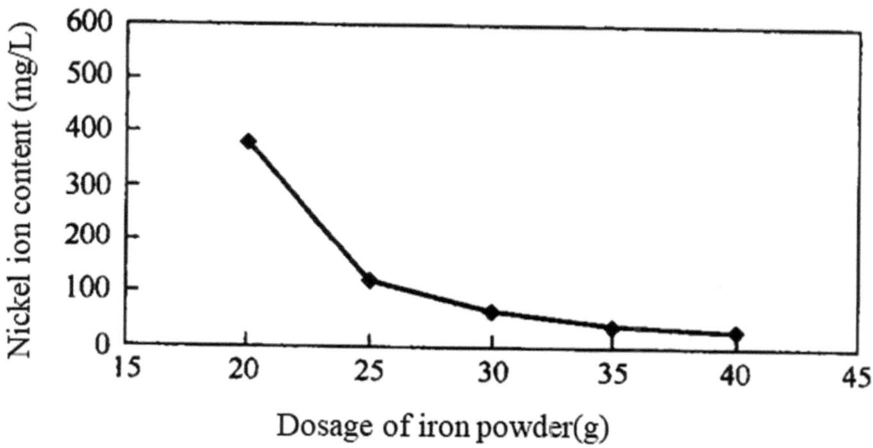

Fig. 2.6 Effect of iron powder dosage on the removal of nickel ions

the amount of iron powder added. Add 20 g, 25 g, 30 g, 35 g, 40 g of iron powder respectively. The effect of iron powder dosage on the removal of nickel ions is shown in Fig. 2.6.

As the amount of iron powder added increases, the removal rate of nickel ions increases. According to the reaction mechanism, when the theoretical value of the added iron powder is slightly excessive, under the above reaction conditions, the concentration of nickel ions has significantly decreased, and the excess iron powder cannot fully participate in the reaction, depositing in the precipitate, making separation difficult.

⑤ Effect of Iron Powder Addition Method: Under the above selected reaction conditions, a small amount of iron powder is added to the mother liquor after nickel separation, stirred at constant temperature, and reacted for 1 h. After filtration and separation, the concentration of nickel ions in the mother liquor is only 3.8 mg/L, which is lower than the standard nickel content of ferric chloride coagulant. Therefore, it is believed that adding iron powder in batches is better than adding iron powder at once for nickel removal.

Based on the experimental results of the above two methods, the following conclusions can be drawn:

① the ferric chloride solution treated by the concentration crystallization method can be used again for etching, but the separated mother liquor needs to react with iron powder again to further reduce and deposit nickel to meet the requirements of the general industrial water treatment agent heavy metal content standard. The advantage of using the crystallization method is that the precipitation of nickel salt and ferric chloride during the crystallization process reduces the amount of subsequent waste liquid treatment.

② the iron powder reduction method is used to treat ferric chloride solution containing nickel. Then the mother liquor contains less than 40 mg/L of nickel, meeting the standard requirements for water treatment agent ferric chloride to treat sewage and industrial water. Its characteristic is that the reaction process is easy to control, and the cost is slightly lower compared to the crystallization method. However, if 5 tons of waste liquid are treated, 7–7.5 tons of ferrous chloride solution will be recovered, and chlorine gas oxidation is required to produce nickel-free ferric chloride coagulant.

③ the method of using iron powder and other reducing agents to displace and separate heavy metals, nickel and other metals are mixed in the precipitate, making the subsequent separation of precipitated metals very complex. Currently, the price of nickel salt on the market is relatively expensive. If nickel can be separately separated and purified, it will have high economic value. The subsequent treatment of metal precipitates and the separation of precipitated metals can be a valuable research direction.

④ the iron powder reduction method is used to treat ferric chloride waste liquid containing nickel. Excessive iron powder is added in stages compared to the theoretical value, the reaction temperature is 40 ~ 60 °C, and the reaction time under stirring is 4 h, which can achieve the best effect of removing nickel ions.

⑤ the above methods are suitable for the removal of copper in the solution and are suitable for large-scale production.

2.3.2 *Main Properties of Ferrous Chloride*

Ferrous chloride (liquid): the appearance is a light green liquid.

Purpose: It is used for water treatment. The product quality is stable and the cost is low. It is currently a more economical and practical industrial wastewater treatment agent on the market, especially suitable for the treatment of printing and dyeing wastewater, chromium-containing wastewater, and cadmium-containing wastewater. It has significant effects on precipitating heavy metals and sulfides, decolorization, deodorization, and oil removal.

Compared with other wastewater treatment agents, the main features are as follows:

① Low cost, compared with the use of ferrous sulfate, it can save more than 30% of the cost of the drug;

② It is a water solution itself, eliminating the operation of preparing a solid solution and the problem of incomplete dissolution;

③ The amount of sludge produced after treatment is greatly reduced, which can save the cost of sludge treatment;

④ The content of free acid is low, and the corrosiveness is small.

⑤ The current control of heavy metals is relatively relaxed compared to previous standards, and the index item requirements of "water treatment agent—ferrous chloride" (HG/T 4538-2022) are shown in Table 2.8.

Table 2.8 National Standard in China "water treatment agent-ferrous chloride" (HG/T 4538-2022)

Item		Index	
		Solid	Liquid
Mass fraction of ferrous chloride (calculated as $FeCl_2$)/%	≥	60.2	22.7
Mass fraction of acid insoluble matter/%	≤	0.50	0.20
Mass fraction of sulfate (SO_4^{2-})/%	≤	0.5	0.5
Mass fraction of free acid (calculated as HCl)/%	≤	0.5	3.0[a]
Mass fraction of ammonia nitrogen (calculated as N)/%	≤	0.01	0.05
Mass fraction of iron (Fe) (III)/%	≤	0.60	0.40
Mass fraction of arsenic (As)/%	≤	0.0005	
Mass fraction of lead (Pb)/%	≤	0.002	
Mass fraction of mercury (Hg)/%	≤	0.00002	
Mass fraction of cadmium (Cd)/%	≤	0.0005	
Mass fraction of chromium (Cr)/%	≤	0.005	
Mass fraction of zinc (Zn)/%	≤	0.05	
Mass fraction of copper (Cu) (%)	≤	0.01	
Mass fraction of nickel (Ni) (%)	≤	0.005	
Total organic carbon (TOC)/(mg/L)	≤	400	

[a]According to supply and demand negotiation, special requirements can be no more than 5%

In summary, the production technology of coagulants in China is constantly improving, but the related theory of coagulation has not undergone major changes. The influence of engineering experience on the development of the coagulant industry is more realistic and important than theory, and theoretical research needs further development. Also, the standardization of the hazardous waste industry will guide iron-aluminum waste acid and slag, as well as a large amount of by-product hydrochloric acid, into the water treatment agent industry, resulting in overcapacity and fierce competition. Many traditional enterprises are increasingly struggling to survive, even going bankrupt.

The recent upgrading of the sewage and sludge industry also provides a new opportunity for the water treatment agent industry. The development of inorganic-inorganic composites, inorganic–organic composites, and even the combination of particles and inorganic and organic flocculants, all have great potential and can better meet different needs. For water treatment with various complex components, inorganic polymer coagulants are highly applicable, but they have large dosage, small flocs, and large sludge production. Organic polymer coagulants can fill these defects, coupled with the high adsorption performance of adsorbents, the effect of using the three together is much better than using them alone. Therefore, this is also a direction for the future development of coagulants. In addition, water treatment users should avoid a deeply rooted misconception that the higher the purity of the drug, the better the water purification effect.

2.4 Ferric Chloride

2.4.1 Main Properties of Ferric Chloride

Ferric chloride is a very widely used water treatment agent. For drinking water treatment plants, ferric chloride is not a new reagent. Because the density of iron salt flocs is greater than that of aluminum salt, it is often used in combination with aluminum salt and iron salt in the water treatment process, especially in the low-temperature and low-turbidity water treatment process. Over the past period, whether it is drinking water treatment or sewage sludge treatment, everyone has gradually paid attention to it, largely due to the great improvement in the economy, quality, and usability of the product.

After the 1990s, people's interest in ferric chloride and polyferric sulfate has increased. It can not only be used to treat turbidity, but also remove color, natural organic matter, and arsenic in raw water. Its effect on removing COD and phosphorus from industrial wastewater is far better than polyaluminum chloride, so the market share of iron salt coagulants in water treatment agents is increasing year by year. From the perspective of the coagulation effect of ferric chloride, its comprehensive performance should be better than polyaluminum chloride, but the biggest problem is corrosion. At present, the main raw materials for producing ferric chloride are the etching waste liquid of steel and circuit boards. Some regions often use fluorine-containing waste hydrochloric acid or fluorine-containing by-product hydrochloric acid. After use, this part of the fluoride ions is still in the etching or pickling waste liquid, which puts more stringent requirements on subsequent production equipment and use equipment. Moreover, during the etching process, zinc, copper, nickel, lead, etc. on the surface of metal components, depending on the composition, enter the etching waste liquid. Therefore, it is necessary to separate the above-mentioned heavy metal ions during the production of ferric chloride to achieve the purpose of purification.

In recent years, the demand for sludge dewatering has been very large, and the method of combining iron chloride with lime is widely used. This is the most efficient method of sludge dewatering. Although all-organic based on polyamines, diene propylene homopolymers and copolymers are gradually expanding the market, purely from the perspective of coagulation effect, the combination of iron chloride and lime is ideal. Because chloride ions form hydrochloric acid during the incineration process, its corrosiveness far exceeds that of sulfuric acid, so many waste incineration companies limit the use of chlorinated coagulants, including iron chloride and polyaluminum chloride, and prefer to use composite or all-organic sludge dewatering agents.

The Zeta potential distribution of ferric chloride solution at different concentrations is shown in Fig. 2.7. Figure 2.8 shows the pH values of ferric chloride solution at different concentrations.

Another feature of iron chloride is that it can form flocs in a wide range of pH. Figures 2.7 and 2.8 also shows that compared with aluminum hydroxide, iron

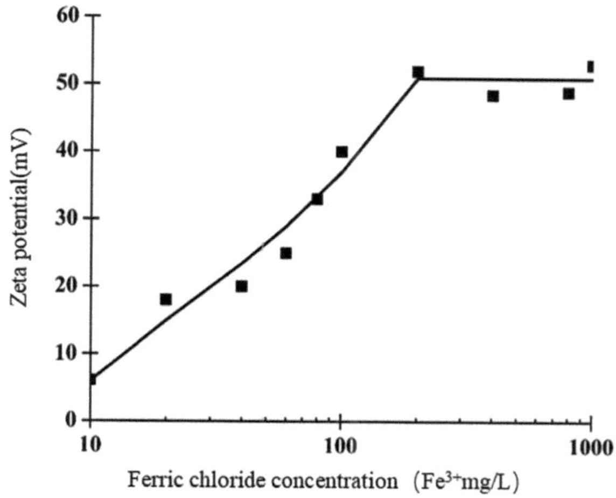

Fig. 2.7 Zeta potential distribution of iron chloride solution at different concentrations

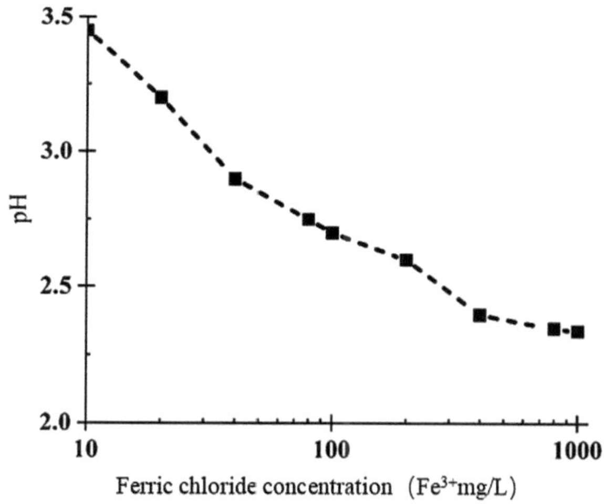

Fig. 2.8 pH values of iron chloride solution at different concentrations

hydroxide has a lower solubility, which allows iron chloride to function in a very wide range of pH values. Iron salts, including iron chloride, have an unavoidable defect, which is the color problem. Whether it is a production company or a user company, if the coagulant used is mainly iron, there are color problems caused by splashing on the facilities and environment. If too much is added to the treatment water body, there are even more color and iron taste problems. For sewage and sludge treatment, this problem is not very serious.

2.4.2 Synthesis Method of Iron Chloride

The oxidation of ferrous chloride can use various oxidants, such as hydrogen peroxide, sodium aluminate, potassium chlorate, sodium hypochlorite, nitric acid, ozone, etc. Through nearly thirty years of research and actual operation in the industry, the currently widely accepted are chlorine and oxygen oxidation methods. These two oxidation processes have simple production processes and fast oxidation speeds. Especially when using oxygen oxidation, its oxidation speed is fast, and its safety is very high. It has been widely used in many production companies.

2.4.2.1 Principle of Chlorine Oxidation

$$2FeCl_2 + Cl_2 = 2FeCl_3 \tag{2.17}$$

After chlorine gas is dissolved in water, part of it forms hydrochloric acid and hypochlorous acid, and the other part still exists in the form of chlorine molecules. Cl_2 exists in the following equilibrium in water:

$$Cl_2 + H_2OHClO + HCl \tag{2.18}$$

The solubility of chlorine in water is 0.091 mol/L, of which 0.061 mol/L exists in the form of Cl_2, and 0.03 mol/L exists in the form of HClO and HCl, so the free acid content is (calculated as HCl) 2×0.03 mol/L $\times 36.5 = 2.171$ g/L.

(1) When Cl_2 is in excess to saturation, the content of free acid

During production, the amount of $FeCl_2$ fed into the oxidation kettle is 1 m^3, the concentration is 35%, the density is 1.375, and Cl_2 is added 390 kg to obtain the product $FeCl_3$, with a density of 1.44 g/cm^3 and a concentration of 41%. Assuming that Cl_2 is in excess and reaches saturation, the content of free acid in the finished product is:

$2.171 \times 1 \times 10^3/1.44 \times 106 = 0.15\%$ reaching the first-grade product standard.

Excess $Cl_2 = 0.091 \times 1 \times 103 \times 36.5 \times 10^{-3} = 3.3$ kg.

(2) The maximum limit of excess chlorine allowed by product indicators

The first-grade products require a free acid content of $\leq 0.4\%$, the product contains HClO, HCl as

$$\frac{1 \times 1.44 \times 0.4\% \times 10^3}{36.5 \times 10^{-3} \times 10^3 \times 2} = 0.0789\,\text{mol/L}$$

Equivalent to an excess of 0.2393 mol/L of chlorine gas.

Oxidizing 1m^3 of ferrous chloride can produce an excess of $0.239 \times 10^3 \times 36.5 \times 10^{-3} = 8.73$ kg of chlorine gas.

In fact, when there is an excess of 8.73 kg of chlorine gas, only 3.3 kg dissolves in water, the rest escapes from the reaction solution in the form of gas.

The above calculation is based on a temperature of 25 °C. In actual production, the temperature of the reaction solution is 65 °C, and less chlorine gas will dissolve in the reaction solution.

(3) According to the product index, the maximum allowable deficiency of chlorine gas

When the amount of chlorine gas is insufficient, some $FeCl_2$ cannot be fully oxidized, affecting the final product index.

The first-grade $FeCl_3$ requires $FeCl_2$ content $\leq 0.30\%$.

If 1 m^3 of $FeCl_2$ is added,

$$2FeCl_2 + Cl_2 = 2FeCl_3 \tag{2.19}$$

$$\begin{array}{ccc} 2 \times 127 & 71 & 2 \times 162.5 \\ 1 \times 1.375 \times 35\% \times 10^3 & 134 & 615 \end{array}$$

The product obtained is 1506 kg, allowing $FeCl_2$: $1506 \times 0.30\% = 4.518$ kg, Cl_2 can be reduced by 1.26 kg, that is, the amount of chlorine gas cannot be less than $1.26/134 = 0.9\%$ of the chemical calculation amount.

Production calculation using ferrous chloride as raw material: Take 1000 L of ferrous chloride solution, with a density of 1.33 and a concentration of 31%. Put it into the reaction kettle and heat it to 60 ~ 65 °C. React with chlorine gas. After the reaction, the concentration of the ferric chloride solution is 36%, and the acidity is not detectable. By this analogy, when the concentration of the ferrous chloride solution is 35%, the concentration of the final product ferric chloride solution is 41%.

Alternatively, iron filings can be used as the main raw material. The iron filings are soaked in water, and chlorine gas is passed into the water to directly oxidize the iron to ferric chloride.

$$Fe + Cl_2 = FeCl_2 \tag{2.20}$$

$$Cl_2 + 2FeCl_2 = 2FeCl_3 \tag{2.21}$$

Many companies in China still use this process. The product is used for water and wastewater treatment. If iron filings are used for production, the biggest drawback is that the heavy metal content seriously exceeds the standard. For methods to remove heavy metals, please see the ferrous sulfate section.

2.4.2.2 Principle of Oxygen Oxidation

Chlorine gas oxidation is a relatively fast method, but in some industrial parks and factories, chlorine gas cannot be used due to concerns about leakage. In the 1990s, Professor Li Fengting's teamfirst proposed the method of oxygen catalyzed oxidation of ferrous chloride, this method draws on the method of nitrogen oxide catalyzed ferrous sulfate, its oxidation mechanism is as follows:

$$HCl + NaNO_2 = HNO_2 + NaCl \tag{2.22}$$

$$HCl + HNO_2 + FeCl_2 = FeCl_3 + NO + H_2O \tag{2.23}$$

$$2NO + O_2 = 2NO_2 \tag{2.24}$$

$$2FeCl_2 + 2HCl + NO_2 = 2FeCl_3 + NO \uparrow + H_2O \tag{2.25}$$

The same mechanism applies to the production of ferric chloride. Because oxygen is used as an oxidant, if the treatment of nitrogen oxides is well controlled during the production process, high-speed oxidation reactions can be achieved under low or even negative pressure. A 25 cubic meter reaction kettle was used, which can achieve an oxidation cycle of 1–2 h under low pressure, and each batch can produce about 20 tons. It can also achieve continuous feeding and discharging, realize the recycling of nitrogen oxides, avoid waste and secondary treatment and pollution of catalyst gas. Since oxygen is converted into water during the oxidation process, it must be concentrated when producing high-concentration products to increase the iron content. For the production of polyferric chloride and 38% ferric chloride, no concentration process is needed.

2.5 Ferric Chlorosulfuric Acid

2.5.1 Synthesis Method of Ferric Chlorosulfuric Acid

In the 1950s, ferrous sulfate was commonly used as a coagulant in China. Because the hydrolysis tendency of ferrous sulfate itself is very low, it is necessary to further oxidize the ferrous ions into trivalent iron with chlorine gas after addition to play a coagulating role. The production of ferric chlorosulfuric acid is generally based on ferrous sulfate as the raw material, and chlorine gas is filled into the ferrous sulfate solution. When the concentration of ferrous ions reaches the required level, the product can be obtained. This production process is very simple, and often uses reaction kettles or reaction towers as production equipment.

$$2FeSO_4 + Cl_2 = 2FeSO_4Cl \tag{2.26}$$

The above reaction can appropriately control the ratio of acid to iron, and polyferric chlorosulfuric acid can be prepared. Alternatively, ferrous sulfate can be used as the main raw material, supplemented with hydrochloric acid, and catalyzed by oxygen to avoid the use of oxygen, and ferric chlorosulfuric acid can also be obtained.

2.5.2 *Main Properties of Ferric Chlorosulfuric Acid*

Chlorinated iron sulfate is a reddish-brown liquid or solid product. Compared with ordinary polyferric, it has high iron content, fast settling speed, dense sludge, easy sludge concentration and dehydration, low treatment cost, and no toxicity. It is widely used in water supply treatment, urban sewage and industrial wastewater treatment, especially suitable for desulfurization wastewater treatment in power plants.

Chapter 3
Preparation and Quality Control of Inorganic Polymer Coagulants

3.1 Poly Aluminum Chloride

China began to study polyaluminum chloride in the 1970s. Academicians Tang Hongxiao and Qu Jiuhui of the Chinese Academy of Sciences, and Mr. Li Runsheng of Shenzhen Zhongrun CompanyWater Industry Technology Development Co., LTD are pioneers and founders in this field. They have done a lot of research on the structure, properties and synthesis process of this polymer, promoting the progress of this industry.

Poly aluminum chloride (PAC) is a highly efficient coagulant widely used at home and abroad. It has the advantages of small dosage, less sludge, and small impact on the pH value of the effluent. It is widely used in various water treatment coagulation processes, with a market share of over 95% in the drinking water market.

3.1.1 Distribution of Al(III) in Solution

Polymeric aluminum salts are a type of efficient inorganic coagulant, generally referring to a series of quasi-stable substances between Al^{3+} and $Al(OH)_3$, that is, hydroxy complexes between dialuminum and aluminum hydroxide, which may exhibit structural features of shared hydroxy complexes or shared oxygen base coordination. There are many studies on the hydrolysis of aluminum salts. Evidence suggests that, based on aluminum chloride as the calculation basis, the aqueous solution of simple aluminum salts can be accurately described by hexahydrate ions. When the aluminum salt solution lacks acid, the basic salt is formed, and a more complex system, that is, multiple multinuclear polymers, can be formed. The properties and structure of this multinuclear body have been the focus of debate for a long time. There are two completely different models for its structure.

L. Fengting, *Water Treatment Chemicals: Coagulants and Flocculants*,
https://doi.org/10.1007/978-981-96-8221-8_3

(1) **"Core links" model or "gibbsite fragment" model (hexagonal ring continuous distribution model)**

In such models, multinuclear aggregates are composed of one or more hexagonal rings, which are composed of octahedral aluminum atoms with hydroxyl groups. In the case of high hydrolysis, these ring structures further aggregate and grow, which can explain the final gibbsite surface or related crystalline solid phase. The interesting part of this model is that it maintains the double octahedral structure of gibbsite throughout the multinuclear compression process, which can fully explain the reduced activity of the multinuclear with acid or complex ligands when the solution matures. However, it is worth noting that researchers have not obtained definitive evidence of the existence of this multinuclear body, especially in dilute solutions at room temperature.

The traditional research methods on the state of aluminum compounds are chemical analysis and potentiometric titration. In addition to monomers such as Al^{3+}, $Al(OH)_2{}^{+}$, $Al(OH)^{2+}$, $Al(OH)_3$, $Al(OH)_4{}^{-}$, different authors have successively proposed $Al_2(OH)_2{}^{4+}$, $Al_2(OH)_5{}^{+}$, $Al_3(OH)_4{}^{5+}$, $Al_4(OH)_8{}^{4+}$, $Al_6(OH)_{15}{}^{3+}$, $Al_7(OH)_{16}{}^{5+}$, $Al_8(OH)_{20}{}^{4+}$, $Al_{10}(OH)_{22}{}^{8+}$ and other aggregated forms, and the dominant forms will evolve with conditions. Especially, the continuous change distribution series of hydrolysis forms is proposed, which is believed to develop from monomers to aggregates according to the hexagonal ring mode. Some people propose that its aqueous form can reach $Al_{54}(OH)_{144}{}^{18+}$, until the precipitate $Al_x(OH)_{3x}$ is formed.

$$Al^{3+} + H_2O \rightleftharpoons Al(OH)^{2+} + H^{+} \tag{3.1}$$

$$Al(OH)^{2+} + H_2O \rightarrow 1/6\ Al_6(OH)_{12}^{6+} + H^{+} \tag{3.2}$$

$$1/6\ Al_6(OH)_{12}^{6+} + H_2O \rightarrow Al(OH)_3 + H^{+} \tag{3.3}$$

(2) **Al_{13} poly-cation model**

This model believes that only a few forms of aluminum exist in the aluminum solution, and they can directly transform each other. In addition to monomers and oligomers, there are only a few forms such as $Al_{13}O_4(OH)_{24}{}^{7+}$ (Al_{13}) and high polymers. This is mainly identified and inferred by the "^{27}Al NMR method" and small-angle X-ray diffraction method. Al_{13} acts as the dominant hydrolysis body.Johansson first clarified the existence of the dimer $[Al_2(OH)_2(H_2O)_8]^{4+}$ and the poly-cation $[Al_{13}O_4(OH)_{24}(H_2O)_{12}]^{7+}$ with Keggin structure. The former is composed of two parallel octahedra centered on aluminum atoms and connected by a common edge; the latter's structure is described by special resonance shifts in the Al-NMR spectrum. The central aluminum atom is a tetrahedral structure, surrounded by 12 octahedral aluminum atoms. Its Keggin structure of Al_{13} is shown in Fig. 3.1.

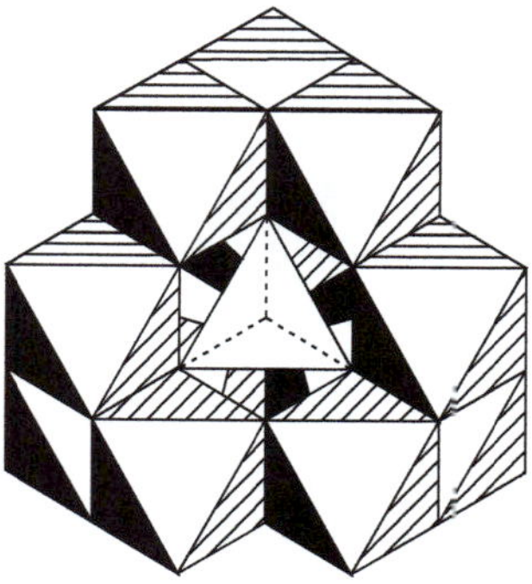

Fig. 3.1 Keggin structure of Al_{13}

Furthermore, up to now, the largest cationic polymer discovered is $Al_{30}O_8(OH)_{56}(H_2O)_{24}{}^{18+}$, known as δ-Tauelle^{18+}. Its structure consists of two δ-Al_{13}Keggin units, which are bonded by four AlO_6 octahedra. The structure of the cationic polymer $[Al_{13}(OH)_{24}(H_2O)_{24}]^{15+}$ is shown in Fig. 3.2. Al_{30} can be formed through the following process: during the heat treatment of Al_{13}, some monomers separated from AlO_6 are added to the unseparated ε-Al_{13}. These ε-Al_{13} with one to three AlO_6 units isomerize with the δ form. Once the isomerization is complete, Al_{30} can fully stabilize the polymer.

It is well known that $Al(H_2O)_6{}^{3+}$ only exists at low pH values in aluminum salt solutions. When pH > 3, aluminum ions hydrolyze and produce more or less water-soluble multinuclear formations, known as hydroxy complexes, cationic polymers, or hydroxy complexes. These hydrolysis species can be prepared and characterized in various ways:

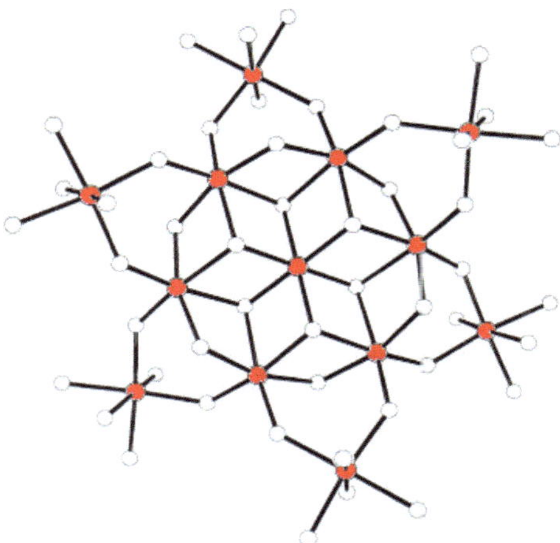

Fig. 3.2 Structure of the polymeric cation $[Al_{13}(OH)_{24}(H_2O)_{24}]^{15+}$ (• represents aluminum atoms, o represents oxygen)

① Potentiometric titration method. Brosset et al. used 0.5 mol/L aluminum chloride and NaOH solution to obtain a solution with an aluminum concentration of 0.1 mol/L. The specific process is as follows: under the condition of 20 °C and purging with nitrogen, 50 mL of 0.5 mol/L $AlCl_3$ solution is added to a container, and NaOH solution is added under continuous stirring to control the ratio of (OH)/(Al). Based on the study of the prepared solution, the following variants are proposed: $Al(Al_2(OH)_5)_n{}^{(3+n)+}$ or simpler $Al_6(OH)_{15}{}^{3+}$ and $(Al_8(OH)_{20}{}^{4+}$, with the ratio r = (OH)/(Al) limited to 2.5. The properties of these complexes are consistent with Sillen's "core links" theory. Van Cauwelaert et al. discussed the series from $Al_4(OH)_8{}^{4+}$ to $Al_{13}(OH)_{32}{}^{7+}$. Recently, Vermeulen and Stol emphasized the importance of kinetics in the hydrolysis process.

② The method of coagulation with silver halide colloid. Matijevic et al. believe that the main existing formation is $Al_8(OH)_{20}{}^{4+}$.

③ Chemical method. After a series of experiments (by adding alkali to control the ratio r = (OH)/(Al) between 1.5 and 2.5), Hsu and Bates proposed a continuous series of polymers, the basic unit of which is the cyclic $(Al_6(OH)_{12}(H_2O)_{12})^{6+}$. As r increases, the rings bind at the edges (two rings: $(Al)_{10}(OH)_{22}(H_2O)_{16})^{8+}$; seven rings: $(Al)_{24}(OH)_{60}(H_2O)_{24})^{12+}$; ...), thus deriving the double-enclosed structure of crystalline hydroxides. Hem et al. also reached similar conclusions from Ferron reagent complexation experiments. The Ferron reagent reacts at different rates with monomers and different polymers in Al(III), thus using the Ferron stepwise complexation spectrophotometric method can divide them into three categories, namely: Ala includes monomers and initial polymers ($Al_{1\text{-}3}$); Al_b includes low polymers (such as $Al_{6\text{–}8}$) and medium polymers (Al_{13}); Alc includes high polymers (Al > 13) and sol–gel state.

④ Small-angle X-ray scattering. After heating the hydrolysis solution at 70 °C for 1 h, under the condition of r = (OH)/(Al) between 1.5 and 2.25, Raush and Bale found a polymer with a rotation radius of 4.3 angstroms, they speculated that it is $(Al_{13}O_4(OH)_{24}(H_2O)_{12})^{7+}$, Johanson also proposed the same structure.

⑤ Nuclear magnetic resonance. Akitt et al. believe that the following forms exist as the pH value changes: $(Al(H_2O)_6)^{3+}$, $(Al_2(OH)_2(H_2O)_8)^{4+}$, $(Al_{13}O_4(OH)_{24}(H_2O)_{12})^{7+}$, there may be $(Al_8(OH)_{20}(H_2O)_{12})^{4+}$.

3.1.2 *Forms of Al_{13}*

The hydrolysis products in aluminum salt solution are polyhydroxy complexes and polymers. Some scholars believe that Polythirteen Aluminum (Al_{13}) is the best coagulation component in polyaluminum, and its content can reflect the effectiveness of the product. However, there is still a lot of controversy about this, and many researchers have conducted detailed studies on Al_{13}.

In $AlCl_3$ solution, in addition to the previously mentioned monomers, dimers, and trimers, there are also non-polymers and polymeric cations. So far, there are only vague descriptions about their size and shape. After very slowly evaporating water from a concentrated $AlCl_3$ solution (Al/Cl = 0.7), in addition to forming large $AlCl_3 \cdot 6H_2O$ crystals, other crystalline phases are also formed. The lattice parameters of the formed single crystals can be determined by single crystal X-ray diffraction. Based on the obtained lattice parameters, Breuil reported the data of $5AlCl_3 \cdot 8Al(OH)_3 \cdot 37.5H_2O$ powder. Wilhelm et al. confirmed the existence of this crystal phase through X-ray structure analysis, although not all positions of H atoms could be determined. However, in the crystal cell, it was found that its water content was slightly lower than reported, being $5AlCl_3 \cdot 8Al(OH)_3 \cdot 37.5H_2O$. The crystal cell contains four units, and the atomic positions determined in the structural analysis indicate two similar types of units, composed of AlO_6 octahedra.

Each type has two existing in the crystal cell unit, and Cl atoms are located outside this complex unit. Considering the charge carried (Al^{3+}, Cl^-), this large unit is considered to be a polymeric cation. Each polymeric cation is composed of 13 octahedra (Fig. 3.2).

One is located at the center, six are surrounded by common edges, and finally, the six outer aluminum atoms each connect to the inner ring through common bonds formed by two oxygen atoms. The Al^{3+} in the central octahedron is located at the transformation center of the spatial group. Therefore, the polymeric cation is centrally symmetric. The central Al and the 6 Al on the ring are on the same plane, forming a hexagon. The position of O corresponds to the octahedron with the degree of deformation. In the central octahedron, the distance of Al-O is almost equal (188–189 nm), and the O–Al-O bond angle is also almost equal (97°), but different from the bond angle of the ideal octahedron (90°). The octahedral ring has a certain degree of deformation, and the outer octahedra occupies replaceable positions above or below the hexagonal ring. Therefore, they also have a certain degree of deformation; the distance from Al-O to the free angle is obviously longer than the distance to the common angle. Since not all positions of H atoms can be determined, the issue of the distribution of hydrogen atoms and oxygen atoms to form OH^- and H_2O still exists. The charge number can clarify this issue. In the polymeric cation, 13 Al^{3+} carry 39 positive charges. On the other hand, there are 15 Cl^-, carrying 15 negative charges, so there are still 24 negative charges left in the polymeric cation, which corresponds to the number of oxygen atom bonds. Therefore, it can be guessed that the oxygen atoms belong to OH^-. Then there must be water molecules on the 24 free angles of the outer octahedra. Another 13 water molecules are located outside the polymeric cation. Therefore, the molecular formula of the hydrolysis product of aluminum chloride is $Al_{13}(OH)_{24}(H_2O)_{24}]Cl_{15} \cdot 13H_2O$. However, this model is also different from the aforementioned Keggin model.

Some progress has been made in the study of the morphology of polyaluminum chloride, such as the relationship between the structure and properties of polymer cationic aggregates, and flocculation morphology. The coagulation mechanism of aluminum aggregates is between traditional coagulants and cationic organic flocculants, belonging to the surface complexation, surface hydrolysis, and surface precipitation process of multinuclear hydroxy complexes. It has a unique mechanism model and quantitative calculation model, which needs further research.

3.1.3 Environmental Effects of Polymeric Aluminum

In groundwater and surface water systems, there is a dissolution equilibrium of aluminum. The transformation between dissolved and fixed aluminum mainly depends on the pH value. In acidic or weak alkaline environments, the equilibrium moves, and dissolution increases. When the pH value is below 5, aluminum exists as a trivalent ion, and when the pH value is above 6.2, it is a hydrated cation, which is also the highest pH value at which aluminum can dissolve. The concentration of aluminum in natural water bodies is a function of pH, and its maximum value can be obtained through thermodynamic calculations.

However, the chemical migration of aluminum is very complex, and no equilibrium can explain the morphological changes and migration process of aluminum. The chemical properties of aluminum in acidic soils and rivers in inorganic form can be considered from several aspects: mineral solubility, ion exchange, water mixing process, and solid-phase organic matter, etc.

Aluminum in the soil can easily dissolve and enter the river. The chemical properties of aluminum in acidic soils, such as solubility, are similar to those in surface water. However, it is not feasible to infer that the chemical properties of aluminum in water are similar to those in soil, because the influence of water conditions and chemical reactions in water mixing cannot be ignored. Natural acidification leads to an increase in dissolved aluminum. In a neutral environment, aluminum can undergo ion exchange, such as aluminum ions can exchange calcium ions. There are a large number of aluminum ions in acidic soils, which can undergo ion exchange with soluble ions that have been soil-filtered. When the pH value is between 5 and 7, chloride ions or organic anions can form complexes with aluminum ions, making the aluminum content in the solution far higher than expected. The migration of aluminum in an acidic environment provides a source for plant absorption of aluminum. Plants can enrich aluminum in nutrient-rich solutions, especially in the roots, where the enrichment of aluminum is large. People have studied the situation of aquatic plants enriching aluminum, of course, biological enrichment is related to the aluminum available to the organism.

The concentration of aluminum in different environments is related to the form and migration of aluminum. Aluminum is an important element in soil particles and coal ash particles. The concentration of aluminum in surface water and soil water systems is very stable and is related to physicochemical and geological factors. Aluminum in water exists in several different forms, either suspended or dissolved, or it can form complexes with organic or inorganic ligands or exist as free aluminum ions. Aluminum salts can exist in water in monomer form but will form aggregates over time. The form of aluminum is determined by pH, dissolved organic carbon, chloride ions, sulfate ions, and suspended particles. Usually, there is very little dissolved aluminum in the water body, with a concentration of 1–50 μg/L, and under acidic conditions, the concentration can reach 500–1000 μg/L. Under extremely acidic conditions, the concentration of dissolved aluminum can reach 90 mg/L. Under neutral conditions, aluminum is enriched in sediments due to precipitation or sedimentation, with a concentration of 20,000–80,000 mg/kg. Aluminum is one of the most abundant elements in the soil, and the concentration varies greatly (700–100,000 mg/kg), this range has been cited by the American Geographical Survey Magazine. Therefore, there is a considerable amount of aluminum in the environment, however, very little exists in the ecosystem, because under general conditions, the form of aluminum is unabsorbable to organisms.

The standards set for toxic substances in drinking water, such as nitrates, fluorides, and lead, are often controversial. In fact, the process of establishing these standards is a complex one that encompasses technical judgments, public and standard setters, and economic levels. If people have always believed that the bioavailability of aluminum is indeed very low, the presence of aluminum in drinking water would not cause much concern. The establishment of standards for aluminum is quite complex. Aluminum is widely present in the environment, and as the number of aluminum-containing containers, food additives, and drugs increases, people are exposed to more and more aluminum. For animal experiments, the lethal dose (LD50) of aluminum ranges from a few hundred to 1000 mg/kg. Although the form of aluminum-DNA complex is complex and chromosomal mutations have appeared in mice in toxicity tests, there has been no sign that aluminum has carcinogenic effects. There is still no conclusion about the harm to the human body and the relationship between aluminum and Alzheimer's disease, which is also a hot spot in the toxicological study of aluminum.

The impact of aluminum exposure in the environment on organisms is very complex. Because the concentration of aluminum in the environment is affected by thermodynamic factors and pH values, it is often infeasible to infer the real situation in the environment based on laboratory research. It can be affirmed that organisms will become diseased or die after being affected by different forms of aluminum for a long time. Aluminum is different from other chemical pollutants. Environmental exposure not only refers to the release and concentration of aluminum but also includes its bioavailability and factors affecting its existing form.

Before the early 1980s, the standard for drinking water did not include aluminum content as a quality parameter. Currently, countries generally adopt the WHO standard, limiting total aluminum to 0.2 mg/L.

The toxicity of inorganic mononuclear aluminum complexes to aquatic plants and terrestrial plants is greater than that of organic complexes. In acidic or poorly buffered environments, aluminum has multiple effects on both aquatic and terrestrial plants. Calcium and organic acids can reduce the biological effects caused by the increase in inorganic aluminum concentration. The increase in toxic aluminum in acidic water bodies reduces the number of biological species in the water, even under nutrient-rich conditions.

3.1.4 Synthesis Method of Poly Aluminum Chloride

There are many raw materials for the production of polyaluminum chloride, including aluminum hydroxide, aluminum scraps, aluminum ash, bauxite, and aluminum calcium ore powder.

3.1.4.1 Aluminum Hydroxide Acid Dissolution Method

(1) Synthesis Method

The production of polyaluminum chloride from aluminum hydroxide is a process commonly used in China in the 1980s and currently abroad. Aluminum hydroxide is generally obtained through the Bayer process, using sodium hydroxide to dissolve bauxite, obtaining sodium aluminate solution, and then using carbon dioxide to neutralize and precipitate, drying to obtain white aluminum hydroxide crystals. Since many heavy metals are not alkali-soluble, the alkali dissolution process can separate the heavy metals in the ore. The industry standard divides the aluminum hydroxide used for water treatment into two types, Type I and Type II, the main difference being the different dissolution rates of alumina (Table 3.1). The dissolution rate of alumina in Type I products should be higher than 94%. Both can be used for the production of polyaluminum chloride for drinking water (Table 3.1).

Polymeric aluminum chloride aluminum salt, also known as easily soluble aluminum hydroxide, is widely used in various aluminum salts, catalysts, filler production, etc., due to its superior solubility. It is a good raw material for the above products, especially in the polyaluminum water purifier industry, it has incomparable advantages over ordinary aluminum hydroxide: under the same conditions, it consumes less raw materials, reacts faster, has a longer equipment service life, and produces products with high whiteness and transparency, and low production cost. Therefore, the emergence of aluminum salt-specific aluminum hydroxide has brought

Table 3.1 Industry standard in China "aluminum hydroxide for water treatment" (HG/T 5567–2019)

Item		Class I	Class II
Aluminum oxide (calculated as Al_2O_3) mass fraction/%	≥	64.0	
Aluminum oxide dissolution rate/%	≥	95.0	–
Moisture mass fraction/%	≤	12.0	6.0
Total iron (calculated as Fe_2O_3) mass fraction/%	≤	0.01	
Mass fraction of Lead (Pb)/%	≤	0.001	
Mass fraction of Mercury (Hg)/%	≤	0.00002	
Mass fraction of Chromium (Cr)/%	≤	0.001	
Mass fraction of Arsenic (As)/%	≤	0.0002	
Mass fraction of Cadmium (Cd)/%	≤	0.0002	

a new technological revolution to the water purifier industry. Under normal pressure and 110 °C, the acid dissolution rate of aluminum hydroxide can be increased to more than 97%, solving the problem of high temperature and high-pressure corrosion damage to the enamel reaction kettle in the "one-step method", and shortening the chemical reaction process and time, improving productivity and ensuring the qualification rate of PAC products. As shown in Table 3.2, the enterprise standard for water treatment agent-specific aluminum hydroxide of CHALCO Shandong Co., Ltd.

Through the comparison and analysis of the electron microscope images of aluminum salt-specific aluminum hydroxide and ordinary aluminum hydroxide (Fig. 3.3), the aluminum salt-specific aluminum hydroxide has loose crystals, a large specific surface area, and fully contacts with hydrochloric acid, accelerating the reaction speed of hydrochloric acid and aluminum hydroxide under the same conditions, with an acid dissolution rate as high as 99%. Ordinary aluminum hydroxide has dense crystals, a small specific surface area, and needs to be pressurized to react, with a long reaction time and an acid dissolution rate of only 65% under the same conditions.

Aluminum salt-specific aluminum hydroxide can reduce the production cost of polyaluminum chloride from four aspects: ① Environmental protection: it does not boil at 90 °C, and can be completely dissolved, and emits less acid gas; ② Energy: it uses less new steam, reducing the amount of new steam used; ③ Production: high

Table 3.2 CHALCO Shandong Co., Ltd's enterprise standard for water treatment agent-specific aluminum hydroxide

Grade	Al_2O_3%	Na_2O %	Fe_2O_3%	SiO_2%	Moisture %	Acid dissolution rate%
	≥	≤	≤	≤	≤	≥
HY-01	64.5	0.50	0.02	0.02	20	96

(a) Aluminum salt-specific aluminum hydroxide (b) Ordinary aluminum hydroxide

Fig. 3.3 Comparison of electron microscope images of aluminum salt-specific aluminum hydroxide (**a**) and ordinary aluminum hydroxide (**b**)

acid dissolution rate, shortening reaction time; ④ Equipment: no need for a reaction kettle, and it can be produced without pressurization.

The heavy metal content of aluminum salt-specific aluminum hydroxide is zero, it is currently an internationally trusted raw material for polyaluminum chloride, and its heavy metal indicators meet food hygiene standards.

Aluminum hydroxide has a high purity, and the heavy metal content of the products produced is very low. However, due to the poor acid solubility of aluminum hydroxide, a production process of heating and pressurizing acid dissolution is generally adopted. This process is relatively simple, but the salt basicity of the produced polyaluminum chloride is often relatively low, generally in the range of 30–50%. This is the main reason why the lower limit of salt basicity in foreign standards is 40%. In recent years, Japanese companies have continued to use this method in patents applied for in China. Many patents and articles have reported methods to increase salt basicity, generally by adding aluminum scraps, sodium aluminate, calcium carbonate, aluminum carbonate, aluminum hydroxide gel, lime, etc. to low salt basicity products. In order to ensure the low heavy metal content of this product and not introduce other impurities, the method of adding sodium aluminate is generally adopted. The cost of adding sodium aluminate is higher than that of adding calcium aluminate. At present, companies in China that have strict requirements for drinking water coagulants all use the method of aluminum hydroxide acid dissolution and calcium aluminate adjustment. Due to the high cost of aluminum hydroxide, this process is rarely used for non-drinking water treatment in China.

Using aluminum hydroxide heating and pressurizing or normal pressure acid dissolution and calcium aluminate ore powder neutralization and polymerization in two steps, the process flow chart of aluminum hydroxide two-step production of polyaluminum chloride is shown in Fig. 3.4.

When aluminum chloride is acid-dissolved under high temperature and high pressure (or normal pressure), low salt basicity aluminum chloride is generated, with the reaction formula as follows:

$$2Al(OH)_3+nHCl = Al_2(OH)_{6-n}Cl_n+nH_2O \tag{3.4}$$

Generally, the n value is less than 3, which means the salt basicity is less than 50%. When adjusting the salt basicity with calcium aluminate and aluminum hydroxide gel, the salt basicity can be controlled even lower, which can enhance the dissolution of aluminum hydroxide and reduce the amount of insoluble aluminum hydroxide. The salt basicity can be controlled at 10–20% during the production process. This can also enhance the dissolution rate of aluminum in the adjustment agent. The reaction conditions are as follows: temperature 110–130 °C; pressure 0.4 MPa; reaction time 3–5 h. The consumption and reaction of materials are basically carried out according to the stoichiometric value. If the reaction is carried out under normal pressure, an aluminum chloride solution or an intermediate product with a salt basicity of less than 20% is often obtained, and the research degree is adjusted later.

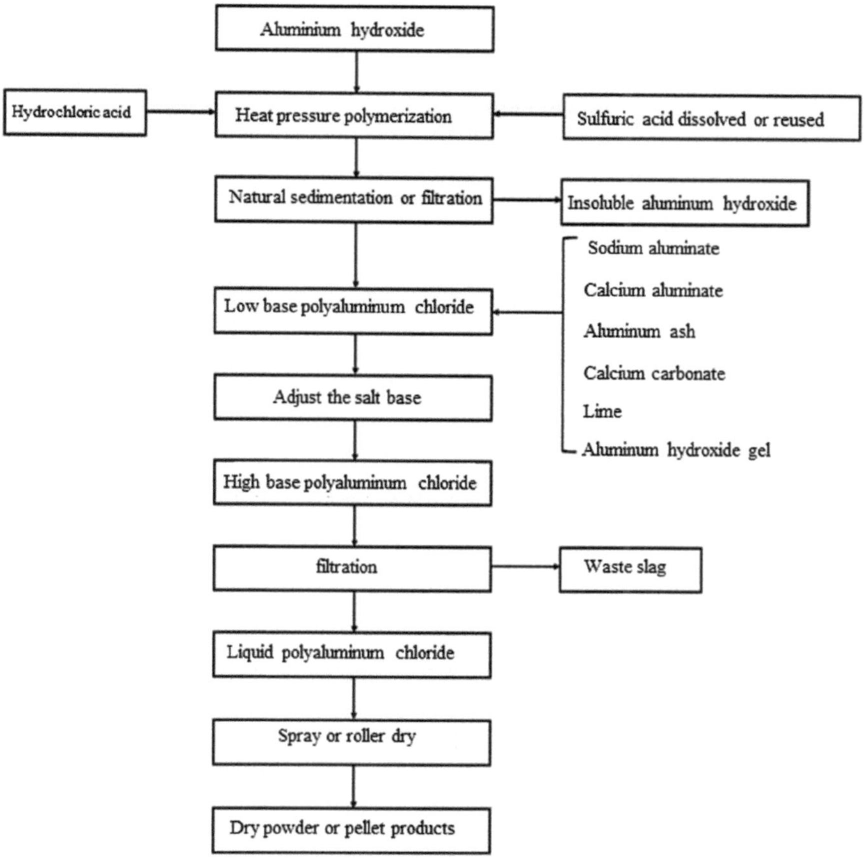

Fig. 3.4 Process flow chart of aluminum hydroxide two-step production of polyaluminum chloride

In the above reaction, the concentration of hydrochloric acid can be above 20%, and the higher the acid concentration, the better the dissolution effect of aluminum hydroxide. In the actual production process, the salt basicity can be controlled even lower (10–30%), which can not only fully dissolve aluminum hydroxide, but also easily adjust the salt basicity. The reaction ingredient ratio is $A1(OH)_3$: calcium aluminate ore powder = 0.6–0.9:1 (weight ratio). Specific operation is as follows: Add calcium aluminate ore powder to the conventional aluminum hydroxide acid solution in proportion, react at normal pressure at 80–105 °C for 3–4 h, filter, and obtain liquid polyaluminum chloride product, dry to obtain solid product. If soluble aluminum hydroxide is used, aluminum chloride solution can be easily obtained under normal pressure, and the production cost can be greatly reduced, and the corrosion of the acid dissolution process on the equipment can be reduced. The dissolution rate of soluble aluminum hydroxide can reach about 95%, and the utilization rate of raw materials is already very high.

(2) Adjustment of salt basicity

The relationship between the conventional aluminum hydroxide acid dissolution ratio and the dissolution rate and salt basicity is shown in Table 3.3.

In the actual production process, the first step does not need to pursue a high salt basicity. The salt basicity is controlled to zero, and the salt basicity is increased by adjusting the subsequent aluminum acid calcium. In this way, the highest possible leaching rate of aluminum hydroxide can be achieved, reaching 90% or even 95%. And the reaction conditions can be not very harsh, which can be carried out under normal pressure. The relationship between alumina leaching rate and reaction pressure is shown in Table 3.4.

In order to increase the salt basicity of the first leaching product of aluminum hydroxide, it is necessary to react under high temperature and high pressure, and the salt basicity can reach close to 50%. For papermaking chemicals, the iron ion requirement of polyaluminum chloride is less than 50 mg/L, so this method must be adopted. For products required for general drinking water, the reaction can be carried out under normal pressure, which can make the reaction vessel volume very large, such as 50 m^3 or even 300 m^3, can be produced, thereby greatly improving production efficiency.

Currently, the tap water industry is almost all using solutions with an alumina content of 10–15%.

Domestic companies produce a colorless liquid product with an alumina content of 22%, which is mainly used as an additive for cosmetics. This product can be obtained by the one-step method of aluminum foil; it can also be obtained by reacting aluminum hydroxide with acid to obtain a moderately concentrated low-salt basicity polyaluminum chloride and then reacting with aluminum foil to obtain the final product.

(3) Material consumption

The material consumption for producing 1 ton of solid polyaluminum chloride with regular aluminum hydroxide is shown in Table 3.5.

Table 3.3 Relationship between the conventional aluminum hydroxide acid dissolution ratio and the dissolution rate and salt basicity

Item	Al_2O_3 conversion rate (%)	Salt basicity (%)	pH value
$Al(OH)_3$:HCl = 0.6:1	88.8	$\leq$ 20.0	1.8
$Al(OH)_3$:HCl = 0.9:1	70.8	$\geq$ 20.0	2.2

Table 3.4 Relationship between alumina leaching rate and reaction pressure

Reaction pressure (MPa)	Al_2O_3 leaching rate (%)	Salt basicity (%)	pH value
0.3	80.0	15	1.8
0.4	88.0	20.0	2.2
0.45	90.5	21.5	1.5

Table 3.5 Material consumption for producing 1 ton of solid polyaluminum chloride with aluminum hydroxide

Aluminum hydroxide (anhydrous)	Calcium aluminate powder	Industrial hydrochloric acid	Industrial sulfuric acid	Precipitant
0.18–0.2 tons	0.30 tons	1.0 tons	0.05 tons	2–3 kg

(4) Preparation of polyaluminum chloride with anthraquinone acid

In the sulfonation process of benzene or polyaromatic hydrocarbons, anhydrous aluminum chloride is often used as a catalyst. The by-product of the reaction contains a solution of aluminum chloride, sometimes with a small amount of sulfuric acid. This type of acid is commonly referred to in the industry as anthraquinone acid. Its alumina content is between 3 and 5%, free acid is less than 5%, and it has good activity, which can be used to produce polyaluminum chloride. Generally, a certain amount of calcium aluminate is added to the waste acid solution, the reaction temperature is controlled between 90 and 100 °C and stirred for 2 h. Hot filtration can yield liquid polyaluminum chloride, or flocculants can be added to separate the supernatant and sediment. During the reaction process, the amount of calcium aluminate added can be determined based on the free hydrochloric acid, aluminum, and the basicity of the final product. After the implementation of the new national standard for polyaluminum chloride in China, the requirements for heavy metals are extremely strict. Heavy metals are often brought into the product after the dissolution of calcium aluminate acid. In order to control the content of heavy metals, most companies have to control the use of calcium aluminate, which also affects the increase in the basicity of the final product and the water treatment effect in some areas. If the anthraquinone acid contains benzene organic matter, it must be pretreated before use, adsorbing the organic matter in it, or using advanced oxidation methods to degrade the organic matter.

3.1.4.2 Preparation of High Purity Polyaluminum Chloride by Alkali Dissolution Method

(1) Synthesis Method

Firstly, calcium aluminate ore powder reacts with pure alkali solution to obtain sodium aluminate solution. Carbon dioxide gas is introduced into the sodium aluminate solution to produce aluminum hydroxide precipitate. Then, an appropriate amount of hydrochloric acid is added to the obtained aluminum hydroxide precipitate to produce a liquid product of polyaluminum chloride, and after drying, a solid product of polyaluminum chloride is obtained. The specific reactions are as follows:

$$Ca(AlO_2)_2 + Na_2CO_3 = 2NaAlO_2 + CaCO_3 \downarrow \tag{3.5}$$

$$2NaAlO_2 + CO_2 + 3H_2O = 2Al(OH)_3 \downarrow + Na_2CO_3 \tag{3.6}$$

$$2mAl(OH)_3 + (6-n)HCl = \left[Al_2(OH)_nCl_{6-n}\right]_m + (6-n)mH_2O \tag{3.7}$$

① Generation of Sodium Aluminate. Sodium carbonate is prepared into a solution of appropriate concentration and reacts with Al_2O_3 in calcium aluminate ore powder in equimolar amounts to obtain a sodium aluminate solution. The reaction temperature is 100–110°C, and after reaching the required temperature, the reaction continues for 3–4 h.

② Generation of Aluminum Hydroxide. Carbon dioxide gas is introduced into the obtained sodium aluminate solution. When the pH value of the reaction system is 6–8, a large amount of aluminum hydroxide gel is formed, and the reaction is stopped. The temperature of the entire reaction system should not exceed 40 °C, otherwise, an aged insoluble gel will form. The aluminum hydroxide obtained is washed, separated, and filtered.

③ Generation of Polyaluminum Chloride. A certain concentration of hydrochloric acid is added to the aluminum hydroxide generated and heated to dissolve, obtaining a colorless, transparent, viscous liquid polyaluminum chloride. After drying, solid polyaluminum chloride is obtained.

The method of alkali dissolution can effectively improve the extraction rate of alumina, and non-amphoteric heavy metal ions cannot enter the leaching solution, so the heavy metal content in the product is very low. However, due to the large use of sodium carbonate or sodium hydroxide (caustic soda) in the production process, and then neutralizing the excess alkali with hydrochloric acid, a large amount of by-product sodium chloride is produced. At the same time, the viscosity of aluminum hydroxide gel is very large and difficult to filter, these factors lead to higher production costs. However, it is an ideal method for producing high-purity polyaluminum chloride, especially for the production of cosmetic-grade polyaluminum chloride, which can meet the quality requirements of Japanese Patent No. 1990–4534 “Preparation of Pharmaceutical and Daily Cosmetic Grade Polyaluminum Chloride Solution”.

(2) Adjustment of Salt Basicity

In order to increase the content of polyaluminum ions (salt basicity), it is necessary to reduce the hydrochloric acid content in polyaluminum chloride, so it is necessary to add alkaline substances or aluminum scraps. Simply adding alkaline substances, such as lime, caustic soda, etc., can only increase the salt basicity and cannot increase the aluminum content of the solution. However, using aluminum scraps, calcium aluminate, sodium aluminate, etc. can achieve both purposes.

Sodium aluminate and calcium aluminate are weak alkaline aluminates, which have very good reactivity with low salt basicity polyaluminum chloride.

$$Al_2(OH)_2Cl_4 + 0.4NaAl(OH)_4 = 1.2Al_2(OH)_3Cl_3 + 0.4NaCl \tag{3.8}$$

$$Al_2(OH)_2Cl_4 + 0.2Ca\left[Al(OH)_4\right]_2 = 1.2Al_2(OH)_3Cl_3 + 0.2CaCl_2 \tag{3.9}$$

As can be seen from the above equations, the amount of alkali consumed when adjusting with aluminate is lower, and the yield of polyaluminum chloride can be increased.

The current cost of calcium aluminate is very low, similar to high-alumina cement, especially in Gongyi of Henan Province, Kaili of Guizhou Province, etc., the price of calcium aluminate is around 1200 yuan/ton. Therefore, the cost of adjusting the salt basicity with calcium aluminate is lower than that of using sodium aluminate. However, sodium aluminate is produced by alkali dissolution and will not introduce heavy metal ions during the adjustment process. If calcium aluminate is used for adjustment, a certain number of heavy metals will be introduced. This is also an important reason why polyaluminum chloride for drinking water currently uses aluminum hydroxide acid solution-calcium aluminate adjustment.

(3) Calculation of the amount of sodium aluminate solution or calcium aluminate added during the adjustment process

Sodium aluminate solution is a mixture of sodium aluminate and sodium hydroxide. In the process of mixing with low salt basicity polyaluminum, it can be considered that the hydroxide in the sodium aluminate solution replaces the chloride ion of polyaluminum chloride to form a high salt basicity product. The salt basicity of the adjusted polyaluminum chloride is equal to the molar ratio of the sum of the hydroxide content in the low salt basicity polyaluminum chloride and the hydroxide content in the sodium aluminate solution to three times the total aluminum content in the sodium aluminate and low salt basicity polyaluminum. The expression for the adjusted salt basicity is:

$$B = \frac{6B_0G_3n_3/M_{Al_2O_3} + G_1n_1/M_{NaOH} + 4G_1n_2/M_{NaAl(OH)_4}}{3[G_1n_2/M_{NaAl(OH)_4} + 2G_3n_3/M_{Al_2O_3}]} \tag{3.10}$$

where G_1 is the amount of calcium aluminate solution added, kg;

G_3 is the amount of polyaluminum chloride to be adjusted, kg;

n_1 is the percentage content of sodium hydroxide in the sodium aluminate solution, %;

n_2 is the percentage content of sodium aluminate in the sodium aluminate solution, %;

n_3 is the percentage of aluminum oxide in the polyaluminum chloride to be adjusted, %;

B_0 is the basicity of the polyaluminum chloride before adjustment, %;

B is the basicity of the polyaluminum chloride after adjustment, %;

$M_{Al_2O_3}$ is the molar mass of Al_2O_3, g/mol;

M_{NaOH} is the molar mass of NaOH, g/mol;

$M_{NaAl(OH)_4}$ is the molar mass of $NaAl(OH)_4$, g/mol.

The amount of sodium aluminate required in production, G1, is:

$$G_1 = \frac{6G_3n_3(B - B_0)/M_{NaAl(OH)_4}}{n_1/M_{NaOH} + n_2(4 - 3B)/M_{NaAl(OH)_4}} \tag{3.11}$$

The polyaluminum chloride solution with adjusted basicity can be directly added to the modified polyacrylamide precipitant, the supernatant is separated, or it can be directly filtered.

3.1.4.3 One-Step Method of Aluminum Hydroxide and Two-Step Method of Alum Soil-Aluminum Hydroxide

The one-step acid dissolution method of aluminum hydroxide uses high-alumina cement, i.e., aluminum hydroxide, to produce polyaluminum chloride, which appeared in the Gongyi area of Zhengzhou in the early 1990s. Since aluminum hydroxide itself is a highly active substance, it can react at room temperature by adjusting the concentration of hydrochloric acid and the amount of aluminum hydroxide added, accompanied by a large amount of heat release, which further accelerates the reaction. This technology is a huge advancement in China's coagulant industry. Because the reaction conditions have become relaxed, the volume of the reactor can be expanded to 100 m^3 or 300 m^3, greatly improving production efficiency. However, the disadvantage of this method is the obvious consumption of hydrochloric acid, and the introduction of impurity calcium chloride into the product, resulting in a low content of aluminum oxide in the target product, easy water absorption, and excessive total heavy metals.

In the late 1980s and late 1990s, the two-step method of alum soil-aluminum hydroxide appeared in industrial production. In order to reduce the consumption of hydrochloric acid and fully utilize the technical advantages of aluminum hydroxide raw material production, some enterprises use roasted aluminum vanadium soil to react with hydrochloric acid to produce aluminum chloride solution, which is diluted and then reacted with aluminum hydroxide. This method has become the process route used by most production enterprises in China, because it can not only fully utilize the advantage of China's rich aluminum ore resources but also improve economic benefits. At present, 90% of the polyaluminum chloride products in Gongyi of Henan Province and Zibo of Shandong Province are produced by this method, which can well meet the treatment of industrial water and drinking water. However, with the implementation of new national standards, this process cannot meet the new requirements for polyaluminum chloride in drinking water standards in terms of heavy metals and therefore has to withdraw from the drinking water industry.

$$Al_2O_3 + 6HCl = 2AlCl_3 + 3H_2O \tag{3.12}$$

$$AlCl_3 + Ca(AlO_2)_2 \rightarrow Al_2(OH)_nCl_{6-n} + xH_2O + CaCl_2 \tag{3.13}$$

In the above equation, bauxite reacts with hydrochloric acid to form aluminum chloride, which then reacts with calcium aluminate to produce a solution of polyaluminum chloride. The calcium aluminate used above was previously known as high alumina cement. Due to the rapid development and huge demand in the water purification industry, calcium aluminate has now become a separate product for this industry. Due to the nature of the ore, calcium aluminate often has a problem with heavy metals. Heavy metal ions enter the polyaluminum chloride solution with the dissolution of calcium aluminate, often resulting in excessive heavy metals. In response to this problem, Mr. Ouyang Weihua from used the tunnel kiln roasting method in 2010, using alumina and high-purity bluestone as raw materials, to produce white calcium aluminate, thereby greatly reducing the content of heavy metals in calcium aluminate. Moreover, this calcium aluminate maintains the activity of conventional calcium aluminate and can react with hydrochloric acid under conditions below 100 °C and normal pressure to obtain a colorless transparent solution, and the salt basicity can exceed 70%. Unfortunately, the quality standard achieved by this technology far exceeds the quality requirements of the national standard in China "polyaluminum chloride for drinking water" (GB/T 15892-2020). The application of white calcium aluminate products will be a milestone in this industry, greatly improving the product quality of polyaluminum chloride in China and simplifying the production process.

3.1.4.4 One-Step and Two-Step Methods of Aluminum Hydroxide Solution

In the 1980s, the teams of Academician Tang Hongxiao and Professor Li Runsheng independently conducted a large amount of research work, further realizing technology industrialization. Academician Tang Hongxiao first applied spray drying technology to the production of solid polyaluminum chloride in Tangshan of Hebei Province, and obtained white solid dry powder products. This technology laid the foundation for the widely used polyaluminum chloride drying technology in this industry.

The process of the one-step method of crystalline aluminum hydroxide solution is relatively simple. Hydrochloric acid and aluminum hydroxide react in a closed container to obtain a product with a salt basicity of less than 50%. The process conditions can be achieved at 0.4PMa and 150 °C. The widely used bauxite alkali Bayer method in the industry produces sodium aluminate and aluminum hydroxide, so the heavy metal content is relatively low, and the product meets the requirements of drinking water. Due to the harsh reaction conditions and poor operational safety, this method was used less in the late 1990s, but with increased supervision of waste in recent years in China and the improvement of the quality requirements of polyaluminum chloride for drinking water, this method has gradually been used more widely.

$$2Al(OH)_3 + (6 - n)HCl = Al_2(OH)_nCl_{6-n} + (6 - n)H_2O \quad (3.14)$$

The salt basicity of the polyaluminum chloride obtained by the above one-step method is generally difficult to exceed 50% (salt basicity B% = 40–50%), so it is necessary to adjust the salt basicity to above 80%.

$$2Al(OH)_3 + HCl = 2AlCl_3 + 3H_2O \quad (3.15)$$

$$AlCl_3 + Ca(AlO_2)_2 \rightarrow Al_2(OH)_nCl_{6-n} + xH_2O + CaCl_2 \quad (3.16)$$

After 2010, CHALCO Shandong Co., Ltd introduced a soluble alumina hydroxide product. The so-called solubility refers to the reaction of alumina hydroxide with hydrochloric acid, which can dissolve alumina hydroxide under conditions below 100 °C and normal pressure, resulting in aluminum chloride or low-salt basicity polyaluminum chloride. Soluble alumina hydroxide has been the biggest advancement in this industry over the past 20 years, making the production process simple and safe.

The reaction of alumina hydroxide and aluminum acid calcium usually consists of three steps. ① The first is the acid-matching reaction. 1500 kg of acid and 1000 kg of water are added to the reaction kettle, making the concentration of hydrochloric acid between 20 and 22%; ② the second is the acid-dissolving alumina hydroxide reaction. 600 kg of fine alumina hydroxide is added to the reaction kettle containing acid. The reaction of the two compounds was conducted at 110–130 °C and 0.1 MPa for 2 h (at present, the reaction can be carried out at normal pressure). The alumina content of the product is 8–9%, and the base degree is 19–21%.; ③ the third is the polymerization reaction. After the aluminum dissolution reaction is completed, 250 kg of aluminum acid calcium powder is added to the reaction kettle, and the various mixtures continue to react at 105 °C, 0.25PMa for 1.5 h. The product's alumina content is 13–14%, and the salt basicity is 45–55%. The product contains 3–5% slag. To improve the product's salt basicity, water and aluminum acid calcium can be added during the secondary polymerization process, which not only improves the product's flocculation effect but also facilitates drying the product and improving drying efficiency.

The process of producing polyaluminum chloride by the alumina hydroxide-aluminum acid calcium two-step method is shown in Table 3.6. The quality indicators of alumina hydroxide one-step and two-step products are compared in Table 3.7.

If active alumina hydroxide is used for production, all reactions can be carried out under normal pressure. If the salt basicity of the first step dissolution product is controlled to zero, it is more conducive to the dissolution and polymerization of aluminum acid calcium.

Table 3.6 Alumina hydroxide-aluminum acid calcium two-step method for producing polyaluminum chloride process

	Acid matching		First polymerization		Second polymerization	
	Hydrochloric acid	Water	Alumina hydroxide	Indicator	Water	Aluminum acid calcium
Process 1	1500	1000	600	B:20–30% Al_2O_3:8–9%	0	250
Process 2	1050	520	400	B:20–30% Al_2O_3:8–9%	975	300

Table 3.7 Comparison of quality indicators of alumina hydroxide one-step and two-step products

	Al_2O_3 (%)	Salt basicity (%)	pH value	Insoluble matter	Density (g/ cm^3)	Slag content quality percentage (%)
One-step method	11.92	50.29	3.5	0.3	1.25	9.5
	12.20	48.50	3.5	0.3	1.26	9.7
	11.88	51.29	3.5	0.3	1.25	9.6
	12.18	50.21	3.5	0.3	1.25	9.6
Two-step method	11.72	74.68	4.5	0.3	1.24	6.6
	12.07	72.49	4.5	0.3	1.24	6.3
	11.68	76.52	4.5	0.3	1.24	6.5
	11.36	77.25	4.5	0.3	1.24	6.5

3.1.4.5 Removal of Heavy Metal Ions in Poly Aluminum Chloride

For polyaluminum chloride solutions with high heavy metal content, sulfides, such as sodium sulfide, can be added to form precipitates and remove elements such as lead, copper, tin, and chromium. Alternatively, methods such as aluminum chip displacement, activated carbon adsorption can be used to remove heavy metals, or the method of aluminum carbon micro-battery enhanced reduction can be used to enhance the reduction of heavy metal ions. The method of sulfide precipitation can be used in conjunction with modified polyacrylamide flocculants to enhance the removal effect of precipitates.

3.1.5 Resource Utilization of Hazardous Aluminum Ash and Production of Poly Aluminum Chloride

Aluminum ash refers to the solid waste generated in the aluminum industry through processes such as aluminum electrolysis, processing, and regeneration. Depending on the way and form in which the aluminum ash is produced, it can be divided into primary and secondary aluminum ash. Primary aluminum ash is usually the insoluble slag produced in the process of aluminum electrolysis, which is usually white and also known as white ash. Secondary aluminum ash, also known as black ash, is the slag produced by recycling and remelting primary aluminum ash or other aluminum-containing metals and their processing, and it is black in color.

In recent years, with the continuous increase in primary aluminum production, the production of aluminum ash has also been increasing. In 2020, China's primary aluminum production was 37.337 million tons, and the production of recycled aluminum was 7.25 million tons, accounting for 57% and 23% of the world's total, respectively. Under normal circumstances, the direct production of 1000 tons of metallic aluminum from bauxite will produce 10–20 tons of aluminum ash, and the recycling of secondary aluminum resources will produce 20–50 tons, or even higher, of more complex aluminum ash. It is estimated that China's annual production of aluminum ash exceeds 3 million tons.

The main components of aluminum ash include metallic aluminum, aluminum oxide, aluminum nitride, and salts such as chlorides and fluorides. Depending on the different raw materials and different process flows, the components of aluminum ash from different sources often vary greatly. The specific range of aluminum ash components is roughly as shown in Table 3.8.

Due to the complex composition of aluminum ash, long-term large-scale accumulation can easily cause heavy metal pollution of water bodies and soil, soil salinization, and the production of irritating ammonia gas when damp, causing great environmental pollution and serious harm to human health. In view of the hazards of aluminum ash, it was officially included in the "National Hazardous Waste List" in 2016 in China.

In terms of the treatment and utilization of aluminum ash, it is mainly used for the synthesis of hard materials or the production of coagulants. The synthesized hard materials mainly include building materials, road materials, refractory materials, recycled alumina, composite material desulfurizers, etc. Although they have certain research value, their general application value is not high, and they consume a lot of energy and are prone to cause other pollution.

The coagulants prepared from aluminum ash are mainly polyaluminum chloride and aluminum sulfate. Due to the mild reaction conditions, simple process flow, and low production cost of aluminum ash preparation coagulants, aluminum ash-based coagulants have good economic benefits in production practice. However, the coagulants produced using aluminum ash often exceed the standard in terms of heavy metal content and ammonia nitrogen. According to industry standards,

Table 3.8 Specific components and content range of aluminum ash

Type of aluminum Ash	$w(Al)/\%$	$w(Al_2O_3)/\%$	$w(AlN)/\%$	$w(salts)/\%$
Primary aluminum As	15–80	20–85	< 5	< 5
Secondary aluminum As	5–10	50–65	15–30	30–50

polyaluminum chloride and aluminum sulfate produced from hazardous waste can only be used for various types of reclaimed water and wastewater treatment and are strictly prohibited for use in drinking water treatment. Therefore, how to improve the process of producing coagulants from aluminum ash, reducing the content of heavy metals, and increasing the purity of coagulants is an important part of the aluminum ash coagulant production industry.

3.1.5.1 The Problem and Resource Utilization of Aluminum Ash

Industrial production often accompanies the generation of waste and the consumption of natural resources. Industrial waste accumulates over time, causing serious harm to the environment and public health. Aluminum is the third most abundant element on Earth, and aluminum metal is the second largest category of metals after steel. The aluminum production industry, as one of the most important strategic industries, also brings many hazardous waste problems. Aluminum metal is light in weight, has good conductivity, and has excellent corrosion resistance. These advantages make it a widely applicable material in fields such as aerospace, building structures, and marine industries. In 2021, the world's primary aluminum output was 67.243 million tons, of which China's output was 38.837 million tons, accounting for 57.7%; at the same time, China is also the world's largest producer and consumer of recycled aluminum, recovering more than 10 million tons of metal from waste each year, accounting for about 1/3 of the global annual output.

The production cost of electrolytic aluminum is mainly composed of electricity and alumina costs. As a resource and energy-intensive industry, China's electrolytic aluminum industry distribution has obvious regional focus characteristics. Shanxi Province and Henan Province have rich bauxite and coal resources, and their electrolytic aluminum industry has a first-mover advantage; Shandong Province has risen rapidly with its own power and low-cost imported bauxite, and Xinjiang Uygur Autonomous Region has become a rising star due to low-cost electricity. With the proposal of the "dual carbon" goal, the advantage of coal-fired aluminum is no longer, and provinces with abundant water resources such as Yunnan Province are welcoming the growth of electrolytic aluminum capacity.

In order to control the disorderly expansion of electrolytic aluminum capacity, the state listed it as an overcapacity industry in 2017, implementing supply-side structural reforms, set capacity ceiling, and strictly restricted the scale of new capacity. Under this background, the development of the recycled aluminum industry is a trend. According to the "Non-ferrous Metal Industry Carbon Peak Action Plan" issued by the Ministry of Industry and Information Technology, the National Development and Reform Commission, and the Ministry of Ecology and Environment in China, China's recycled aluminum output will reach 11.5 million tons by 2025. At present, China's recycled aluminum enterprises are mainly distributed in the southeast coast, with the distribution feature being close to the aluminum consumption market, and at the same time, the transportation cost is low, which can conveniently obtain foreign waste aluminum resources. According to data released by the International Aluminum Institute (IAI), the electricity consumption for producing 1 ton of electrolytic aluminum is 13,500 kWh, while the electricity consumption for producing 1 ton of recycled aluminum is only 230 kWh. According to data released by the China Product Life Cycle Greenhouse Gas Emission Factor Library, under the industrial structure of thermal power supply, the greenhouse gas emissions of recycled aluminum are only 1/40 of that of primary aluminum (recycled aluminum 0.72 tCO_2/t, thermal power aluminum 29.04 tCO_2/t). With the improvement of the carbon emission mechanism, the electrolytic aluminum industry will be gradually included in the carbon trading market, which will further make up for the cost disadvantage of recycled aluminum. Under policy promotion, the recycled aluminum industry has begun to show a trend of decreasing the number of enterprises and scale of production.

As the China's emphasis on the ecological environment continues to increase, the emission standards for solid waste are becoming increasingly strict. The National Hazardous Waste List (2021 Edition) published in 2021 has already listed secondary aluminum ash as hazardous waste, and it is not allowed to landfill aluminum ash. Therefore, how to dispose of secondary aluminum ash has become a difficult problem that urgently needs to be solved in the aluminum production industry. On the other hand, during the "the Fourteenth Five-Year Plan" period, China's goal of promoting high-quality development requires us to comprehensively improve resource utilization and encourage the resource utilization of bulk solid waste. Secondary aluminum ash contains rich aluminum resources. The recycling of secondary aluminum ash can not only alleviate environmental pressure but also achieve efficient use of resources and sustainable economic development.

3.1.5.2 The Source, Composition, and Harm of Secondary Aluminum Ash

(1) The source of aluminum ash

As mentioned earlier, aluminum ash is divided into primary aluminum ash and secondary aluminum ash. The former is produced in the production of primary aluminum, and the latter is produced in the process of recycling primary aluminum ash or regenerated aluminum in secondary smelters. Figure 3.5 shows the process of the aluminum industry.

In the production of primary aluminum, bauxite is first processed by the Bayer process to manufacture alumina. Depending on the type of bauxite, the processing conditions are different. Gibbsite can be leached under mild conditions, while diaspore requires a temperature above 145 °C to ensure the leaching rate. China's bauxite is mainly hard diaspore, with higher leaching temperature and alkali concentration, and requires pressure leaching, usually also needs to add lime sintering to increase yield. The alumina extracted from bauxite is electrolyzed into metallic aluminum by the Hall-Héroult process. In order to reduce the melting point of alumina, cryolite (Na_3AlF_6) and aluminum fluoride (AlF_3) are usually added to the electrolytic

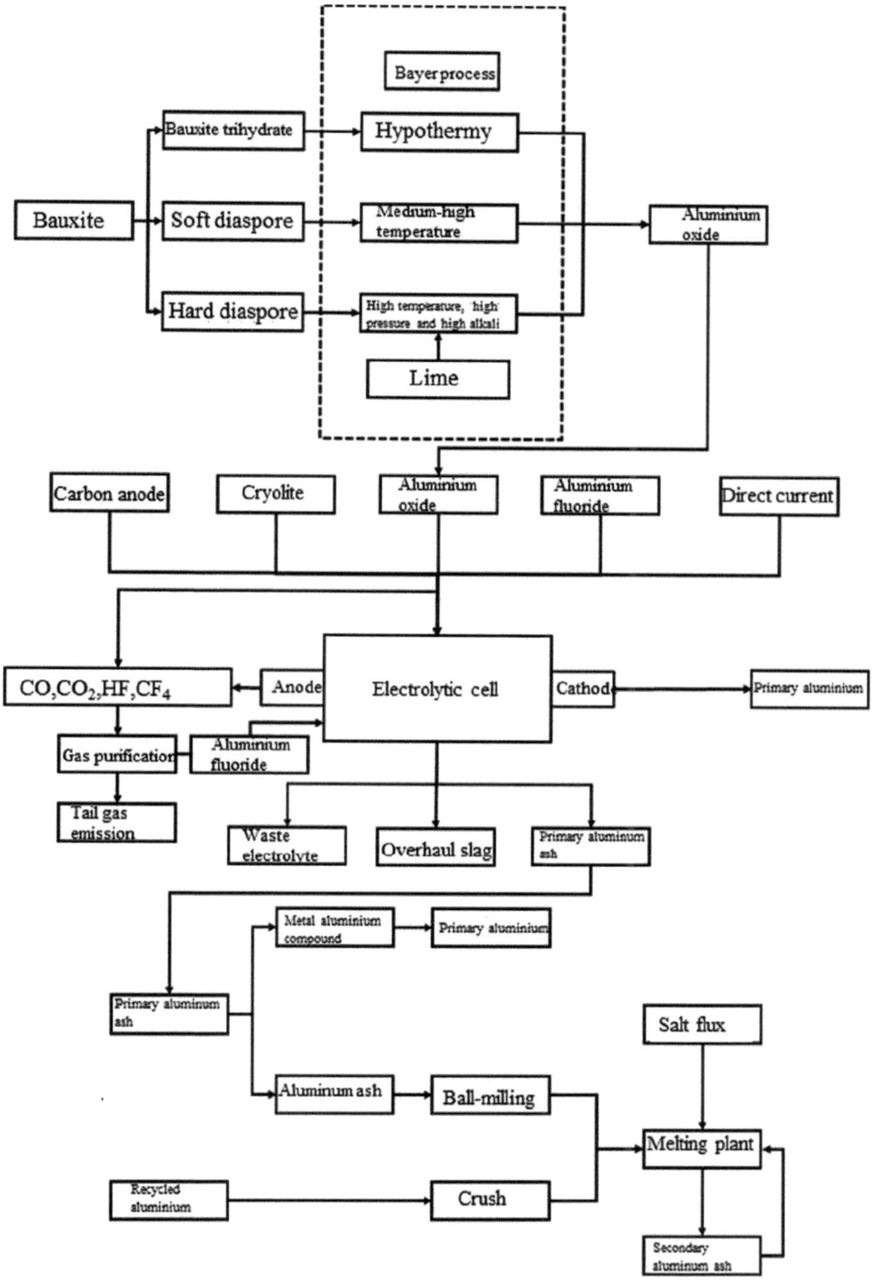

Fig. 3.5 Aluminum industry flow chart

cell. Aluminum ions are reduced at the cathode to concentrate primary aluminum, and the carbon at the anode will react with the oxygen atoms that lose electrons to generate carbon monoxide (CO), carbon dioxide (CO_2), and some hydrogen fluoride (HF), carbon tetrafluoride (CF_4) and other gases. Among them, HF can react with alumina to generate aluminum fluoride and continue to enter the electrolytic cell as raw materials. The electrolytic cell will deform and rupture after being eroded for a long time and will be replaced. The waste generated is called overhaul slag. The overhaul slag mainly includes carbon and fluorides, where carbon can be recycled, and fluorides must be harmlessly treated. The waste electrolyte produced by electrolytic aluminum can be used as an additive in the smelting industry or a supplementary raw material in the cement industry. During the electrolysis process, the surface of the molten aluminum liquid is oxidized, and the oxide film is broken with the movement of the aluminum liquid, and the oxide particles continuously aggregate to form primary aluminum ash. After the primary aluminum ash is sorted, large aluminum particles can be directly recycled, and the remaining aluminum ash is milled and then enters the melting device to recover metallic aluminum. Usually, such aluminum ash contains 80% of metallic aluminum, which is recovered by two or three times of melting. The secondary melting stage is also called "frying ash", and the aluminum ash produced at this time is secondary aluminum ash.

(2) The composition of secondary aluminum ash

The main components of secondary aluminum ash are oxides such as alumina (Al_2O_3), magnesia-alumina spinel ($MgAl_2O_4$), and silica (SiO_2), as well as some unreacted metallic aluminum (Al). The remaining components vary greatly due to different production processes and raw materials. In most cases, salt fluxes are used in the recycling process of aluminum flash to minimize oxidation and promote the coagulation and separation of metals. Salt fluxes can promote the transfer of heat to the metal, protect the metal from atmospheric influences, and some salt fluxes can chemically react with and dissolve aluminum oxide. The commonly used salt fluxes may be chlorides such as sodium chloride (NaCl), potassium chloride (KCl), or fluorides such as calcium fluoride (CaF_2), sodium fluoride (NaF), cryolite (Na_3AlF_6), etc.

In the process of aluminum electrolysis, inexpensive nitrogen (N_2) is usually used as a carrier gas. The high-temperature aluminum liquid reacts with N_2 to form aluminum nitride (AlN) which remains in the aluminum flash; in the recycling process of aluminum flash, molten aluminum will also react with N_2 in the air to produce AlN.

In some cases, various alloy elements are used together with salt fluxes to produce aluminum alloys. Different alloy elements lead to differences in the content and types of metal oxides in different aluminum flash.

Organic pollution caused by aluminum waste products with paint or plastic layers may bring carbon elements. When molten aluminum reacts with dispersed carbon particles, aluminum carbide (Al_4C_3) is formed. During the remelting of aluminum products, organic compound refining agents such as hexachloroethane (C_2Cl_6) are often added to the aluminum liquid. Incomplete combustion of these organics can produce toxic gases that pollute the air around the aluminum plant and form some Al_4C_3 in the aluminum flash.

Some aluminum products are coated with a phosphating layer to protect and improve adhesion. It reacts with molten aluminum to form aluminum phosphide (AlP) which remains in the aluminum flash; AlP may also be formed by the reaction of aluminum with phosphates in the raw materials.

(3) Hazards of Secondary Aluminum Flash

The hazards of secondary aluminum flash are mainly reflected in its reactivity and toxicity. Secondary aluminum flash can react with water or water vapor in the air, producing various gases. Some of these gases are flammable and explosive, such as hydrogen (H_2), methane (CH_4), and some are toxic and foul-smelling, such as ammonia (NH_3), phosphine (PH_3). The related reactions are as follows:

$$2Al + 6H_2O = 2Al(OH)_3 + 3H_2(g) \tag{3.17}$$

$$Al_4C_3 + 12H_2O = 4Al(OH)_3 + 3CH_4(g) \tag{3.18}$$

$$AlN + 3H_2O = Al(OH)_3 + NH_3(g) \tag{3.19}$$

$$AlP + 3H_2O = Al(OH)_3 + PH_3(g) \tag{3.20}$$

The release of NH_3 has a significant impact on the environment. NH_3 is very soluble in water, causing its pH value to reach 9 or even higher, further promoting the hydrolysis of secondary aluminum flash. At the same time, the accumulation of NH_3 in water bodies can easily cause ammonia poisoning in aquatic organisms. Under the action of nitrifying bacteria, NH_3 will be oxidized and produce strong acidic substances, causing serious damage to the soil and lowering the pH value of groundwater. CH_4 is a greenhouse gas that is more potent than carbon dioxide, and PH_3 is a highly toxic gas. The gases released by secondary aluminum flashflash seriously pollute the surrounding environment and pose an explosion risk.

Improper landfilling of secondary aluminum flashflash can increase the aluminum element in the soil and also bring the risk of exceeding the standard of other heavy metals. The chlorides in the aluminum flashflash can increase the risk of soil salinization, and the fluorides can cause excessive fluoride concentration in the soil and groundwater.

Given the risk of secondary aluminum ash polluting the atmosphere, soil, and water sources, China has strictly prohibited the landfilling of secondary aluminum ash. According to the "Environmental Protection Tax Law of the People's Republic of China", a tax of 1000 yuan will be levied on each ton of hazardous waste, and companies also face substantial taxes on air and water pollution emissions. The harmless treatment and resource utilization of aluminum ash has become a focus of industry research.

3.1.5.3 Harmless Treatment of Secondary Aluminum Ash

(1) Removal of AlN in secondary aluminum ash

① Hydrolysis removal of AlN

AlN, due to its excellent thermal conductivity, corrosion resistance, high-temperature resistance, and good chemical stability, is widely used as packaging material, heat dissipation material, and circuit components for large-scale integrated circuits and high-power electronic devices. Since AlN easily absorbs moisture, the research focus is usually on the surface modification of AlN powder and ceramics to prevent its hydrolysis. These two types of AlN have an induction period before the hydrolysis reaction, which shortens with increasing temperature. However, when secondary aluminum ash comes into contact with water, AlN hydrolyzes quickly, releasing ammonia gas. This is because AlN powder and AlN ceramics have a dense structure and relatively poor chemical activity, while secondary aluminum ash is loose, with very fine AlN particle size and a large specific surface area. Also, the heat released during the hydrolysis of secondary aluminum ash further promotes AlN hydrolysis.

Andraz et al. have studied the hydrolysis mechanism of AlN powder. The hydrolysis reaction of AlN powder is accompanied by an increase in pH value and temperature. Increasing the initial reaction temperature will increase the reaction rate; the initial temperature and hydrolysis time significantly affect the reaction products and their morphology. The study confirmed that at room temperature, the main crystalline reaction product is bayerite ($Al(OH)_3$). At high temperatures, the crystalline product becomes boehmite (AlOOH). As the hydrolysis time extends, it will reconvert to bayerite, with the specific reaction as follows:

$$AlN + 2H_2O = AlOOH_{(amorph)} + NH_3(g) \tag{3.21}$$

$$NH_3 + H_2O = NH_4^+ + OH^- \tag{3.22}$$

$$AlOOH_{(amorph)} + H_2O = Al(OH)_{3(xstal)} \tag{3.23}$$

Jiang Lan et al. have studied the hydrolysis behavior of AlN in aluminum ash. The study shows that the reaction temperature and time have a significant impact on the hydrolysis rate and degree of AlN in aluminum ash, stirring can accelerate the reaction, while the particle size has no obvious effect on hydrolysis. The study also points out that compared to AlN powder and ceramics, the AlN in aluminum ash is more active, there is no obvious induction period, and no AlOOH phase was observed in the hydrolysis products.

The hydrolysis removal of AlN in aluminum ash is simple and energy-efficient, making it the preferred process for many companies. Industrial water can be directly used for hydrolysis denitrification, which is low-cost, and the absorbed ammonia gas can be used to produce ammonium hydroxide. The drawback of direct hydrolysis is that the degree of hydrolysis is not thorough. According to Tang Linghong's research, even under optimal reaction conditions, the denitrification efficiency of AlN in distilled water is only 38.12%. Alkaline conditions can effectively improve the hydrolysis efficiency of AlN, with the hydrolysis efficiency of AlN reaching 94.14% in a 5% Na_2CO_3 solution, and 99.76% in a 5% NaOH solution. The flaw in this experiment is that the liquid–solid ratio of the reaction is too high, with a liquid–solid ratio of 10:1 in the NaOH system, and even 40:1 in the Na_2CO_3 system. Although the removal of AlN is very effective, the consumption of alkali is too large, making it not economically viable.

② Oxidation Removal of AlN

AlN has reductive properties. AlN in secondary aluminum flashflash will undergo an oxidation reaction in high-temperature air, ultimately producing α-Al_2O_3 as a product. Li Yong et al. have studied the mechanism of high-temperature roasting and nitrogen removal of aluminum flashflash. AlN particles react with O_2 on the surface to form an Al_2O_3-AlN shell structure, then O_2 diffuses inward through the Al_2O_3 layer and oxidizes AlN, causing the Al_2O_3 layer to increase in thickness. Due to the different thermal expansion coefficients of Al_2O_3 and AlN, when the temperature rises, the shell structure ruptures, exposing AlN to oxygen again and being oxidized.

Su et al. proposed a new method to detoxify, recycle, and apply high-value secondary aluminum flash through calcination roasting in an air atmosphere, that is, co-roasting secondary aluminum flash with CaO in the air can simultaneously achieve nitrogen removal, chlorine salt recovery, and the preparation of calcium aluminate. Under optimal conditions, the molar ratio of Ca to Al is 1.7, the roasting temperature is 1400 °C, and after reacting for 2 h in an air atmosphere, 99.7% of the AlN in the secondary aluminum flash is converted into harmless N_2.

The main components of secondary aluminum flash are similar to cement raw materials. Using cement kilns to co-process aluminum flash can not only detoxify and deactivate aluminum flash but also resource it. High-temperature treatment can convert the aluminum nitride in the aluminum flash into N_2, and the metallic aluminum will also become aluminum oxide, solving the environmental risk of aluminum flash. Research has found that the addition of aluminum flash can also improve the SCR (selective catalytic reduction) nitrogen oxide reduction efficiency in cement kilns, and the AlN in the aluminum flash can act as a reductive nitrogen to react with NO_x to form N_2 under high-temperature conditions.

The energy consumption of the fire method AlN oxidation removal is large, the composition of the produced gas is difficult to determine, and nitrogen oxides may be produced, causing environmental damage. The wet method AlN hydrolysis removal cannot avoid the explosion risk brought by gases such as H_2 and CH_4 during the process. Designing a safe and reliable low-cost AlN removal process is the key to achieving harmless treatment of secondary aluminum flash.

(2) Removal of Salt in Secondary Aluminum Flash

The salts in secondary aluminum flash are divided into chloride salts and fluoride salts. Graziano et al. proposed four potential salt recovery technologies: (a) basic process: leaching in water at 25 °C, crystallizing the salt by evaporation; (b) high-temperature process: leaching in water at 250 °C, precipitating the salt by flash crystallization; (c) solvent/anti-solvent process: leaching in water at 25 °C, precipitating the salt by adding acetone; (d) electrodialysis process: leaching in water at 25 °C, concentrating and recovering the salt by electrodialysis.

Binnaz et al. studied the salt leaching technology of aluminum flash and determined the solubility conditions of aluminum flash in water. To study the effect of temperature, dissolution experiments were conducted at 20 °C, 40 °C, 60 °C, 80 °C, and 100 °C with a solid–liquid ratio of 1:15. The results showed that the temperature had little effect on the solubility, so room temperature of 20 °C is sufficient to achieve the best salt solubility. The test results show that mixing for 30 min is enough to dissolve the salts in the aluminum flash and achieve the maximum recovery rate. The composition of the recovered flux (70% NaCl, 30% KCl) is almost the same as that of the original flux (69% NaCl, 29% KCl, 2% CaF_2). Davies et al. recovered salt at room temperature with a liquid–solid ratio of 7:3 after crushing and grinding the aluminum flash. The results showed that 90% of Cl, 55% of Na, and 45% of K could be recovered.

According to China's Standard for Identification of Hazardous Wastes Leaching Toxicity Identification (GB5085.3-2007), the leaching concentration limit of inorganic fluorides is 100 mg/L, so research on the fixation of fluorine in secondary aluminum ash is necessary. Bao Shanci et al. conducted leaching research on the fluorine and aluminum elements in secondary aluminum ash. The research shows that under the optimal leaching parameters, the maximum leaching rate of fluorine elements is 87.62%, and the maximum leaching rate of chlorine elements is 99.02%. The filtrate can be returned to fluoride salt and chloride salt after evaporation and concentration.

Wan et al. used secondary aluminum ash as raw material, first washed and evaporated to recover soluble chloride salt, and then extracted aluminum elements through an alkaline roasting-leaching process at 500 °C. They introduced CO_2 into the leaching solution to produce cryolite (Na_3AlF_6), with secondary aluminum ash providing the source of aluminum and part of the fluorine. Song Xuefeng et al. used gypsum to fix the fluorine elements in secondary aluminum ash in combination with the calcium salt precipitation method. The calcium in gypsum can react with fluoride ions to form stable calcium fluoride, thus achieving the effect of fluorine fixation. The research shows that desulfurization gypsum doped with some kaolin has a strong fixation effect on the fluoride ions in secondary aluminum ash, and its water resistance is better.

(3) Wet recovery of secondary aluminum ash

The key to the wet treatment technology of aluminum ash lies in the leaching medium and leaching parameters, with the core being the recovery of alumina in the aluminum ash. Secondary aluminum ash contains 50–65% alumina, and the cost of producing the same quality of alumina using secondary aluminum ash as raw material is only 49.54% of the cost of using bauxite technology. According to the different leaching

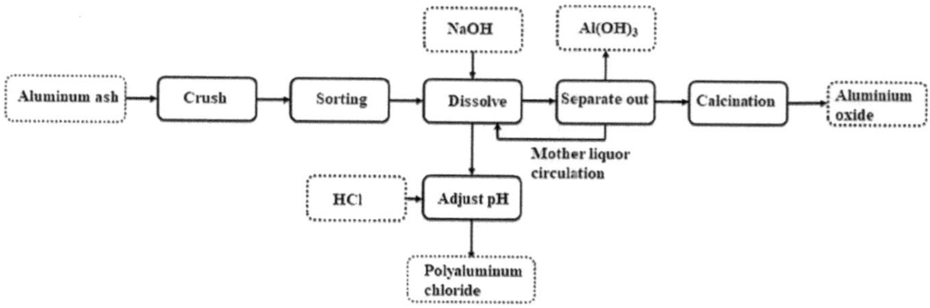

Fig. 3.6 Alkali leaching process flow chart of aluminum ash

media, the wet treatment technology can be divided into alkali leaching process and acid leaching process.

① Alkali leaching method for recycling secondary aluminum ash

Aluminum ash can be leached with alkali solution to recover most of the aluminum elements in it. The commonly used alkali solution is NaOH solution. The alkali leaching process flow chart of aluminum ash is shown in Fig. 3.6. Aluminum and its oxides react with NaOH to obtain a leaching solution rich in sodium aluminate. The leaching solution can adjust by the pH value, adding crystals to precipitate aluminum hydroxide, and then calcine to obtain alumina products. Since other heavy metals cannot be dissolved during alkali leaching, the purity of aluminum hydroxide obtained after dehydration is very high, suitable for the production of various water treatment agents. Depending on the process, the sodium aluminate solution obtained after alkali leaching. It can also adjust the pH value with hydrochloric acid to obtain polyaluminum chloride products.

Many scholars at home and abroad have researched on the alkali leaching of aluminum ash. Jafari et al. found that a 1–2 mol/L NaOH solution can provide enough hydroxide ions to react with the metal aluminum in the aluminum ash. Zauzi et al. studied the characteristics of aluminum ash at different NaOH concentrations, different reaction times, and different reaction temperatures. At a reaction temperature of 40 °C, the specific surface area of aluminum ash increased from 10.1 to 80.0 m^2/g after being wetted with a 1% NaOH solution for 15 min, which makes aluminum ash have the potential to become a catalyst material. The research shows that the alkali leaching of aluminum ash will release a lot of heat, and the system will quickly enter a boiling state; at the same time, the concentration of alkali solution will significantly affect the alkali leaching efficiency. At 100 °C, using 70% alkali solution to wet the aluminum ash with an alkali ash ratio of 70:20, the 0.5 h leaching rate of alumina in the aluminum ash can reach 93%.

Yoldi et al. ground secondary aluminum ash in a ball mill for 3 h, then extracted it with 2 mol/L NaOH for 1 h. The study showed that treating secondary aluminum ash with NaOH can extract a large amount of Al elements, and the obtained aqueous solution is a high-value-added product as an Al source. In addition, since Al extraction is carried out under reflux, NH_3 and all other volatile compounds have been removed. However, the product still faces the problem of excessive salt, which needs to be washed with deionized water.

Li Lingling et al. studied the influence of various parameters of the aluminum ash alkali leaching process on the alkali leaching results. Under the condition of 250 °C, most of the alumina in the secondary aluminum ash will dissolve rapidly within 0.5 h; at the same time, they explored the best conditions for alumina dissolution: solid–liquid ratio is 80 g/L, the mass concentration of NaOH is 248 g/L, the reaction temperature is 250 °C, the reaction time is 3 h, and the alumina dissolution rate is 98.6%.

Thanks to the amphoteric nature of aluminum, the aluminum elements in the aluminum ash can dissolve in the alkali solution, while other impurity metals cannot dissolve, so the product obtained by the alkali leaching process has high purity, and the process can refer to the Bayer method to produce alumina, realizing the recycling of alkali solution. And different characteristics of alumina products can be obtained by adjusting during the production process. However, the alkali leaching process requires a large amount of alkali to ensure the leaching rate of alumina, the production cost is high, and the generation of H_2, NH_3 and other gases needs to be considered during the production process, and it is also accompanied by the difficulty of waste liquid treatment. The alkali leaching process is suitable for aluminum ash production with high aluminum and iron content and low silicon content to avoid the impact of silica and obtain high-purity alumina.

② Acid leaching method for recycling secondary aluminum ash

More than thirty years ago, the acid leaching process has been widely used in the recycling of secondary aluminum ash. With the gradual improvement of the national standard for polyaluminum chloride, people have more understanding of aluminum ash, including safety issues in production and heavy metal issues in products, or the problem of irregular transfer of hazardous waste. The acid leaching process uses an acidic leaching medium, adjusts the leaching conditions to dissolve the aluminum elements in the aluminum ash, and separates them from insoluble substances by filtration. The acid leaching process can very conveniently obtain water treatment agent products such as polyaluminum sulfate and polyaluminum chloride and can also continue to process to obtain alumina products. The acid leaching process also requires the crushing and sorting of aluminum ash first. The process flow chart of wet method and acid leaching is shown in Fig. 3.7.

The reaction activity of aluminum ash and hydrochloric acid is very intense, and it can be acid leached with hydrochloric acid or a mixture of hydrochloric acid and sulfuric acid. Sarker et al. directly dissolved aluminum ash with HCl without any pretreatment. The study showed that as the leaching temperature, time, and acid concentration increased, the leaching rate of alumina continued to increase. Under the best conditions of 4 mol/L HCl, 2 h leaching time, and 100 °C, up to 71% of alumina was extracted from the aluminum slag. By heat treating the dry gel at 1000, 1200, and 1400 °C, θ-Al_2O_3, $(\alpha + \theta) - Al_2O_3$, and α-Al_2O_3 were obtained. XRD analysis showed that the alumina obtained by this method is all nanoscale. This

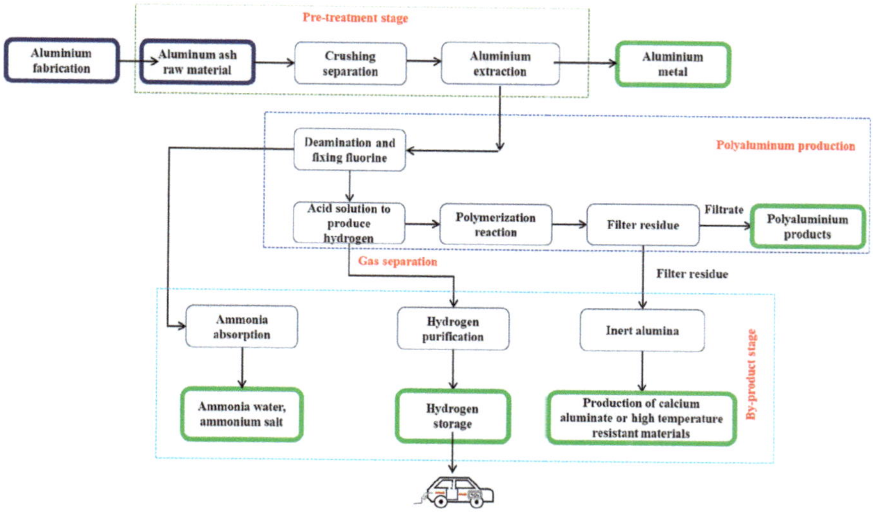

Fig. 3.7 Wet method and acid leaching process flow chart

treatment method combines wet treatment and fire treatment, the product value is higher but the treatment process is more complicated.

The method of producing polyaluminum from aluminum ash as a raw material has been used since the 1970s. It reacts with HCl and aluminum ash, and polyaluminum chloride products are obtained by adjusting the salt base with the addition of aluminum calcium. Chao Xi explored the effects of HCl concentration, reaction temperature, reaction time, rotation speed, and liquid–solid ratio on the leaching of aluminum ash, and gave the best parameters for acid leaching in combination with production costs: HCl 6 mol/L, 85 °C, reaction time 2 h, rotation speed 200r/min, liquid–solid ratio 12:1(mL/g); and the optimal parameters for preparing polyaluminum: the ratio of calcium aluminate to acid leaching solution is 12:80(g/mL), temperature 85 °C, reaction 1.5 h, stirring speed 200r/min. The production of water treatment agents using aluminum ash is currently a relatively mature and simple method of aluminum ash recycling.

The cost of acid leaching process is lower than that of alkali leaching process, mainly because the cost of hydrochloric acid is much lower than that of sodium hydroxide. Moreover, acid leaching requires less pretreatment, has higher leaching efficiency, and more mild leaching conditions. Acid leaching can conveniently obtain water treatment products such as polyaluminum chloride, while the alkali leaching process is more mature if aluminum oxide products are to be obtained. Whether it is acid leaching or alkali leaching, the gases produced during the production process, especially H_2, which can pose an explosion risk, need to be considered. The disadvantage of the acid leaching process is that the purity of the product is greatly affected by the raw materials. If the original aluminum ash contains other metal impurities, a heavy metal removal process is necessary.

The conditions for wet treatment are relatively mild, but the leaching rate is limited by the form of aluminum oxide in the aluminum ash. In production, the removal of nitrogen, fluorine, and heavy metals in aluminum ash needs to be considered. At the same time, a high silicon content in the aluminum ash also affects production. In the alkali leaching process, aluminum ash with a high silicon content will consume more alkali solution, and the requirements for pressure and temperature are also higher; in the acid leaching process, a high silicon content will affect the filtration performance, leading to difficulties in filtration. Therefore, the wet treatment process requires more operating units in actual production to solve problems in each process. For the process of producing solid polyaluminum chloride products, in order to increase the proportion of solid active ingredients, whether it is alkali leaching or acid leaching, desalination can be used before dissolution to reduce the sodium chloride and potassium chloride in the mixture, and recover these resources to the aluminum processing process to save costs.

3.1.5.4 Production of Aluminum Ash and Coagulants

The coagulants produced using aluminum ash are mainly aluminum salts, including polyaluminum chloride, aluminum sulfate, aluminum chloride, etc., the first two are widely used in various types of reclaimed water and wastewater treatment.

The process of producing water treatment agents using aluminum ash has been in use for over 40 years and has been widely applied. In the past decade or so, the industry standards for water treatment agent polyaluminum chloride have become stricter regarding the heavy metal content in the production products, and have added requirements for ammonia nitrogen indicators; without special treatment processes, heavy metals often seriously exceed the standards, and ammonia nitrogen reaches a concentration of 4000–5000 mg/kg, seriously affecting the water quality treatment indicators. At the same time, the reaction of aluminum ash with hydrochloric acid carries certain risks. A large amount of hydrogen gas is released during the reaction process, which can easily ignite and explode; therefore, most companies use open equipment, where hydrogen gas evaporates, self-ignites or burns; or they quickly discharge the mixed gas through air dilution. Currently, with the increasingly strict supervision of chemical companies, the above processes and product quality cannot meet the requirements for cleaner production and higher product quality. The national standard in China "water treatment agent polyaluminum chloride" (GB/T 22627-2022) index item requirements are shown in Table 3.9.

The national standard of polyaluminum chloride has been updated and revised, increasing the requirement for ammonia nitrogen, which is controlled below 0.05%. Due to the increase in technical difficulty of standard production, few enterprises have used aluminum ash for large-scale production in China in recent years. As the Ministry of Ecology and Environment has classified aluminum ash as hazardous waste, many companies producing aluminum ash have no choice but to pay for its disposal by other companies. Even so, there are very few companies that are currently recycling it on a large scale. Aluminum ash can not only be used to refine aluminum

Table 3.9 National standard in China "water treatment agent polyaluminium chloride" (GB/T 22627-2022)

Items	Indicator	
	Liquid	Solid
Mass fraction of aluminum oxide (Al_2O_3)/%	≥ 8.0	≥ 28.0
Density (20 °C)/(g/cm^3)	≥ 1.12	–
Salinity/%	20–98	
Mass fraction of insoluble matter/%	≤ 0.4	
pH value (10 g/L aqueous solution)	3.5–5.0	
Mass fraction of iron (Fe)/%	≤ 1.5	
Mass fraction of ammonia nitrogen (calculated as N)/%	≤ 0.05	
Mass fraction of arsenic (As)/%	≤ 0.000 5	
Mass fraction of lead (Pb)/%	≤ 0.002	
Mass fraction of cadmium (Cd)/%	≤ 0.000 5	
Mass fraction of mercury (Hg)/%	≤ 0.000 05	
Mass fraction of chromium (Cr)/%	≤ 0.005	

Note The indicators of insoluble matter, iron, ammonia nitrogen, arsenic, lead, cadmium, mercury, and chromium listed in the table are all calculated based on a 10% mass fraction of Al_2O_3. When the Al_2O_3 content ≠ 10%, it should be converted into a 10% Al_2O_3 product ratio based on the actual content, and the corresponding mass fraction is calculated

metal but can also be further recycled to produce water treatment agents, providing good economic benefits. Moreover, compared to the production of traditional coagulants, the comprehensive cost of coagulants produced from aluminum ash is lower due to the offset of hazardous waste disposal fees. The product quality is similar, so aluminum ash has a better competitive advantage in the production of water treatment agents.

(1) Acid dissolution method

The main principle of the acid dissolution method is to dissolve the Al_2O_3 and a small amount of Al in the aluminum ash with hydrochloric acid, then add an appropriate amount of water, and through hydrolysis, polymerization reactions, etc., by continuously adjusting the alkalinity with the addition of calcium aluminate, polymers are formed, and a liquid sample is finally obtained. The specific reaction process is as follows:

$$Al_2O_3 + 6HCl + 9H_2O \rightarrow 2AlCl_3 \cdot 6H_2O \tag{3.24}$$

$$2Al + 6HCl + 12H_2O \rightarrow 2AlCl_3 \cdot 6H_2O + 3H_2 \uparrow \tag{3.25}$$

$$\left(2-\frac{n}{4}\right)AlCl_3+\frac{n}{2}H_2O+\frac{n}{8}Ca(AlO_2)_2 \rightarrow Al_2(OH)_nCl_{6-n} + \frac{n}{8}CaCl_2 \tag{3.26}$$

In addition, the reaction between aluminum ash and hydrochloric acid will generate a large amount of heat. If the ratio of hydrochloric acid to water is effectively adjusted and the released heat is used reasonably, the use of external heat sources can be reduced, or even no external heat source is needed, to carry out self-heating reactions, further reducing costs and saving energy.

Many researchers have studied the effects of acid–water ratio, reaction temperature, maturation temperature, and maturation time on the performance of the prepared polyaluminum chloride. For example, Luo Ziqin et al. prepared a liquid polyaluminum chloride product with a salinity of more than 50% and an Al_2O_3 content of more than 9.5% under the conditions of a reaction temperature of about 100 °C, a reaction time of 6–12 h, natural cooling dilution, and maturation at 95 °C for 15–24 h. Lu Xiuguo et al. used the acid dissolution one-step method to prepare polyaluminum chloride, conducted experimental research on various factors, and finally determined the optimal process conditions: 3.4 g aluminum ash, 10 mL water, V(concentrated hydrochloric acid):V(water) is 1:1.5, the reaction time is 2 h, medium speed stirring, the reaction temperature is 85 °C, and the maturation time is 36 h. Liu Xixiang et al. used the aluminum ash slag from a certain aluminum profile processing factory in Nanning as raw material to study the effects of reaction temperature, reaction time, acid–water ratio, etc. on the product through single factor and orthogonal experiments, and determined the optimal process conditions for preparing PAC, and by adding calcium aluminate to adjust, the final product salt basicity reached 85%, and the Al_2O_3 content reached 10.50%. Chao Xi et al. also made detailed adjustments to the amount of calcium aluminate added while determining the best conditions for preparing polyaluminum chloride from aluminum ash through acid leaching experiments, and finally determined that the best polyaluminum product was under the conditions of 12 g of calcium aluminate/80 mL acid leaching solution, temperature 85 °C, and reaction 1.5 h. The above explorations of preparation conditions are all obtained in the laboratory, and in specific large-scale production, the difference from laboratory data is often quite large, and the production process and conditions should be continuously optimized in combination with specific production realities.

However, in the specific production process, the acid dissolution method has certain disadvantages: the unstable composition of aluminum ash leads to unstable product quality, especially the heavy metal content is easy to exceeds the standard; during the reaction of aluminum ash with water and hydrochloric acid, a large amount of ammonia, hydrogen, hydrogen chloride and other gases are easily produced, which may pose certain dangers. Therefore, facing these problems, it is necessary to continuously improve the process, continuously improve the purity of the product, and reduce the danger in the industrial production process. For example, the problem of excessive heavy metal content in the product can be solved by adding some impurity removers, heavy metal removers, etc., such as sulfides, organic metal chelating agents, etc., to fully reduce the heavy metal content in the product and improve product purity. Facing the safety issues in industrial production caused by the reaction of aluminum ash and hydrochloric acid to generate hydrogen, Zhang Yuejun et al. added aluminum ash to hot water to make aluminum ash slurry, continuously stirred in hot water, and alternated stirring and standing to destroy the alumina in the aluminum ash, accelerate the hydration reaction of various active components in the aluminum ash, slowly release ammonia and hydrogen under mild hydration conditions, and greatly reduce or eliminate the generation of ammonia and hydrogen under the severe acidification conditions of high-temperature reaction, greatly reducing the danger in the industrial production process.

(2) Preparation and production of calcium aluminate

By mixing calcium carbonate or calcium oxide with aluminum ash, then heating and calcining to form calcium aluminate, polyaluminum chloride products are produced through acid dissolution, maturation, filtration, and drying processes. This method not only effectively improves the utilization efficiency of aluminum ash, but also effectively alleviates the flammability and explosiveness of hydrogen gas released from the hydrolysis or acidolysis of metallic aluminum in aluminum ash. Jiao Zhanzhong et al. mixed aluminum ash with calcium carbonate or calcium oxide, ground and mixed them, and then calcined them at a temperature of 850–1200 °C in an oxidizing atmosphere. The metallic aluminum in the aluminum ash was oxidized to alumina, the aluminum nitride was oxidized to alumina and nitrogen, and the low-reactivity alumina was reacted to form activated calcium aluminate. This was then reacted with hydrochloric acid and water to produce a slurry, which was separated by maturation and drying to finally produce high-quality polyaluminum products. Zhang Yeling et al. mixed aluminum ash with calcium hydroxide and water, fully mixed and denitrified, and then carried out solid–liquid separation by pressure filtration. The aluminum-containing filter cake was mixed with quicklime for calcination, then cooled and crushed to produce fine calcium aluminate powder. The filtrate was returned to the reaction kettle to react with some of the produced calcium aluminate powder and 30% hydrochloric acid, and after maturation, filtration, and drying, polyaluminum chloride products were finally obtained. This process not only reduces the content of heavy metals in polyaluminum chloride, obtains high-quality polyaluminum chloride products, but also obtains high-quality intermediate product calcium aluminate powder, some of which can be sold externally, improving economic benefits. In addition, mixing aluminum ash and bauxite and calcining with calcium carbonate to prepare bauxite is also a good method. During the calcination of aluminum ash, salts such as sodium chloride will evaporate, which will affect the operation of the exhaust port at the back end of the rotary kiln. If there is a salt separation process in the early stage, the effect will be better, and the alumina content of the product will also increase accordingly. The alumina content of calcium aluminate prepared from pure aluminum ash is generally 40–45%, which is correspondingly lower than the product prepared from bauxite. Several domestic companies have already adopted the production of calcium aluminate from aluminum ash.

3.1.5.5 The Impact of the Aluminum Ash Industry on the Traditional Coagulant Industry

At present, the main production method of polyaluminum chloride in China is to use bauxite and hydrochloric acid reactions and calcium aluminate and hydrochloric acid reactions to produce polyaluminum. Moreover, most of the products produced are relatively single, the scale of enterprise production is small, the technical content is low, the product effective component alumina content is low, impurities are many, and there are very few efficient, cheap composite polyaluminum salts and high-purity polyaluminum chloride products, which are difficult to meet market demand.

The comprehensive utilization of aluminum ash raw materials has brought a huge challenge to the traditional coagulant industry. China's current annual production of aluminum ash exceeds 3.7 million tons. Because the annual production of aluminum ash is large enough, once it can be fully utilized, it can not only improve the environmental problems caused by aluminum ash accumulation but also effectively meet the coagulant market. With the continuous improvement of the process, the production cost is significantly lower than the traditional coagulant, the quality of polyaluminum products manufactured by the aluminum ash acid dissolution method is continuously improved, the impurity content is reduced, and the purity is also continuously improved. For example, the indicators of polyaluminum chloride produced by the acid dissolution method using aluminum ash from a certain place in the south are compared with the indicators of "Water Treatment Agent Polyaluminum Chloride" GB/T 22627-2022 as shown in Table 3.10, and the products all meet the national standard requirements.

In the process of producing polyaluminum chloride by reacting aluminum ash with hydrochloric acid, a large amount of hydrogen gas is generated, which brings the risk of flammability and explosion, posing a safety hazard to the factory and personnel. To eliminate the safety hazards brought by the acid dissolution of aluminum ash, the hydrogen gas can be separated and purified to form a highly valuable by-product,

Table 3.10 Comparison of the indicators of polyaluminum chloride produced by aluminum ash in a certain place in the south with the national standard

Index	1#	2#	3#	"Water Treatment Agent Poly Aluminum Chloride" GB/T 22627-2022
$w(Al_2O_3)$/%	15.33	15.54	17.90	≥ 8.0
$w(Fe)$/%	0.0907	0.0581	0.0451	≤ 1.5
$w(As)$/%	0.0002	0.0003	[a]	≤ 0.0005
$w(Pb)$/%	0.0006	[a]	[a]	≤ 0.002
$w(Cd)$/%	[a]	[a]	[a]	≤ 0.0005
$w(Hg)$/%	[a]	[a]	[a]	≤ 0.00005
$w(Cr)$/%	0.0037	0.0047	0.0024	≤ 0.005

[a] Indicates not detected by the instrument

hydrogen gas. This not only greatly reduces the danger in the production process but also brings high economic value. Moreover, with the increasing emphasis on the hydrogen industry in coastal areas and the layout of the hydrogen industry, the hydrogen gas produced by this method will have a very high market prospect, and its potential is huge.

The acid dissolution method requires a large amount of hydrochloric acid. To further reduce costs and improve economic benefits, aluminum ash can often be combined with industrial waste hydrochloric acid for "double waste" combination, fully realizing the resource utilization and recycling of waste. Shandong Province, Jiangsu Province, and other places have a large number of chlor-alkali industries and hydrochloric acid or hydrochloric acid used in other processes after the reaction of chlorine gas, which provides a guarantee for the production of industrial-grade polyaluminum chloride, greatly reducing the cost of raw materials.

Since aluminum ash is currently a hazardous waste on the market, companies that produce aluminum ash must pay disposal companies funds to handle aluminum ash to achieve safe use or "detoxification" before being used as raw materials for the production of aluminum salts. The price of entrusting the disposal of one ton of aluminum ash varies by region, generally between 500 yuan/ton and 2000 yuan/ton. However, if the chosen process is safe and reliable, it can be used for the production of water treatment agents, not only can detoxify aluminum ash, but also resource treatment, generating very good economic benefits. Moreover, aluminum ash can be combined with waste hydrochloric acid, which is also a hazardous waste, to further reduce production costs, effectively solve the environmental problems brought by aluminum ash and waste acid, fully realize the effective resource utilization and recycling of waste and bring considerable economic benefits.

In summary, the comprehensive utilization of aluminum ash has brought a huge impact on the traditional coagulant industry. The process technology of producing polyaluminum chloride from aluminum ash is constantly improving, the quality of the produced polyaluminum chloride products is constantly improving, and the treatment effect on wastewater and sludge is also constantly improving. Moreover, during the production of polyaluminum products, the hydrogen gas produced during the acid dissolution process can be effectively utilized to form valuable by-products. Not only can it effectively reduce the environmental pollution caused by aluminum ash and waste acid, but it also brings high economic benefits. Therefore, traditional water treatment agent production must change its thinking, make full use of aluminum ash and waste hydrochloric acid, combine with the hazardous waste treatment industry, and connect the hazardous waste industry with new energy and traditional water treatment agent industries to effectively improve market benefits.

3.1.6 Drying Methods of Poly Aluminum Chloride

3.1.6.1 Drum Drying

(1) Principles and Characteristics of Drum Drying

The most common method currently used in the production of solid polyaluminum chloride is drum drying, where steam at 0.4–0.6 MPa is introduced into the drum. During the rotation process, the polyaluminum chloride liquid adheres to the outer wall of the drum, and the water and a small amount of hydrochloric acid in the adhering liquid evaporate. After one rotation, the liquid dries to form a flaky substance, which is scraped off by a scraper to obtain the solid granular finished product.

The drying process of the steam drum dryer is mainly achieved through drum drying equipment heated by steam. The outstanding advantage of this process is that the drying equipment is simple, the investment is small, and the production cost is low. It has been widely used in many medium and small-scale production factories. On the other hand, this process also has many shortcomings.

① The heat utilization rate is not high. Since the indirect heating of steam is an open heating system, the heat of the hot water and secondary steam discharged from its drainage system is generally difficult to reuse; coupled with the widespread existence of the phenomenon of steam traps, therefore, it is difficult to improve the heat utilization rate of the steam drum drying process.

② Pressure operation, poor operational safety. Since the pressure of the heating steam in the drying drum in this process generally needs to be maintained at about 0.4–0.6 MPa, the equipment is always in a pressurized working state. If there are quality issues with the equipment itself or improper operation, it can easily cause safety accidents.

③ The service life of the drum is short. Since the drum operates under strong corrosion and wear conditions (the scraper constantly wears the drum surface), for commonly used materials (such as 16MnR) of the drum, the wall thickness will continuously thin during use. According to the design specifications of the internal pressure drum, under a working pressure of 0.6 MPa, if the drum wall reaches a certain thickness and is not replaced in time, it can easily cause safety accidents. To extend the service life of the drum, the wall thickness of the drum currently produced is generally about 20 mm. Even so, the utilization rate of the drum wall thickness is only about 50%. Each drum can only be used to produce 200–300 tons of solids. This not only affects continuous production but also significantly increases the proportion of equipment costs in production costs.

④ The flaky polyaluminum chloride produced has poorer dispersibility and solubility in water than the powdered products of spray drying.

(2) Process Flow of Conductive Oil Drum Drying PAC

In order to fully utilize the simple structure of drum equipment and overcome its low heat utilization, enterprises in China have already used a new process of conductive

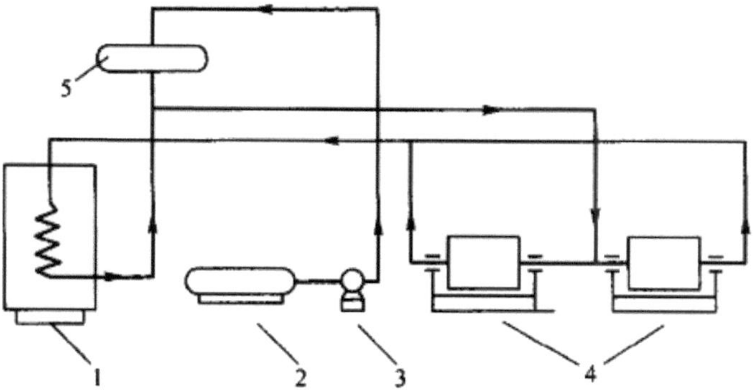

1-Heating furnace; 2-Low position trough; 3-Heat oil pump;
4-Drum dryer; 5-Expansion trough

Fig. 3.8 The PAC drying process in a drum dryer

oil drum drying with conductive oil as the heating medium. Figure 3.8 shows the PAC drying process in a drum dryer using thermal oil as heating carrier.

The heat-conducting oil in the heating system is always in the liquid phase, with almost zero vapor pressure, so the heat carrier itself does not generate transport driving force. Instead, it relies on the heat oil pump to transport it to the heat-conducting oil heating furnace, where it is heated and raised in temperature as it flows through the furnace tube, then flows to the drum dryer. In the drum dryer, the high-temperature heat-conducting oil releases part of its heat as it flows, drying the PAC liquid film on the drum surface. Finally, this cooled low-temperature heat-conducting oil is forcibly circulated back to the heating furnace by the heat oil pump for reheating. This cycle repeats continuously, achieving the heating process of the process. During the operation of the drum dryer, the drum slowly rotates while being heated, and a layer of liquid film adheres to the drum surface through contact with the liquid PAC material through the hot drum wall. The moisture in the film quickly vaporizes due to the indirect heating of the high-temperature heat-conducting oil in the drum, and the PAC material changes from a liquid film to a solid shell. The scraper then scrapes it off the drum surface in time to obtain the flaky solid PAC product.

(3) Comparison of heat-conducting oil heating and steam heating

① The heat-conducting oil heating system is a closed heat circulation system, there is no heat loss caused by the loss and emission of the heat carrier, the heat utilization rate is high, and the energy-saving effect is significant. The heat carrier (heat-conducting oil) in the system does not vaporize, does not generate pressure, and only relies on the forced circulation of the heat oil pump to transfer heat. Therefore, the entire system operates at low pressure, with the maximum pressure at the outlet of the heat oil pump (equal to the sum of the fluid resistance of the heating furnace, dryer, and pipeline) only being 0.25–0.35 MPa, and the working pressure in the dryer is below 0.08 MPa. In this way, the pressure of the entire system is very low, which not only saves construction and maintenance costs, but also minimizes safety risks in the process.

② Since the drum dryer works almost at atmospheric pressure, on the one hand, the dryer drum can be processed as a normal pressure device, eliminating safety risks in production and use; on the other hand, the safe use wall thickness of the drum can be reduced to 2–3 mm, thereby increasing the service life by nearly 80%.

③ The system temperature is stable, which is convenient for automatic control.

④ The auxiliary equipment of the system is reduced, such as the elimination of soft water treatment equipment and sewage disposal facilities, etc. This simplifies the process and saves equipment investment. Compared with steam heating, heat-conducting oil heating requires a relatively larger investment, while the latter requires a smaller investment.

3.1.6.2 Spray Drying

Spray drying has many advantages: it can speed up the drying process and improve the efficiency of heat utilization; it effectively solves the problem of denaturation of heat-sensitive materials due to uneven heating; there is no danger of explosion due to the thinning of the cylinder wall caused by corrosion, enhancing safety; the product has good water solubility, maintains good medicinal effects due to uniform heating; the entire system operates in a closed manner, avoiding environmental pollution caused by the evaporation of hydrochloric acid or sulfuric acid gas during the drying process. Common spray drying methods include centrifugal spray and pressure spray, the difference between the two lies in the atomization process.

Clean hot air directly contacts the liquid droplets, causing the liquid droplets formed by centrifugal atomization or pressure atomization to quickly evaporate moisture and form powder during the falling process. Figure 3.9 is a flow chart of the spray drying process.

The clean hot air that has been filtered and heat exchanged enters the main drying tower tangentially from the top. The material enters from the center of the top of the main tower and is dispersed into extremely small droplets by the high-speed rotating (15,000 r/min) nozzle, directly contacting the hot air. During the falling process of the droplets, the contained moisture quickly evaporates over a short distance and falls to the conical bottom to form a powder product. The tower maintains a negative pressure, and a small amount of solid dust is sucked out and enters the cyclone separator. Using the principle of centrifugal sedimentation, solid particles are separated from the airflow and collected in the collection barrel for regular packaging. The rising gas in the middle of the centrifugal separator carries evaporated HCl gas, water vapor, and a small amount of solid dust into the wet dust removal system, is wet and absorbed by water, and then recirculated into the liquid material production process. The remaining trace amounts of HCl gas and water vapor are discharged high into the air.

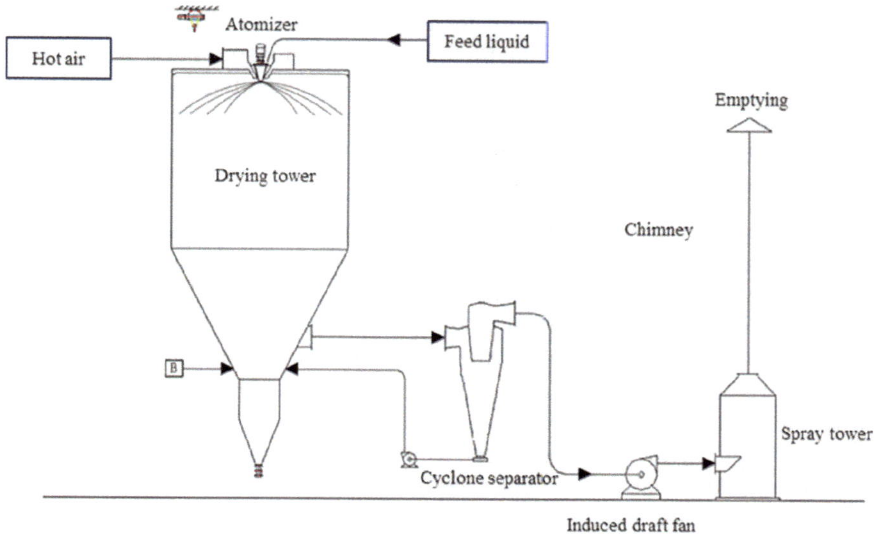

Fig. 3.9 Schematic diagram of a centrifugal spray drying device

During the spray process, the polyaluminum chloride droplets quickly vaporize, forming spherical hollow particles, so their density is relatively low, which is what people often refer to as "throwing".

3.1.7 Main Properties of Polyaluminum Chloride

The fundamental difference between polyaluminum chloride and traditional inorganic coagulants is that it has already formed aluminum polymers of different molecular weights that have partially hydrolyzed during the production process, so it has a higher hydrolysis and bridging starting point than traditional inorganic coagulants with low molecular weights, thus the coagulation speed is fast. Compared with small molecule aluminum salts, it has a wide applicable pH range and less corrosiveness to pipeline equipment. Its coagulation action is as follows:

① Strong neutralization of colloidal substances in water;
② Very good bridging adsorption of hydrolysis products to suspended matter in water;
③ Under normal circumstances, compared with aluminum sulfate, the cost of water purification is 15–30% lower;
④ The consumption of alkalinity in water is lower than all kinds of inorganic small molecule coagulants, so it can be used without or with less alkali addition;
⑤ It can coagulate in the pH range of 5–9 of the source water, and the optimal pH range is 6–8;

⑥ The increase in salt content in the treated water is small, which is beneficial for ion exchange treatment and high-purity water production; but it is still lower than organic coagulants such as polyamine and polydiallyldimethylammonium chloride.
⑦ Its adaptability to the temperature of the source water is better than inorganic coagulants such as aluminum sulfate.

Due to the completely different uses of drinking water polyaluminum chloride and industrial grade polyaluminum chloride, these products are currently divided into two categories, namely drinking water purifier polyaluminum chloride ("Drinking Water Poly Aluminum Chloride" GB15892-2020) and water treatment agent polyaluminum chloride ("Water Treatment Agent Poly Aluminum Chloride" GB/T 22,627–2022). Polyaluminum chloride and polymeric aluminum chloride, which were previously referred to as polymeric aluminum chloride, are completely equivalent in this context. In terms of inorganic polymers, their molecular weights are far smaller than other water-soluble polymers, such as various polyacrylamides, which often exceed 10 million molecular weight. There is currently no suitable method to accurately measure the molecular weight of inorganic polymer compounds.

Polyaluminum chloride technical requirements: In appearance, the liquid is colorless or yellow, brown liquid, and the solid is white or yellow, brown granules or powder. The depth of color is mainly formed by different iron contents. The lower the iron content, the lighter the color, up to white. However, color and purity do not necessarily correlate with coagulation effect. The current dosage of coagulant in coagulation is based on the efficiency of removing various colloids and pollutants. There is no strict quantitative relationship between the drug and the pollutants removed, which must be determined by beaker experiments.

3.1.8 Preparation Method of Polyphosphoric Aluminum Chloride (PPAC)

Strictly speaking, this name is not appropriate because its main component is still polyaluminum chloride. We all know that phosphates, including sulfates, are multivalent salts with better coordination properties than chloride ions. The phosphate and sulfate can coordinate three aluminum ions and two aluminum ions, respectively, which can multiply or increase the molecular weight in polyaluminum chloride, thereby improving the coagulation effect of polyaluminum chloride. Therefore, adding various phosphates or phosphoric acid to the polyaluminum chloride solution can improve the coagulation effect of water, especially for the treatment of low-temperature and low-turbidity water.

Firstly, the newly prepared aluminum chloride solution is slowly dripped with alkali under intense stirring to prepare a solution with a set alkalinity and aluminum concentration, then a certain amount of NaH_2PO_4 solution is dripped, refluxed for several hours in an 80 °C water bath, and matured for 24 h to prepare a PPAC solution.

Phosphate has a polymerizing effect on polyaluminum, making PPAC contain higher charge, larger molecular weight polynuclear complexes, and its long-term stability is good. Using PPAC for coagulation treatment of oily sewage, organic wastewater, wool dyeing wastewater and polluted river water, the removal rate of turbidity, oil content, Cr^{3+} is more than 90% at a lower dosage, and the removal rate of soluble organic matter is also more than 60%, and it adapts to a higher pH value. Its efficacy is significantly better than ordinary PAC and does not increase the phosphorus content in the treated water, which has a broad development and application prospect.

3.1.9 Implementation of the New Standard and Industrial Impact

3.1.9.1 Multi-department Coordination to Solve Water Resource Management

Currently, in China, the national standard and industry standard for drinking water purifiers are formulated by the National Chemical Standardization Technical Committee Water Treatment Chemical Branch, while the production license and quality supervision of coagulants are the responsibility of the health administrative department. The multi-departmental policy leads to contradictions in the current national norms and standards, which need further research and discussion. For example, the "Hygienic Safety Evaluation Standards for Chemicals Used in Drinking Water Treatment (2001)" in China (hereinafter referred to as the "Evaluation Standards") proposes the following health requirements for water treatment chemicals: when chemicals used in drinking water treatment are used at the prescribed dosage, the harmful substances brought into the drinking water by the chemicals are substances specified in the "Hygienic Standards for Drinking Water Quality (2001)", the allowable limit of this substance is 10% of the corresponding specified limit, specifically including metals (arsenic, cadmium, chromium, lead, silver, selenium, and mercury), inorganic substances, organic substances, and radioactive substances. For polyaluminium chloride used in drinking water, the "Evaluation Standards" stipulate that the evaluation dosage is 25.0 mg/L (expressed as Al), converted to aluminum oxide 55 mg/L, then it is necessary to add liquid polyaluminium 550 mg/L. The heavy metal evaluation of polyaluminium chloride is shown in Table 3.11.Taking lead as an example, the control concentration of lead in China's drinking water is 0.01 mg/L, the lead introduction amount under the evaluation dosage reaches 0.0055 mg/L, which leads to the proportion of lead introduction under the current national standard conditions reaching 55%, which does not meet the requirements of the "Evaluation Standards". Based on the calculation of heavy metals in "polyaluminum chloride for Drinking Water" (GB15892-2020), the introduced relative amount is: $0.00275/0.01 = 0.275 = 27.5\% > 10\%$, still not reaching the standard of drinking water treatment agents.

Table 3.11 Heavy metal evaluation of polyaluminium chloride

Item	Drinking water limit (mg/L)	Product quality limit (mg/L)		Heavy metals brought in by the evaluation dosage			
		GB2009	GB2020	GB2009/ (mg/L)	Introduced heavy metals/ %	GB2020/ (mg/L)	Introduced heavy metals/ %
As	0.01	2	2	0.0011	11	0.0011	11
Cd	0.005	2	1	0.0011	22	0.00055	11
Cr	0.05	5 (Cr^{6+})	5 (total)	0.00275	5.5	0.00275	27.5
Pb	0.01	10	5	0.0055	55	0.00275	27.5
Hg	0.001	0.1	0.1	0.000055	5.5	0.000055	5.5

Currently, the "Polyaluminium Chloride for Drinking Water" (GB15892-2020_ 's requirements for heavy metals are basically consistent with Japan's industry standards and far exceed the corresponding standards of the European Union and the United States. Table 3.12 shows the comparison of the quality standards of polyaluminum chloride. Even so, according to the "Evaluation Standards", the products after the implementation of the new standard still cannot meet the requirements of the "Evaluation Standards". That is to say, even the polyaluminium chloride products produced according to the strictest water purifier standards at home and abroad cannot be used for the treatment of drinking water in China, which is obviously unreasonable.

The main reasons for the above contradictions include the definition of residual amounts and the setting of evaluation doses.

(1) the definition of residual amounts

All the raw water we use for drinking contains colloidal particles with anions. commonly used aluminum salt and iron salt coagulants will undergo neutralization and colloidal destabilization in water, eventually forming hydroxides and adhesions or aggregates of colloidal particles in water, and separating from the water body. In this process, colloidal particles can be removed along with some soluble organic matter and heavy metals, including those found in the water body and those introduced by external sources. Failing to acknowledge the important function of heavy metal separation may result in a significant increase in the number of residual heavy metals. The "Evaluation Specification" measures the residual number of heavy metals in the actual water treatment process based on the heavy metal content in the pure water-dissolved water treatment agent and the pure solution, which is completely inappropriate because the prepared polyaluminum chloride solution and the actual environment of polyaluminum chloride are completely different.

(2) the setting of evaluation doses

The evaluation doseis the problem of medication addition. Calculated based on liquid polyaluminum chloride, the addition of 550 mg/L far exceeds the actual addition

Table 3.12 Comparison of polyaluminium chloride quality standards

Index	European Union EN883-2004	United States ANSL/ awwab408-201	Japan JIS/k147-2006	China GB15892-2020	China GB/T22627-2022	Aluminum hydroxide and calcium aluminate process
Appearance	Colorless to yellow	Colorless to light yellow	Colorless to light brown	Colorless to light yellow transparent liquid	Colorless to yellow or yellow–brown	Light yellow–brown
Density (20 °C) g/ cm^3	1.16–1.40	–	≥ 1.19	≥ 1.12	≥ 1.12	1.224
Al_2O_3%≥	7.9–17.9	5.0–25.0	10–11	10	8	10.6
Salinity: %	> 35	10–83	45–65	45–90	20–98	79.24
Insoluble matter: %≤	–	–	–	0.1	0.4	–
Turbidity (NTU)≤	–	50	–	–	–	78
pH value (10 g/L aqueous solution)	–	–	3.5–5.0	3.5–5.0	3.5–5.0	4.3
Fe mg/L	–	–	100	2000	15,000	760
Mn mg/L	–	–	15	–	–	–
Ammonia Nitrogen (CN) mg/L	–	–	100	–	500	–
Pb mg/L	2.12–42.34	–	5	5	20	4.57
Cd mg/L	0.16–5.29	–	1	1	5	0.013
Cr mg/L	1.59–52.93	–	5	5	50	6.09
Cr^{6+}mg/L	–	–	–	–	–	–
Hg mg/L	0.21–1.06	–	0.1	0.1	0.5	0.028

(continued)

Table 3.12 (continued)

Index	European Union EN883-2004	United States ANSL/ awwab408-201	Japan JIS/k147-2006	China GB15892-2020	China GB/T22627-2022	Aluminum hydroxide and calcium aluminate process
As mg/L	0.74–5.29	–	1	1	5	0.53
Sulfate ions (SO_4^{2-}) mg/L	–	–	3.5	–	–	–

Table 3.13 The amount of polyaluminum chloride added when typical rivers are used as drinking water sources (July 2018)

Water source	Shanghai section of the Yangtze River	Wuhan section of the Yangtze River	Harbin section of the Songhua River	Guangzhou section of the Dongjiang River	Lanzhou section of the Yellcw Rive	Zhengzhou section of the Yellow River	Jinan section of the Yellow River	Wuxi section of the Tai Lake
Liquid PAC dosage/ mg/L	15–30	15–60	25–60	15–30	30–60	30–80	60–100	10–30

range. Table 3.13 presents the additional amount of polyaluminum chloride required when utilizing river water as a drinking water source in the typical basins of China. It can be found that under normal circumstances, the addition amount of liquid polyaluminum chloride will not exceed 100 mg/L, and the actual addition amount of high-quality polyaluminum chloride is often 10–30 mg/L, which is far lower than the evaluation dose in the specification. Therefore, it is very unreasonable to measure the residual amount of polyaluminum chloride or other coagulants by using a large dose of medicine and all dissolved. When formulating national standards and specifications, health departments should consult with the national standard setting departments of products to form a more reasonable and effective national standard system.

3.1.9.2 The Impact of the Implementation of the New National Standard on Industry Production

The new standard requires that the raw material hydrochloric acid for the production of polyaluminium chloride should be industrial synthetic hydrochloric acid; the aluminum-containing raw material should be industrial aluminum hydroxide as the main raw material. The implementation of the new standard will have a significant impact on the entire industry's production operations, including raw materials, technological routes, and adjustments to production equipment. Especially the previous national standard controlled hexavalent chromium not to exceed 5 mg/L, now it is modified to total chromium not to exceed 5 mg/L, this requires major adjustments to the entire production materials and equipment to meet the new national standards, production enterprises will significantly increase equipment investment and cope with the sharp rise in raw material prices. The production process and raw material sources of drinking water grade products and industrial water products, as well as production costs, will be significantly widened.

3.1.9.3 The Impact of the Implementation of the New National Standard on Production Technology

The implementation of the 2020 version of the new standard is imminent, 95% of enterprises using alumina two-step production technology will have to use aluminum hydroxide and calcium aluminate acid adjusted two-step method, requiring major adjustments to raw materials, technological routes, and production equipment.

Reaction equations:

$$2AlCl_3 + Ca(AlO_2) \rightarrow Al_2(OH)_{5.4}Cl_{0.6} + CaCl_2 \tag{3.27}$$

$$2AlCl_3 + NaAlO_2 \rightarrow Al_2(OH)_{0.3}Cl_{0.6-3} + CaCl_2 \tag{3.28}$$

$$2AlCl_3 + CaCO_3 \rightarrow Al_2(OH)_{3-5.4}Cl_{0.6-3} + CaCl_2 \tag{3.29}$$

Currently, there are two types of aluminum hydroxide, one is conventional aluminum hydroxide, and the other is easily soluble aluminum hydroxide recently developed by CHALCO Shandong Co., Ltd and CHALCO Zhongzhou Co., Ltd. Conventional aluminum hydroxide can react with excess hydrochloric acid, the temperature is at 100 °C, if low-iron hydrochloric acid is used, a colorless aluminum chloride solution is obtained. This solution can increase the salt base in several ways, the most typical is to adjust the salt base with conventional calcium aluminate, but conventional calcium aluminate often contains iron ions, after increasing the salt base, the concentration of iron ions is higher than 1000 mg/L, often making the solution pale yellow, or the concentration is higher to make the liquid brown–red.

Using conventional aluminate is an easy way to adjust the salt alkalinity, but the choice of aluminate must be made carefully. Aluminate is made by sintering limestone (calcium carbonate ore) and alumina, then ball milling to obtain a powder of 100–200 mesh. Both the alumina ore and limestone contain a certain number of heavy metals, such as lead and copper, and the grinding process can also introduce heavy metals like chromium. Therefore, the choice of aluminate must be very careful, otherwise, the resulting polyaluminum chloride solution may exceed the standard levels of lead, chromium, etc.; even if some products can be completely colorless and transparent, there can sometimes be issues with exceeding heavy metal standards. Currently, some companies use sodium aluminate solution to adjust the salt alkalinity, formula 3.28 is expressed as 90% salt alkalinity, formula 3.29 is the same. Since sodium aluminate is produced from aluminum hydroxide and sodium hydroxide, it does not introduce heavy metals and can keep the product colorless and transparent. If light calcium carbonate is used to adjust the salt alkalinity, a colorless and transparent polyaluminum chloride solution can also be obtained, as shown in formula 3.29.

Somecompanies in China have tried to use white alumina and light calcium carbonate as raw materials to produce aluminate. This aluminate can directly react with hydrochloric acid to obtain a colorless and transparent polyaluminum chloride liquid, and its salt alkalinity can be adjusted arbitrarily. The salt alkalinity of the aluminum chloride solution formed by the reaction of aluminum hydroxide (either conventional or soluble aluminum hydroxide) with hydrochloric acid can also be adjusted to obtain a transparent solution. The iron ion content of the resulting polyaluminum chloride is within the range of 50–100 mg/L, and the heavy metal content is around 5 mg/L, fully meeting the new standard for drinking water polyaluminum chloride. This process can eliminate large amounts of waste residue and prevent secondary pollution problems. Currently, the price of high-purity aluminate is more than twice that of conventional aluminate, but with the increasingly strict standards and regulatory efforts for drinking water polyaluminum chloride, the use of soluble aluminum hydroxide and high-purity aluminate will become a trend.

3.1.9.4 The Impact of the Implementation of the New National Standard on the Local Environment

In addition to the heavy metal issue, the production of polyaluminum chloride using lightly calcined alumina will generate a large amount of waste residue. The total content of alumina and iron oxide in the lightly calcined material generally does not exceed 40%, so after extracting alumina, there is still 50–65% waste residue. The surrounding provinces and cities of Shanghai City are becoming increasingly strict in environmental protection, and the cost of waste residue treatment is unbearable, which is also an important reason this region has taken the lead in using aluminum hydroxide for production.

The regions of Gongyi City in Henan Province and Zibo City in Shandong Province have abundant low-grade bauxite resources, which can produce a large amount of lightly burned material. At the same time, these two regions have a large amount of by-product acid resources, and the cost of waste disposal is relatively low, which is an important reason for the local use of a large amount of lightly burned bauxite. In the past, these wastes were mainly disposed of by landfills. The Gongyi area has recently prohibited landfilling and now requires the recycling of waste materials. Under the premise of increasingly strict environmental supervision and positioning of Gongyi city as a tourist city, scattered polyaluminium chloride production enterprises have to enter industrial parks to strengthen pollution control. Recently, the local environmental protection department's requirement to dispose of and recycle this part of industrial waste has greatly affected production in recent years, causing domestic coagulant prices in China to soar and supplies to be tight. The use of industrial aluminum hydroxide and calcium aluminate in this atmospheric pressure production process not only facilitates mass production but also reduces the pressure of waste disposal on production enterprises, reducing the generation of solid waste by 90%. Even if a small amount of waste is generated, it can be further processed into new water treatment agents, thus completely avoiding the generation of waste.

Gongyi City is the largest production base of polyaluminium chloride in China, exporting about 2 million tons of products to the whole country every year. There are more than 70 coagulant production enterprises and more than 4000 sales personnel or agent teams. Coagulants are also an important local industry, but not yet a pillar industry. Low-price competition, excessive resource consumption, high energy consumption, and high pollution have restricted the further development of the entire industry. While Gongyi has made outstanding contributions to the green development of the country, especially the development of the water treatment industry, it has paid a high price for environmental pollution. Recently, thanks to the national strict control of air pollution and the policy adjustment of "coal to gas", the Gongyi government hopes to change this unsustainable development model as soon as possible with the implementation of new national standards, while improving the technical level of the industry, guiding enterprises to develop upstream and downstream of water treatment, and thus using the advantages of water treatment agents to gain greater service and value space. In this regard, the successful transformation of several water treatment agent enterprises in Jiangsu Province and Guangdong Province is a very good case.

3.1.9.5 National Standards and Supervision

(1) Formulation of National Standards

At present, the national standard for coagulants in China is formulated by the National Chemical Standardization Technical Committee. The production license for coagulants is supervised and issued by the provincial health and epidemic prevention station. The health and epidemic prevention department oversees product quality, but they often do not strictly supervise the entire production process. They only determine that the coagulant sample meets national standards to consider it qualified. The current national standard for polyaluminium chloride, "Polyaluminium Chloride for Domestic Drinking Water" GB15892-2020 has 11 indicators. If the appropriate raw materials and processes are selected, the new national standard can be fully met. According to the previous production process analysis, everyone understands the restrictive meaning of the stipulation in the new standard that "the hydrochloric acid for producing polyaluminium chloride should be industrial synthetic hydrochloric acid, and the aluminum-containing material should be industrial aluminum hydroxide as the main aluminum-containing material".

The formulators of coagulant standards often come from production enterprises and researchers who are more familiar with the production process. Therefore, they are more familiar with the quality and process control during the production process. Therefore, in future product supervision, it is recommended that the health and epidemic prevention department and the standard-setting department cooperate to guide the healthy development of this industry.

(2) Product Quality Supervision

The new national standard for acid stipulates that the production of polyaluminum chloride for drinking water must use industrial synthetic hydrochloric acid as raw material, and the use of any other by-product acid is illegal. The purpose of this restriction is to strictly prohibit products produced from by-product acids from entering the drinking water industry. Therefore, it is recommended that companies purchasing polyaluminum chloride for drinking water, while inspecting product quality, should ask manufacturers to provide the batch numbers and quality inspection forms of hydrochloric acid and aluminum hydroxide used in the production of polyaluminum chloride, to ensure that the synthetic hydrochloric acid and aluminum hydroxide of that batch are used for the corresponding batch of polyaluminum chloride production, especially for companies that purchase through intermediaries. This ensures the traceability of product quality and prevents counterfeit products from entering the drinking water industry. In this regard, the Shanghai Water Authority has done a very good job, being able to strictly supervise production companies to ensure the quality of chemicals used for drinking water treatment.

Many domestic companies in China do not produce polyaluminum chloride for drinking water themselves, but these companies also have licenses for polyaluminum chloride for drinking water, which poses a significant risk for companies to purchase substandard products and enter the drinking water industry. Some products may meet national standard requirements according to test results, but they may be produced from raw materials that do not meet national standard regulations. Currently, the proportion of companies selling products on the market as agents for drinking water coagulants is very high. Especially many small and medium-sized water plants often purchase through agents, rather than directly from manufacturers. On the other hand, health and epidemic prevention departments also lack supervision of agents, which also brings great risks.

The proportion of coagulants in water treatment costs is relatively small, generally less than 5%, but their role is irreplaceable. It should be noted that the above calculations do not include the intangible costs of the company, such as sales costs, management costs, research and development costs, and various fees charged to companies in some places. At the same time, it does not include transportation costs from the production company to the user. In addition, due to regional differences, the cost of raw materials, especially the cost of hydrochloric acid, varies greatly. Recently, the prices of hydrochloric acid and aluminum hydroxide in our country are at historical highs. Therefore, the price of polyaluminum chloride is also relatively high. Currently, the purchase price of liquid polyaluminum chloride by Beijing and Shanghai Water Companies is above 1100 yuan/ton. This ensures a certain profit for the production companies, which is beneficial for the healthy and sustainable development of the company. Many small and medium-sized enterprises in China, in order to reduce the cost of water treatment agents, blindly depressing the price of polyaluminum chloride is completely detrimental to the development of enterprises. If the company's profits or survival cannot be guaranteed, the production company will substitute inferior for good, and counterfeit qualified products with unqualified products. Substitute by-product acid for synthetic acid, and substitute low-quality standard calcium aluminate for high-quality products. As a result, the products entering the drinking water industry cannot meet the quality level required by the new national standard. Some companies deliberately substitute inferior products for better ones to increase profits, and there should be greater supervision and punishment.

3.1.9.6 Trends and Evaluation of Polyaluminum Chloride Development

Polyaluminum chloride is a characteristic product developed in China over the past forty years. Compared with other countries, polyaluminum chloride has great advantages in technology and raw materials, basically meeting the requirements of drinking water and wastewater treatment in our country. For the healthy and sustainable development of this industry, whether it is producers, sellers, or end users, they should all take up their social responsibilities and promote the healthy development of this industry.

Currently, China's water treatment industry is undergoing a rapid process of integration and development, with some large, listed companies also seeking to extend into the water treatment chemical industry. This provides a good opportunity for mergers, reorganizations, and improvements in technology and market in the water treatment reagent industry. Therefore, the next few years will be a stage of great change in this industry. Only by seizing this opportunity can the industry and businesses experience a rebirth akin to a phoenix and pursue sustainable development.

Regarding the development of China's coagulant industry, the state should establish a good cooperation mechanism between the two main departments of industry and health, jointly promote the formulation of national standards for water purification materials related to drinking water safety; at the same time, it should adapt to local conditions and formulate national and industry standards suitable for China's development situation. This will ensure the safety of drinking water for the people and lead to the rapid and healthy development of the industry.

3.2 Poly Aluminum Sulfate

Based on aluminum sulfate, some domestic scholars in China have done a lot of research on the preparation and performance of poly aluminum sulfate and poly silico aluminum sulfate, including their preparation methods, molecular structure, and usage performance. The coagulation effect of these two types of polymers is better than that of aluminum sulfate, with a general salt base of 20–50%, and the alkalinity consumed is lower than that of aluminum sulfate. Polysilicon aluminum sulfate is a new type of inorganic high molecular weight coagulant, which has both electro neutralization and adsorption bridging functions.

Synthesis Method

The research on poly aluminum sulfate (PAS) started in the 1970s. American patents (4,284,611 and 4,536,665) and Canadian patents (1,203,364, 1,203,664 and

1,203,665) introduced specific synthesis methods. In the above patent methods, aluminum sulfate first reacts with sodium carbonate or sodium hydroxide to form insoluble gelatinous hydroxy aluminum. The gel obtained is re-dissolved in an aluminum sulfate solution, allowing the aluminum sulfate molecules to further hydrolyze and form poly aluminum sulfate (PAS).

The reaction can be summarized as follows:

$$Al_2(SO_4)_3 \cdot (14.3H_2O) + 6NaOH + 42H_2O \rightarrow 2Al(OH)_3 + 6Na^+ + 3SO_4^{2-} + 56.3H_2O \quad (3.30)$$

$$2Al_2(SO_4)_3 \cdot (14.3H_2O) + 2Al(OH)_3 + 50H_2O \rightarrow 3Al_2(OH)_2^{4+} + 6SO_4^{2-} + 78.6H_2O \quad (3.31)$$

When using this method, when the alkalinity is greater than 15%, the poly aluminum sulfate solution will have stability problems. Some solutions can be stable for several months, but some can only last for a few days. Since this stability problem has not been resolved, this technology has not yet reached a fully commercialized level.

U.S. Patent No. 4,877,597 introduces another production process for poly aluminum sulfate. This process abandons the initial step in the formation of hydroxy aluminum gel, using the preparation reaction process of aluminum sulfate and sodium aluminate to obtain poly aluminum sulfate with a salt base of about 33%. This reaction is sensitive to temperature reaction, and stirring conditions, so it is quite difficult to obtain a product with a stable alkalinity higher than 33%. A range of 20–25% alkalinity is the most stable.

$$Al_2(SO_4)_3 \cdot (14.3H_2O) + 2NaAl(OH)4 + 83H_2O \rightarrow 4Al_2(OH)_2^{4+} + 2Na^+ + 9SO_4^{2-} + 125.9H_2O \quad (3.32)$$

U.S. Patent No. 3,544,476 introduces a method for producing polyaluminum chlorosulfate (PACS). The process involves first preparing aluminum chloride and aluminum sulfate solutions, then adding calcium carbonate or lime alkali solution to alkalinize the product, resulting in PACS and a byproduct of calcium sulfate precipitation. After removing this precipitation, a pure and stable polyaluminum chlorosulfate solution is obtained, with an alkalinity of 50% and an Al_2O_3 content of about 10%.

The solution produced by this reaction is relatively stable, and its product is an effective coagulant. The efficacy of PACS has been proven by its successful application in replacing aluminum salts in several fields.

$$18Al_2O_3 \cdot (3H_2O) + 33H_2SO_4 + 42HCl + 867H_2O \rightarrow 11Al_2(SO_4)_3 + 14AlCl_3 + 975H_2O \quad (3.33)$$

$$27Ca(OH)_2 + 11Al_2(SO_4)_3 + 14AlCl_3 + 975H_2O \rightarrow 18Al_2(OH)_3^{3+} + 42Cl^- + 6SO_4^{2-} + (27CaSO_4 \cdot 2H_2O + 86H_2O) \downarrow + 835H_2O \quad (3.34)$$

U.S. Patents No. 3,909,439 and 4,082,685 introduce a method for producing polyaluminum chloride (PAC). The method involves reacting trihydrate alumina with hydrochloric acid under high temperature and high-pressure conditions:

$$Al_2O_3 \cdot (3H_2O) + 3HCl + 42H_2O \rightarrow Al_2(OH)_3^{3+} + 3Cl^- + 45H_2O \quad (3.35)$$

The product of this reaction is also relatively stable, with an alkalinity of up to 66% and an Al_2O_3 concentration of up to 18%. Its commercially available products generally have a salt base of 50% and about 10% Al_2O_3. This method is the same as the widely used method of aluminum hydroxide-hydrochloric acid polymerization.

Subsequently, another polymeric metal salt, hydroxyaluminum chloride (ACH), has also been proven to be an effective product. It is produced by the reaction of metallic aluminum with hydrochloric acid or aluminum chloride:

$$5Al + AlCl_3 + 62H_2O \rightarrow 3Al_2(OH)_5^+ + 3Cl^- + 47H_2O + 7.5H_2(\text{gas}) \quad (3.36)$$

This reaction can also produce a stable product, with an alkalinity of up to 83% and an Al_2O_3 concentration of up to 24%. However, if aluminum ash is used for production, it is difficult to obtain a product with a high salt base due to its large inertness.

U.S. Patents No. 4,981,675, 5,069,893, and 5,149,400 introduce various production processes for polysilicic aluminum sulfate (PASS). The reaction occurs under high-speed stirring conditions, and the product is produced by the interaction of solution-state aluminum sulfate, alkali metal silicate, and aluminum salt. The typical reaction formula is as follows:

$$1.25Al_2(SO_4)_3 + 0.0062(Na_2O \cdot 0.322SiO_2) + 0.75Na_2Al_2O_4 + 3H_2O \rightarrow 4Al(OH)_{1.50}(SO_4)_{0.735}(SiO_{2.311})_{0.05} + 0.812Na_2SO_4 \quad (3.37)$$

The product obtained by this method has an alkalinity ranging from 25 to 66%, generally about 50%, and the equilibrium content of Al_2O_3 is 8.3%.

In common use, the salt base is defined as [OH]/(3[M] + 2[N]), where M is a trivalent metal and N is a divalent metal. If there is a + 4-valence metal in the formula, the salt base formula becomes [OH]/(4[M1] + 3[M2] + 2[N]), where M2 is a trivalent metal. The concentration in the formula refers to the molar concentration.

To prepare polynuclear metal hydrates, the reaction process consists of trivalent metal salt solutions such as aluminum sulfate, iron sulfate or zirconium sulfate and divalent metals, metal oxides, metal carbonates, metal hydroxides, metal aluminate salts or other alkaline divalent metal raw materials such as metal magnesium, magnesium oxide, magnesium hydroxide, magnesium carbonate, zinc oxide, metal zinc or copper oxide. In the reaction, the divalent metal reactant is added to the trivalent or multivalent metal salt solution.

(1) Example 1

Add polyaluminum chloride (basicity of salt: 41%, Al_2O_3: 10.6%) to a 1500 mL beaker, and dilute Al_2O_3 to below 8.3% with 225 parts of water. Then add 130 parts of water and aluminum sulfate dilution (containing 8.3% of Al_2O_3). Heat the mixture to near 100 °C, for 3 h, then cool to 25 °C.

Add 25 parts of magnesium oxide slurry (containing 98.5% MgO) to a second 250 mL beaker and dilute with 30 parts of water. Add the MgO slurry to the first beaker and stir for more than 15 min. Stir this mixture for another 4 h, then filter to remove unreacted or other precipitates.

The reaction occurs according to the following equation:

$$\begin{aligned} &6MgO + Al_2(SO_4)_3 \cdot 14.3H_2O + 3Al_4(OH)_5^{7+} \\ &\quad + 21Cl^- + 411H_2O \rightarrow 7Al_2(OH)_3^{3+} + 6Mg(OH)^+ \\ &\quad + 21Cl^- + 3SO_4^{2-} + 419.3H_2O \end{aligned} \tag{3.38}$$

The final solution composition is:

Al_2O_3: 7.32%, MgO: 2.48%, Cl: 7.61%, SO_4: 3.05%, Mg/Al: 0.43 (molar ratio), 0.38 (mass ratio), Cl/SO4: 6.90 (molar ratio), 2.49 (mass ratio). Salt basicity: 48.7%.

(2) Example 2

Add 760 parts of aluminum sulfate (containing 8.3% Al_2O_3) to a 1500 mL beaker and dilute with 155 parts of water. Stir the mixture for 30 min. Add 38 parts of magnesium oxide (containing 98.5% MgO) to a second 250 mL beaker, dilute with 47 parts of water. Next, add the stirred dilute aluminum sulfate to the diluted mixture for 15 min. Finally, stir the mixture continuously for 4 h and filter.

The reaction equation is as follows:

$$\begin{aligned} &3MgO + 2Al_2(SO_4)_3 \cdot 14.3H_2O + 109H_2O \rightarrow 2Al_2(OH)_2 \\ &\quad + Mg_3(OH)_2^{4+} + 6SO_4^{2-} + 134.6H_2O \end{aligned} \tag{3.39}$$

The final solution composition is:

Al_2O_3: 6.31%, MgO: 3.75%, SO_4: 17.70%, Mg/Al: 0.75(molar ratio), 0.38(mass ratio). Salt basicity: 34.9%

(3) Example 3

Add 760 parts of aluminum sulfate (containing 8.3% Al_2O_3) to a 1500 mL beaker and mix with 110 parts of water solution. Stir the mixed solution for 30 min. Add 76 parts of divalent copper oxide (containing 98.5% CuO) to a second 250 mL beaker and stir evenly with 92 parts of water. Add the stirred dilute aluminum sulfate to this mixed solution for 15 min. Heat the mixture to 60 °C, stir continuously for 4 h and cool to 25 °C and filter.

The reaction occurs as shown in the following equation:

$$\begin{aligned} & 3CuO + 2Al_2(SO_4)_3 14.3H_2O + 109H_2O \\ & \rightarrow 2Al_2(OH)_2^{4+} + Cu_3(OH)_2^{4+} + 6SO_4^{2-} + 134.6H_2O \end{aligned} \quad (3.40)$$

The final solution composition is as follows:

Al_2O_3: 6.25%, CuO: 7.52%, SO_4: 17.8%, Cu/Al: 0.75 (molar ratio), 1.8 (mass ratio). Salt basicity: 31.6%

(4) Example 4

Add 760 parts of aluminum sulfate (containing 8.3% Al_2O_3) to a 1500 mL beaker and dilute with 107 parts of water. Stir the mixture for 30 min. Add 78 parts of zinc oxide (containing 95.0% ZnO) to a second 1L beaker and add 95 parts of water. Then mix this mixture into the diluted aluminum sulfate for 15 min, heat the mixture to 50 °C and stir continuously for 4 h, then cool to 25 °C and filter.

The product indicators are as follows: Al_2O_3: 6.35%, CuO: 7.72%, SO_4: 17.9%, Zn/Al: 0.75 (molar ratio), 1.94 (mass ratio). Salt basicity: 31.6%

(5) Example 5

Add 300 parts of polyalanine chloride (containing 10.6%) to a 1500 mL beaker and dilute with 235 parts of water to make Al_2O_3 less than 8.3%. Then add 380 parts of aluminum sulfate (containing 8.3%) and stir for 3 h. Add 38 parts of magnesium oxide (containing 98.5% MgO) to a 250 mL beaker, mix with 47 parts of water. Mix this mixture into the aluminum chloride mixture, aluminum sulfate, and water for 15 min. Stir the final mixture for 4 h and filter.

The reaction occurs as shown in the following equation:

$$\begin{aligned} & 3MgO + Al_2(SO_4)_3 \cdot 14.3H_2O + 2AlCl_3 + 126H_2O \\ & \rightarrow 2Al_2(OH)_2^{4+} + Mg_3(OH)_2^{4+} \\ & + 3SO_4^{2-} + 6Cl^- + 137.3H_2O \end{aligned} \quad (3.41)$$

3.3 Polyferric Sulfate

3.3.1 Properties of Polyferric Sulfate

Research on polyferric sulfate has been conducted domestically and internationally for over 40 years. In recent years, the production technology and scale of polyferric sulfate have greatly improved, making polyferric sulfate the second largest variety after polyaluminum chloride. It is estimated that the annual production of polyferric sulfate will be between 4 and 5 million tons in 2022. Due to the advantages of iron salts in phosphorus removal and COD removal, their application range is constantly expanding.

Polyferric sulfate (PFS) is an intermediate product of the hydrolysis-flocculation process of ferrous sulfate. Liquid polyferric sulfate itself contains a large number of polyvalent cations, such as: $Fe_2(OH)^{5+}$, $Fe_3(OH)_6{}^{3+}$, etc. Therefore, polyferric sulfate can play a role in charge neutralization and bridging flocculation through polyvalent cations, unlike ferrous sulfate and ferric chloride which first hydrolyze and then play a flocculation role after forming polyvalent cations. Polyferric sulfate has the following characteristics compared to the several coagulants mentioned earlier:

(1) The flocculation body forms quickly, the particles are compact, the density is large, and the settling speed is fast;
(2) It has a good removal effect on the chemical oxygen demand (COD) and biological oxygen demand (BOD) in various wastewaters;
(3) Wide range of adaptability;
(4) It has good dewatering properties for sludge;
(5) Excellent decolorization;
(6) The water contains less residual iron after treatment;
(7) The coagulation process is minimally affected by water temperature;
(8) The raw materials for synthesis are widely available, easy to obtain, and the product is inexpensive.

Polyferric sulfate also has all the defects of iron salts. In cases where there is a lot of residual iron, the water body has a larger color after treatment. Compared with aluminum salts, the flocs or residues formed by iron salt treatment, such as the sludge formed in the membrane separation process, are more likely to breed microorganisms.

3.3.2 Synthesis Method of Polyferric Sulfate

China began to develop polyferric sulfate in the mid-1980s, which was also the period with the most published papers. With the maturity and promotion of technology, the research on iron salt coagulants has relatively weakened. Some units used to produce liquid polyferric sulfate, most of which used sodium nitrite as a catalyst, and used enamel reaction kettles or stainless-steel reaction kettles as synthesis reactors. Under heating and stirring conditions, oxidants and catalysts were added for reaction. The characteristics of this process are: insufficient gas–liquid reaction, long production cycle, some need to be carried out under pressure, low utilization rate of raw materials, high energy consumption, long reaction cycle, and the waste gas produced during the reaction process needs to be treated, the equipment has a short service life, daily maintenance tasks are heavy, the investment in the entire set of equipment is large, and the production efficiency is low. Given the defects of existing production technology, and according to the characteristics of the reaction, many production processes have been proposed, such as the circulating jet method, the reaction tower method, the pressure reaction method, etc. In addition, a lot of research has been done on the selection of raw materials and the screening of oxidants. Currently, the advanced domestic production technology of polyferric sulfate in China has reached a level comparable to that of foreign countries. However, there remains a significant gap in terms of heavy metal removal and production scale.

There are many methods for producing polyferric sulfate, such as air (oxygen) catalytic oxidation method, pyrite ash pressure acid melting method, potassium (sodium) chlorate oxidation method, tetrairon tetroxide ore acid dissolution oxidation method, etc. In addition, air is also used to directly oxidize solid ferrous sulfate to produce polyferric sulfate. Regardless of the raw materials used for production, the SO_4^{2+}/Fe^{3+} molar ratio is controlled to be less than 3/2, that is, oxidation is carried out under acid-deficient conditions, relying on the hydrolysis of iron salts to release hydrogen ions, to produce basic salts containing different numbers of hydroxyl ions, that is, polyferric sulfate with different salt basicity. When pyrite ash is used as the raw material to produce polyferric sulfate, due to the presence of a small number of ferrous ions, polyferric sulfate can be produced by simply dissolving ferric ions under pressure. When other raw materials are used for production, it is often necessary to oxidize ferrous ions to form polyferric sulfate. The oxidation of ferrous ions requires a fast oxidant that is easy to prepare and inexpensive; only then can it be widely used in production. At present, many oxidants are used, such as hydrogen peroxide, chlorate, oxygen or air, hypochlorite, nitric acid, etc. Air or oxygen oxidation methods are more commonly used because of cost constraints. The oxygen oxidation method includes the nitrous oxide catalytic method and the manganese dioxide catalytic method.

Ferrous sulfate is oxidized in sulfuric acid solution. If the ratio of sulfuric acid to ferrous sulfate is less than 1/2, trivalent iron ions undergo hydrolysis, producing some hydrogen ions and various hydroxy iron ions. The hydrogen ions produced by hydrolysis can in turn ensure the hydrogen ions needed for the redox reaction. By adjusting the above ratio, polyferric sulfate with the required salt basicity can be produced. If the above ratio is controlled to be > 1/2, a ferric sulfate solution containing free acid can be produced. At present, the domestic industry in China uses the chlorate oxidation method and the oxygen oxidation method. The former uses a more expensive oxidant, and although the production process is very simple, the overall production cost is relatively higher than the latter. The production principle of using air oxidation method and oxygen oxidation method with nitrogen oxides as catalysts is the same. There are many processes for producing polyferric sulfate using this principle, and the technical level varies greatly. Professor Li Fengting's team combined with their own research, focusing on the more practical pyrite method, iron ore acid dissolution oxidation method, chlorate oxidation method, chlorine oxide catalytic oxidation method and so on.

3.3.2.1 Pyrite Method

During the production of sulfuric acid, crushed pyrite is oxidized in the air at high temperatures to produce sulfur dioxide and ore ash. The ore ash contains a large amount of ferric oxide (40–80%), silica (10–30%), and small amounts of calcium and magnesium sulfates, etc. In the past, this kind of ore ash was used for cement manufacturing or discarded. Since it contains a large amount of trivalent iron ions, it can be fully used for the production of polyferric sulfate. The synthesis method is to mix sulfuric acid with a mass concentration of 35–45% and ferric oxide (pyrite ore ash, particle size 0.01–0.05, Fe_2O_3 content 80%) according to 110% of the chemical stoichiometry, and react at 105–110 °C until 90–100% of the sulfuric acid is consumed. Filter out the remaining solids, and dilute the filtrate with water to a ferric sulfate content of 40–45%. This method can prepare a stable polyferric sulfate solution. The specific method is to mix 10 kg of water and 15 kg of concentrated sulfuric acid to form a suspension with an excess of pyrite ore ash and a sulfuric acid concentration of 38%, react at 107 °C for 22 h, and then filter and dilute to obtain the product. Although many years have passed, the above technology, whether using ore as raw material or using red mud as raw material to produce polyferric sulfate, its process is worth learning from. Red mud is a valuable resource. After alkali dissolution of alumina, iron ions and a small amount of aluminum ions remain in the red mud. Table 3.14 is the index of red mud produced by an alumina company in Inner Mongolia.

When using red mud with a higher iron content, some alumina may also form aluminum sulfate, eventually obtaining polyaluminum ferric sulfate, but its coagulation performance still retains the characteristics of polyferric sulfate. In some special occasions, the performance even exceeds that of polyferric sulfate. If it is dissolved with hydrochloric acid, a ferric aluminum chloride solution is obtained. Due to the high content of sodium aluminate in red mud, it forms sodium sulfate or sodium chloride after acid dissolution, resulting in a high salt content in the product liquid, which appears to have a high density. If further dried to make solid products, it is often difficult to reach the corresponding standard of iron content, but it does not

Table 3.14 Indicators of red mud from an alumina enterprise in inner Mongolia

Index	Moisture content	Burn	Si	Al	Fe	Ti	Na
Percentage, %	29.88	10.66	16.97	20.73	34.80	4.29	7.5

Table 3.15 Relationship between the dissolution rate of iron ions and the molar ratio of Fe^{3+}/Fe^{2+}

Fe^{3+}/Fe^{2+}	Fe leaching rate/%	Fe^{3+}/Fe^{2+}	Fe leaching rate/%
2	98	1/3	30
1	95	1/4	15
1/2	45	1/5	10

affect the use effect. Therefore, in areas with red mud resources, this resource is also a high-quality raw material for the production of water treatment agents.

3.3.2.2 Acid Dissolution Oxidation Method of Iron Ore

This type of ore mainly contains magnetite. 1000 g of ore containing 58% Fe^{3+} and 11% Fe^{2+} is mixed with sulfuric acid with a concentration of 420 g/L, the SO_4^{2-}/Fe^{2+} ratio is adjusted to 1.3–1.4, and the oxide is dissolved at 90 °C for 30 min, 98–99% of the iron oxide can be dissolved. Then, the above reaction liquid can be made into polyferric sulfate under the catalysis of nitrogen oxides, using air, oxygen, or hydrogen peroxide for oxidation. The specific index is 155 g/L Fe^{3+}. During the synthesis of iron oxide, the content of trivalent and divalent iron in the ore is different. Under different temperatures, the dissolution rate of iron ions is different when dissolved with sulfuric acid. The relationship between the dissolution rate of iron ions and the molar ratio of Fe^{3+}/Fe^{2+} is shown in Table 3.15.

3.3.2.3 Chlorate Oxidation Method

Chlorate or perchlorate is added in solid form to a suspension with a ferrous sulfate content of 45–80% (mass percentage). This method can increase the concentration of trivalent iron ions in the solution and effectively remove colloids in the water together with chlorate. The specific synthesis method is: 600 parts of crystalline ferrous sulfate ($FeSO_4 \cdot 7H_2O$) are mixed with 150 parts of water and 105 parts of concentrated sulfuric acid, and then 80 parts of sodium chlorate are added. The reaction can yield a product containing 200 parts Fe(III) and 50 parts chlorate. This solution is an effective coagulant and disinfectant. When treating water, if added at a dosage of 10 parts, each cubic meter of purified water contains 2.5 parts of chlorate, which is sufficient for disinfection.

This method combines the production of iron salt and disinfectant, enabling the product to have both coagulation and disinfection effects, but the product produced by this method has zero basicity. This method is also widely used in China. The difference is that The difference is that the production of polyferric sulfate does not contain chlorate, and there must be a certain base degree. This method reacts quickly and safely, but compared with the nitrogen oxide catalysis method, the cost is higher. The cost difference between the two is due to the cost of consuming oxygen and chlorate. The residual chlorate in polyferric sulfate still has oxidizing properties, which will interfere with the COD measurement of the treated water, resulting in a lower post-test COD. In recent years, the environmental inspection department has arrested several cases of using chlorate to mask COD tests, all of which use chlorate to mask COD tests. The apparent test results meet the standard, but the actual discharge does not meet the standard.

3.3.2.4 Nitrogen Oxide Catalyzed Oxidation Method

Ferrous sulfate is difficult to be oxidized by oxygen in an acidic solution, making it difficult to achieve the purpose of rapid oxidation of ferrous iron. A stronger oxidant must be used.

Oxygen is a strong oxidizing agent [$V(O_2/H_2O) + O_2 = 1.23$], which can directly oxidize solid ferrous sulfate into basic ferric sulfate. The reaction equation is as follows:

$$4FeSO_4 + 2H_2O + O_2 = 4Fe(OH)SO_4 \tag{3.42}$$

To accelerate the oxidation of ferrous sulfate by oxygen, nitrite or nitrate is chosen as the catalyst for the catalytic oxidation experiment. The reaction equation is:

$$H_2SO_4 + 2NaNO_2 = 2HNO_2 + Na_2SO_4 \tag{3.43}$$

$$H_2SO_4 + 2HNO_2 + 2FeSO_4 = Fe_2(SO_4)_3 + 2NO + 2H_2O \tag{3.44}$$

$$2NO + O_2 = 2NO_2 \tag{3.45}$$

$$H_2SO_4 + NO_2 + 2FeSO_4 = Fe_2(SO_4)_3 + NO \uparrow + H_2O \tag{3.46}$$

The overall reaction is:

$$\begin{aligned} &4FeSO_4 \cdot 7H_2O + (2 - n)\ H_2SO_4 + O_2 \\ &\quad = 2Fe_2(OH)n(SO_4)_{3-n/2} + (30 - 2n)\ H_2O \end{aligned} \tag{3.47}$$

(Note: In all reactions, $n < 2$).

There are many factors affecting the above reactions, such as reaction temperature, reaction pressure, stirring speed, etc. High oxidation rates can be achieved using either a reaction kettle or a reaction tower.

The reaction mechanism of the synthesis method using ferrous sulfate and sulfuric acid as raw materials and nitric acid as an oxidant is completely different from the synthesis method using ferrous sulfate, sulfuric acid, and oxygen as raw materials and nitric acid as a catalyst. The former uses liquid nitric acid as an oxidant, and the reduction product nitric oxide escapes from the solution without playing any catalytic role. This method is characterized by a large amount of nitric acid, and the ferrous ions are completely oxidized by the pentavalent nitrogen in the nitric acid, which is a liquid–liquid reaction. The latter uses nitric acid as a catalyst. Its characteristic is to add a small amount of nitric acid (not enough to oxidize ferrous ions) to the mixture of sulfuric acid, ferrous sulfate, and water. The reduction product nitric oxide reacts with oxygen to form nitrogen dioxide, which then oxidizes the ferrous ions. This cycle continues until the ferrous ions are completely oxidized. This method is a gas–liquid reaction.

The United States also has patents for the production of polyferric sulfate using similar principles. The production process is at 70–150 °C Under certain pressure and the catalysis of nitrogen oxides, ferrous ions are oxidized to ferric ions with oxygen. The process involves injecting a mixture of ferrous sulfate and sulfuric acid into the reactor in a spray manner, maintaining the reaction liquid at one third of the total volume. During the spraying process of the solution, it reacts with NO, NO_2, and O_2 in the gas phase, finally generating ferric polysulfate. It has been determined that the solution contains a large number of macromolecular complexes. The generated ferric polysulfate contains some nitrogen oxides, which are sent to the denitrifier and oxygenated in the denitrifier. The mixed gas generated is recycled to the reactor for production. The principle of this method is the same as the principle of preparing ferric polysulfate by the reaction tower method. The difference between them is that the former sprays the liquid to atomize, while the latter uses the larger specific surface area of the packing to enhance the absorption process of the gas and increase the reaction speed. Since the raw material ferrous sulfate often contains impurities during production and transportation, higher requirements are put forward for the production equipment of ferric polysulfate. Before the reaction, the material can be filtered to remove particulate impurities. In the twentieth century, everyone thought that the above oxidation reaction was endothermic. Later, Professor Li Fengting's team used the process of the reaction kettle to quickly oxidize, realizing the self-heating oxidation process, and took the lead in realizing the non-heating oxidation process in China, making the production equipment and process simpler. It can be carried out at lower temperatures and normal pressure, and has the characteristics of safety and convenience.

In the catalytic oxidation process, the amount of nitrogen oxides consumed is very small, especially when treated properly, nitrogen oxides can be recycled, which will not cause waste or pollution problems. In 2023, in a northern province, yellow smoke leakage and explosion occurred during the transportation of iron salt coagulant. The primary reason was that a significant amount of nitrogen oxides remained dissolved in the final liquid. During the transportation process, the liquid kept hitting, which easily caused a large amount of nitrogen oxide gas to suddenly release, leading to an explosion.

3.3.2.5 Preparation of Ferric Polysulfate from Steel Pickling Waste Liquid

In the steel and machinery industries, pickling of the steel surface is a very important surface treatment process. The pickling process produces a large amount of sulfuric acid and ferrous sulfate or hydrochloric acid, ferrous chloride solutions. This waste has not been fully utilized. Some factories arbitrarily discharge them. For example, a few years ago, some areas in Hebei Province dumped waste acid on farmland, and some illegal enterprises discharged waste acid into rivers through secret pipes, causing not only great environmental pollution, but also a huge waste. These wastes are good raw materials for the production of ferric polysulfate and polyaluminum chloride. Synthesis Principle.

Due to the variation in iron-containing raw materials, the current process for producing polymeric ferric sulfate also varies. When producing with iron ore, iron filings, iron oxide skin, and pyrite as raw materials, the above substances are first dissolved under certain conditions with sulfuric acid, and then different oxidants are used to oxidize the remaining ferrous ions. If the amount of sulfuric acid in the solution is controlled to be higher than or equal to the amount required to form ferric sulfate before oxidation, a ferric sulfate solution containing free acid or a ferric sulfate solution can be formed. If the amount of sulfuric acid is controlled to be lower than the above value, polymeric ferric sulfate of different salt basicity can be generated. The salt basicity depends on the ratio of sulfate ions to iron ions in the solution. The following equations give the reaction mechanism.

$$H_2SO_4 + Fe = FeSO_4 - H_2 \tag{3.48}$$

$$H_2SO_4 + 2NaNO_2 = 2H\ NO_2 + Na_2SO_4 \tag{3.49}$$

$$H_2SO_4 + 2H\ NO_2 + 2FeSO_4 = Fe_2(SO_4)_3 + 2NO + 2H_2O \tag{3.50}$$

$$2NO + O_2 = 2\ NO_2 \tag{3.51}$$

$$H_2SO_4 + NO_2 + 2FeSO_4 = Fe_2(SO_4)^3 + NO \uparrow + H_2O \tag{3.52}$$

From the above equations, it can be seen that the reaction of nitrogen oxides catalyzing the oxidation of ferrous ions is a typical gas–liquid reaction, so the factors determining the reaction rate are the oxidation rate of NO and the dissolution rate of NO_2, and the oxidation rate of NO is much higher than the latter, so accelerating the dissolution of NO_2 becomes the determining step of the reaction rate. To enhance the dissolution of the gas, two methods are usually used, one is to increase the contact surface area of the gas–liquid, and the other is to increase the pressure of the gas–liquid phase. Both methods are very effective in increasing the reaction rate. The catalysts for the above reactions are not limited to sodium nitrite that can react with

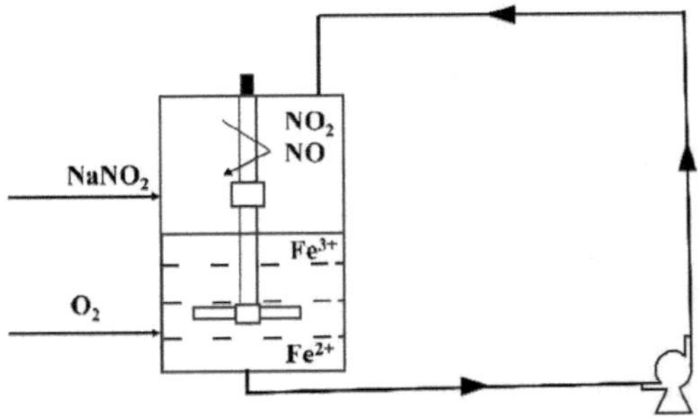

Fig. 3.10 Preparation of polyferric sulfate mechanism and process flow

sulfuric acid to form nitrogen oxides, nitrite salts soluble in a mixture of sulfuric acid and ferrous sulfate, and nitrites, nitrates, and nitric acid all have a catalytic effect on this reaction. At the same time, the above reaction is a self-catalyzed exothermic reaction. If the initial liquid reaction temperature is 20–30 °C, the temperature will naturally rise to 70–80 °C after the reaction is over. If it is summer, the temperature will approach 100 °C, but it will not boil.

(1) Production Process

Add iron filings, concentrate the solution within the range of 60–90 °C. When the solution's density reaches 1.45 g/cm^3 (measured at 60 °C), it can be used to produce polymeric ferric sulfate. The production process can adopt the reaction tower method or the improved reaction kettle method, both of which have very ideal reaction effects. Figure 3.10 is a process flow diagram of the production process using sodium nitrite as a catalyst and a reaction tower as a reactor (this device can also be used to produce polymeric ferric chloride).

During synthesis, a mixture of ferrous sulfate and sulfuric acid is first added to the solution storage tank. The liquid is kept circulating between the reactor and the storage tank by the action of the circulating pump during the reaction process, to achieve the purpose of oxidizing ferrous sulfate. The catalyst is added in the form of a solution at the top of the reaction tower, and oxygen is introduced at the bottom of the reaction. The reaction speed is increased by enhancing gas–liquid contact through the mixing action when the liquid flows over the filler surface. Using a reaction kettle can also achieve a very fast reaction speed, with an average of 1–2 h to produce a batch of products. Depending on the volume of the reaction kettle, a batch can reach 20–30 tons of product. As the proficiency of operation improves, the reaction time can be shortened, but the key is to handle the temperature, pressure, and the balancing process of gas and liquid well. From the introduction above, it is evident that further processing of pickling waste liquid can avoid the pollution and waste generated by pickling by-products, leading to the production of polyferric sulfate coagulant. This not only solves the pollution problem, but also produces a good water treatment agent, creating both economic and social benefits.

3.4 Polyferric Chloride

Currently, many steel processing units in China use hydrochloric acid pickling process, and this part of the waste acid has not been fully utilized. Usually, the waste acid is reacted with iron filings to prepare ferric chloride solution. Although ferric chloride solution is a very good coagulant, because the solution contains a large amount of free acid, it seriously corrodes the equipment during use, thus limiting its application range. Using ferric chloride to produce polyferric chloride is a good choice. Polyferric chloride solution contains a large amount of poly cations, which can play a high level of coagulation effect, and polyferric chloride has a certain salt base, its acidity is lower than ferric chloride solution, and its corrosiveness is relatively weak, so it is an ideal coagulant. Steel hydrochloric acid pickling liquid is a cheap raw material for preparing polyferric chloride, this book introduce the mechanism and process of synthesizing polyferric chloride with it as the main raw material.

3.4.1 *Synthesis Principle*

Due to the different iron-containing raw materials, the current process of producing polyferric chloride is also different. When producing with iron ore, iron filings, iron oxide skin, and pyrite ore as raw materials, the above substances are first dissolved under certain conditions with hydrochloric acid, and then different oxidants are used to oxidize the remaining ferrous ions. If the amount of hydrochloric acid in the solution is controlled to be higher than or equal to the amount required to form ferric chloride before oxidation, a ferric chloride solution or polyferric chloride solution containing free acid can be formed. If the amount of hydrochloric acid is controlled to be lower than the above value, polyferric chloride of different salt bases can be generated. The salt base depends on the ratio of chloride ions to iron ions in the solution. The following equations give the reaction mechanism.

$$HCl + MNO_2 = HNO_2 + MCl \tag{3.53}$$

$$2HCl + HNO_2 + 2FeCl_2 = 2FeCl_3 + 2NO + 2H_2O \tag{3.54}$$

$$2NO + O_2 = 2NO_2 \tag{3.55}$$

$$2HCl + NO_2 + 2FeCl_2 = 2\ FeCl_3 + NO\uparrow + H_2O \tag{3.56}$$

The overall reaction formula is

$$4FeCl_2 + 2(2 - n)\ HCl + O_2 = 2Fe_2(OH)nCl_{6-n} + 2(1 - n)H_2O \tag{3.57}$$

where, M represents alkali metal ions or alkaline earth metal ions, $0 \leq n < 2$.

From the above reaction, the reaction of nitrogen oxides catalyzing the oxidation of ferrous ions is a typical gas–liquid reaction. The determining factors of the reaction speed are the oxidation speed of NO and the dissolution speed of NO_2. However, the oxidation speed of NO is much higher than the latter, so accelerating the dissolution of NO_2 becomes the determining step of the reaction speed. To enhance the dissolution of the gas, two methods are usually adopted: one is to increase the contact surface area of the gas–liquid, and the other is to increase the pressure of the gas–liquid phase. Both domestically and abroad, these two aspects are considered when dealing with the catalytic oxidation of nitrogen oxides and ferrous chloride. In the above reaction, the catalysts are not limited to those that can react with ferrous chloride to form nitrogen oxides. Nitrites and nitric acid, nitrates and nitric acid that are soluble in the mixture of ferrous chloride also have a catalytic effect on this reaction. Polyferric chloride has a strong tendency to hydrolyze, its stability is lower than that of polysulfate iron, and the general salt basicity will not exceed 15%.

3.4.2 *Synthesis Process*

During synthesis, a mixture containing ferrous chloride and hydrochloric acid is first added to the storage tank, and a certain amount of stabilizer is added. Ferric chloride solution and sodium nitrite are introduced at the top of the reaction tower, and oxygen is introduced at the bottom of the reaction tower. When the solution flows over the surface of the filler, the gas and liquid are fully contacted to achieve the purpose of oxidizing ferrous chloride. During the reaction process, the liquid is continuously circulated between the reaction tower and the storage tank through the circulation pump, so that the ferrous iron is continuously oxidized until the ferrous iron is completely oxidized. The operation is similar when using a reaction kettle for production. During the reaction process, the temperature of the solution is kept within the range of 40–90 °C to facilitate the decomposition of the complex. This method is basically the same as the previously introduced method for producing polysulfate iron. Polyferric chloride indicators: $Fe^{3+} = 8\text{–}13\%$; Salt basicity $= 6\text{–}12\%$; $Fe^{2+} \leq 0.1\%$. During the oxidation process, the exothermic reaction of this reaction is more intense, especially in the summer when the temperature of the raw materials is relatively high, the temperature in the later stage of the reaction may exceed 100 °C, so this requires higher corrosion resistance and temperature resistance for the equipment.

3.5 Polyferric Sulfate Chloride and Polyaluminum Ferric Sulfate

Recently, the National Chemical Standardization Technical Committee Water Treatment Chemical Branch in China has been formulating standards for several new products, such as polyaluminum ferric sulfate, polyaluminum ferric chloride, polyferric sulfate chloride, and polyaluminum ferric sulfate, as well as organic–inorganic composite water treatment agents. Given the requirements of health indicators, the metal ions of widely used coagulants are aluminum and iron, and the widely used anions are chloride ions and sulfate ions, or a small amount of silicate ions and phosphate ions. Therefore, from the perspective of structural combination, any composite coagulant can contain aluminum ions, iron ions, or other cations, such as calcium ions, magnesium ions; the anion part can contain chloride ions, sulfate ions, phosphate ions, silicate ions, etc., which can be combined at will.

$$Al_aFe_bCa_cMg_d(OH)_eCl_m(SO_4)_n(SiO_4)_o(PO_4)_p$$

where, a, b, c, d, e, m, n, o, p all represent the number of ions or ion groups in the molecule.

Taking polyaluminium ferric chloride sulfate as an example, if the molar ratio of sulfate and chloride ions is 1:1, and the basicity of the salt is 15%, assuming a = 2, then e = 2 × 15% = 0.3, which is $Al_2(OH)_{0.3}$, the remaining number of anions is 6–0.3 = 5.7. Since the molar ratio of sulfate and chloride ions is 1:1, that is, the equivalent ratio is 1:2, the total equivalent value is 3, so the molar or equivalent value of chloride ions is (5.7/3)/1 = 1.9, the molar value of sulfate is (5.7/3) × 1 = 1.9 or the equivalent value is (5.7/3) × 2 = 3.8, therefore, in the absence of other ions, the molecular formula of the above polyaluminium ferric chloride sulfate can be represented as: $Al_2(OH)_{0.3}Cl_{1.9}(SO_4)_{1.9}$.

The polymer structure above is mainly based on sulfate anions, so the minor chloride ions are added in front of it, which is called polyaluminium ferric chloride sulfate. If the e value is zero, then the hydroxide content in the above molecular formula is zero, which is aluminum chloride sulfate. Table 3.16 is the main indicators of the alkaline waste red mud, a by-product of alumina production by the Bayer process, measured by Professor Li Fengting's team from a certain company.

The general measure of the aluminum content in bauxite and red mud is similar to measure of the indicators of polyaluminium chloride and aluminum sulfate, that is, the effective content of alumina. High-grade bauxite contains more than 60%, even 70% of alumina. There are abundant bauxite resources in Shanxi Province, Shandong

Table 3.16 Chemical composition of alkaline waste red mud used in different implementations

Element		Al	Na	Ca	Si	Fe	Ti	Mg	K	S	P
Content	No.1	21.7	18.7	15.0	13.6	4.86	2.49	0.76	0.73	0.67	0.22
(%)	No.2	17.9	7.90	23.0	7.89	4.83	2.56	1.56	0.28	0.22	0.19

Province, Henan Province, Guangxi Province and other places in China, and China import a large amount of bauxite from Guinea, Ghana, Indonesia and Australia every year. The aluminum content of No.1 red mud is 21.7%, which is 41.9% of alumina content. Calcium, magnesium, iron, etc. all exist in the form of aluminosilicates. In the acid dissolution process, iron, magnesium, calcium, aluminum, sodium, and potassium will all enter the acid solution. If hydrochloric acid is used for dissolution and the basicity is controlled, polyaluminium ferric chloride will be obtained. If sulfuric acid is used for dissolution, a solution with polyaluminium ferric sulfate as the main component will be obtained. In the above two dissolution methods, there are a large amount of calcium, magnesium, sodium, and potassium salts, so it is a very complex multiple salt system. The metal salts that cannot be precipitated in use contribute little to coagulation, and it is mainly iron and aluminum that play an important coagulation role. Since red mud is a widely existing solid waste, it can be used to produce high-quality coagulants in areas with relatively abundant acid resources.

If conventional raw materials, such as aluminum hydroxide, ferrous sulfate, ferric chloride, etc., are used to produce iron-aluminum composite coagulants, their production process is basically the same as that of single metal coagulants.

Chapter 4
Polyacrylamide Polymers

4.1 Flocculation Theory and Application

Artificially synthesized polymer flocculants can significantly improve the effect of solid–liquid separation, especially in precipitation, filtration, and centrifugation processes. This enhancement causes dispersed particles to aggregate into particle sizes that easily form a solid phase. The stability of the suspended matter is disrupted, and then the particle liquid is released from the phase. The vast majority of flocculants and filtration aids currently used are polymers with repeating units of acrylamide and its derivatives, with cationic or anionic groups.

The flocculation process and the coagulation process are often confused, but they are actually two different stages. The coagulation process is based on electrostatic action, which reduces the repulsion of the double electric layer to achieve the purpose of aggregation. Flocculation comes from the Latin "flocculus", which literally means a tuft of wool or a loose fibrous structure. Flocculation works through high molecular weight substances, such as starch and polyelectrolytes, these polymers bridge between two or more particles, causing the solid particles to form a loose, porous, random three-dimensional structure.

4.1.1 The Necessity of Flocculation

All particles have attraction to each other, but the attraction only exists in the separation process where the particle distance is very short. When the particle distance is very close, van der Waals forces (London forces) play a dominant role, this widely existing force causes the coagulation process to occur. However, suspended particles maintain stability through repulsive forces, which prevent the particle distance from decreasing to the extent that electrostatic attraction works. The relationship between interaction potential energy and particle distance is shown in Fig. 4.1.

L. Fengting, *Water Treatment Chemicals: Coagulants and Flocculants*,
https://doi.org/10.1007/978-981-96-8221-8_4

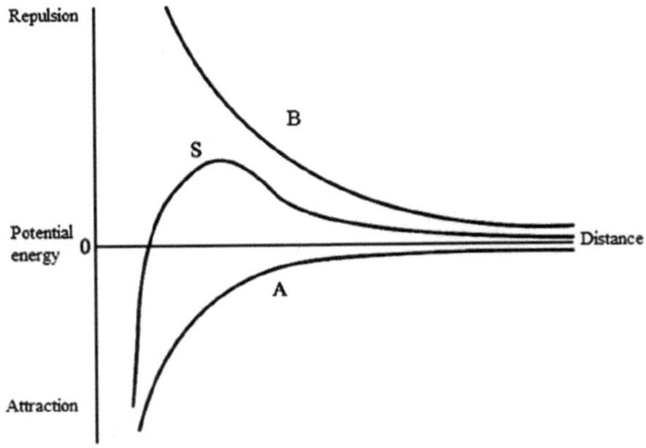

Fig. 4.1 Relationship between interaction potential energy and particle distance

Curve A represents attraction; Curve B represents repulsion; Curve S represents the sum of the two forces, which also represents the electric energy curve when two particles approach each other.

The sources of particle repulsion mainly come from two aspects:

1. ① Water molecules adsorb to the particle surface, forming a solvent layer, causing particles to repel each other.
2. ② Charge action—all particle surfaces carry a charge; its state and intensity depend on the surface properties and the chemical properties of the water suspension medium (Fig. 4.2). Generally, suspended particles carry a negative charge at a pH value of 4 or higher.

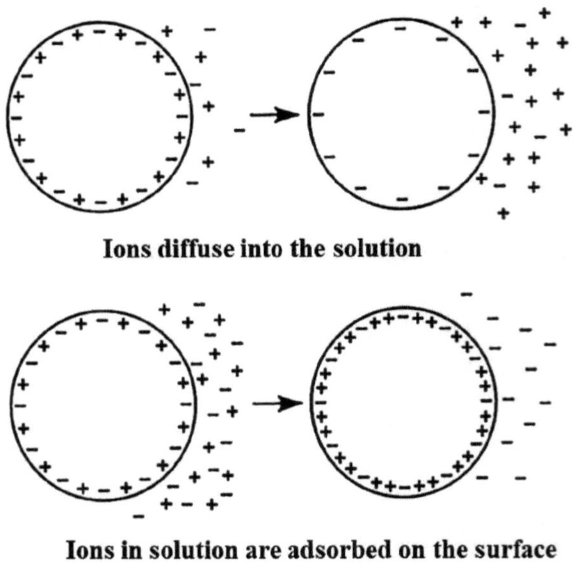

Fig. 4.2 Changes in surface charge of ions in solution

In strong acid solutions, suspended particles carry a positive charge. The charge comes from several aspects: the discontinuity of the lattice surface leads to uneven distribution of cations or anions on the surface; in addition, ions enter the water phase from the surface, causing unsaturation of surface ions, or H^+, Al^{3+}, SO_4^{2-} adsorbed on the crystal surface.

Each particle surface carries a charge, its charge density and intensity depend on the surface properties of the particle and the chemical properties of the water suspension medium.

The repulsive force not only prevents the aggregation of particles, but also keeps the particles in a continuous motion state to prevent sedimentation from occurring. The reason why the charge repulsion of small particles is very obvious is that its area/mass and charge/mass are very large.

4.1.1.1 Main Mechanism of Flocculation

1. **Collision between particles**

When the distance between particles is close enough (0.01 microns or lower), van der Waals forces exceed the repulsive forces. Stirring suspended particles leads to a reduction in the distance between particles and even collisions, resulting in a coagulation (flocculation) process. Because the binding force between particles is weak, excessive stirring can damage the flocs, so an appropriate degree of stirring is necessary. Obviously, the above can only form a certain degree of flocculation process.

2. **Reduction of Surface Charge**

The reduction of surface charge weakens the repulsive forces and promotes the flocculation process. By adding substances with different charges from the particle surface for electrical neutralization, inorganic salts such as $NaCl$, $CaCl_2$, $Al_2(SO_4)_3$, etc., can provide Na^+, Ca^{2+}, Al^{3+}. High-valence cations are more effective in the flocculation process, and Al^{3+} and Fe^{3+} are widely used in practice. When suspended particles are uncharged in the suspension medium, according to most reported flocculation processes, the particle stability is at its lowest. At this time, the Zeta potential is zero.

Zeta Potential: If the particle surface carries a negative charge, cations will be attracted to its surface and form a fixed cation layer. Away from this layer, the counter-ion diffusion layer diffuses outward. When the particle moves in the water phase, shear force is generated along the fixed ion layer. The potential between the shear plane and the bulk solution is the Zeta potential. The changes in the Zeta potential of the particle surface are shown in Fig. 4.3.

Excessive electrolyte concentration can lead to an overabundance of cations, so it is crucial to determine the optimal amount for addition. In addition, different agents have different effective pH ranges, for example, Al^{3+} is 5.5–7.0, Fe^{3+} is 4.0–6.0.

This type of flocculation process is most suitable for well-dispersed suspensions. The resulting flocs are generally small and compact, with a large settling speed.

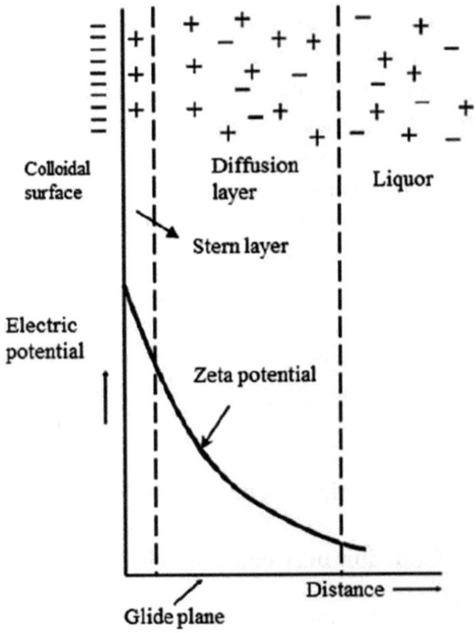

Fig. 4.3 Changes in the Zeta potential of the particle surface

Moreover, the flocs are easily damaged and can be decomposed under the action of high shear forces.

With both 1 and 2, the flocculation process can be effectively improved by adding a "bridging flocculant".

4.2 Artificially Synthesized Bridging Flocculants

These flocculants are all water-soluble high molecular weight organic polymers. The role of bridging flocculants is to strongly adsorb to the particle surface, thereby filling the distance between particles. Synthetic polymer chains are long enough that one end can adsorb to one particle and the other end can adsorb to another particle. High molecular weight polymers can adsorb several particles at once, forming a three-dimensional structure. Generally, the higher the molecular weight, the better the effect of the flocculant. Most synthetic high molecular weight flocculants are polyacrylamides and their derivatives as shown in Fig. 4.4.

Polyacrylamide is non-ionic and forms ionic types when polymerized with other monomers. Anionic polyacrylamide can be obtained by copolymerization of acrylic acid and acrylamide or partial hydrolysis of polyacrylamide, which is the most common synthetic flocculant. Cationic polyacrylamide can be obtained by polymerization of acrylamide and quaternary ammonium salt derivatives of acrylamide. The type of cationic flocculant used varies depending on the application. Most bridging flocculants carry cations or anions.

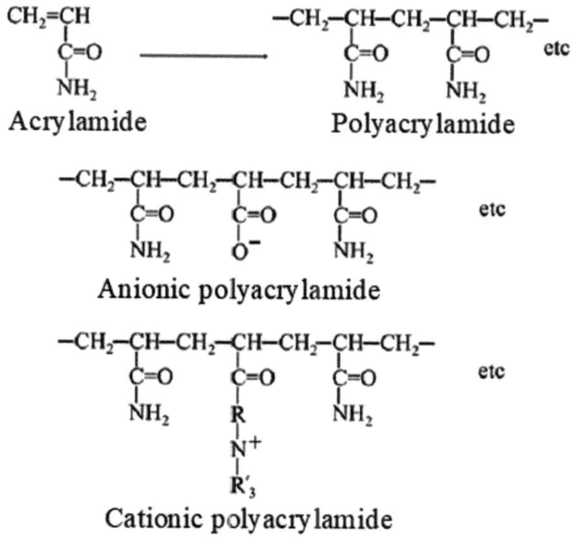

Fig. 4.4 Most polymers have repeated units of acrylamide and its derivatives

The charge of the polymer has two functions:

① Adsorb to the particle surface through electrostatic attraction;
② Cause the polymer molecules to extend and expand under the repulsion of long chains, thereby maintaining the linear shape of the molecule and adsorbing more particles.

Since most suspended particles carry a negative charge, cationic electrolytes are the best choice. This also complies with the purpose of electrical neutralization, allowing the polymer to adsorb to the particle surface, but this is not necessary for bridging. For bridging, the polymer must strongly adsorb to the particle surface, and chemical groups such as amino groups can promote adsorption. Cationic or anionic groups can extend the long chains of high molecular weight polymers for bridging. Most commercial bridging polyelectrolytes used for slurry treatment are anionic because they have a higher molecular weight and are relatively cheaper, providing a cost advantage.

In general, the higher the molecular weight, the better the flocculation effect, and the faster the sedimentation rate. However, during rotary vacuum filtration, the lower the molecular weight, the higher the efficiency. High molecular weight polymers form flocs with a lot of bound water in their structure and increase the moisture content of the filter cake, making the cake easier to remove. Low molecular weight polymers form smaller flocs with higher shear resistance. The resulting filter cake is a homogeneous porous structure that can be quickly dewatered but prevents fine particles from passing through the cake. The study of the flocculation mechanism here can be divided into two steps:

① The drug adsorbs to the particle surface;
② Formation of the polymer body.

Adsorption may occur when the drug and the particle surface carry different charges, causing surface electrostatic forces to not work. Theoretically, any cationic charge should adsorb a polymer carrying the opposite ion, and vice versa. When there is an excess of polymer, the particles are recharged, and stabilization occurs again because the particles can carry a charge different from the original one. At this time, the re-adsorption of the flocculant will stop.

Because the charge distribution on the particle surface is heterogeneous, the Zeta potential of the surface may be high, low, or even have the opposite potential. This means that most surfaces carry a negative charge, and a small part of the surface carries a positive charge. Anionic polymers are adsorbed to this small part of the positive charge, even if the net charge of the particles is the same as the polymer. Because the amount of positive charge carried by the particles is small, the particles are generally located at the end of the long chain, not distributed along the entire long chain. This binding force is a specific electrostatic effect, so there is a threshold of Zeta potential, at which point central adsorption is impossible. After the chemical bridging action is formed, there is a repulsion between the particles, and the floc will be a loose structure.

Hydrogen bonding is also an important adsorption method for non-ionic polymers, such as polyacrylamide. The force of a single hydrogen bond is weak, but the molecular weight of polyacrylamide can reach 10 million, which can form at least 150,000 hydrogen bonds, greatly enhancing its force. The characteristic of hydrogen bonds is that atoms are connected by hydrogen at short distances. Before the formation of hydrogen bonds, the interacting atoms (for example, one located in the polymer, one located in the particle) must be close enough.

Polyelectrolytes can also form in the formation of insoluble salts, such as polyacrylic acid adsorbed on clay, lime, and other calcium-containing minerals to form insoluble calcium polyacrylate.

There are many different hypotheses about the mechanism of floc formation, but very few hypotheses are supported by experiments. One possibility that plays an important role in floc formation is that the Zeta potential is reduced by adsorbing charged polymers, and van der Waals forces bring particles closer together. Another possibility is that the polymer adsorbs particles at both ends at the same time. As the adsorption process continues, and after the polymer molecules are compressed, the distance between the particles decreases. Hypothesis 1-The situation where the polymer connects particles at both ends simultaneously is shown in Fig. 4.5. This theory seems inappropriate because polymer molecules are not linear, and their ends may not be able to bind particles.

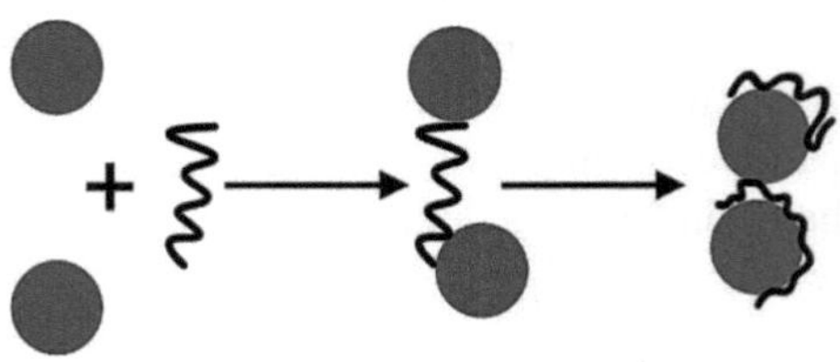

Fig. 4.5 Hypothesis 1 schematic-polymer connects particles at both ends simultaneously

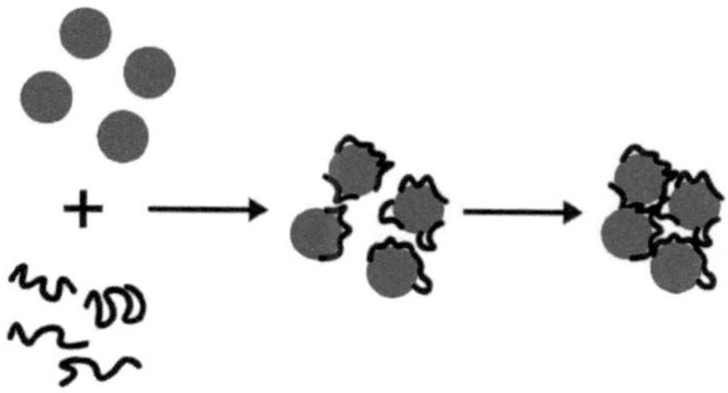

Fig. 4.6 Hypothesis 2 schematic-polymer forms a closed loop with multiple points of contact with the particle surface, connecting each particle

The second hypothesis believes that the shape of the polymer molecule is extremely irregular, so a polymer molecule can adsorb at different positions on the particle surface, thereby forming a closed loop. Depending on the surface, the length of the closed loop is also different. Collision action causes part of the particle surface to be covered, and bridging occurs. Hypothesis 2 is shown in Fig. 4.6. According to this theory, the polymer molecule is completely adsorbed on individual particles before the bridging action occurs. The above occurs at higher polymer concentrations and explains why the flocculation effect decreases after a certain critical concentration. In this case, the flocculation action will be hindered by the repellent force between the polymer molecules. This is also a more credible theory because its explanation is consistent with experimental results.

4.3 Natural Polymers

Natural polymers, such as starch, glue, and alginates, can be used as bridging flocculants, but their molecular weights are generally lower than artificially synthesized polymers, and the degree of flocculation is lower. Polysaccharides (starch, dextrin) work better in neutral or slightly alkaline waters. Organic colloids (glue, gel, protein, casein, etc.), which are composed of large molecules, work better in acidic solutions. These polymers have some shortcomings: high concentration requirements, unstable solubility, easy change of properties, changes in floc strength during the storage process of sediment sludge, but they can be chemically treated to anionize or cationize. Ionic characteristics do not matter much for natural bridging polymers, because they have almost no effect on the Zeta potential, and generally rely on hydrogen bonding.

4.3.1 Factors Affecting Flocculation

(1) Polymer Concentration

For a given treatment system, the flocculation effect does not increase with the increase of electrolyte concentration outside the optimal concentration, and the excessive addition of flocculation effect will decrease instead. The optimal concentration is under experimental conditions, where the maximum amount of polymer can be used for the flocculation process and directly form multi-particle adsorption. Taking silicate clay as an example, at a certain concentration, all polymers are adsorbed on the solid surface. Compared with the optimal concentration, the critical point where the polymer is not completely adsorbed depends on the precipitation rate. However, as the polymer concentration increases, the amount of adsorption will also increase.

The optimal concentration (e.g., maximum precipitation rate) cannot be simply predicted, as it is not only related to the charge characteristics and electricity of the polymer, but also to the molecular weight. The optimal concentration increases with the increase of molecular weight, but at the same time, the precipitation rate is also high.

(2) Shear action on flocs

All flocs, whether formed by coagulation or bridging, will break when over-stirred, and when stirring stops, the flocs cannot reform. If the flocs formed by excessive polymer presence are stirred, their decomposition rate is much higher than that of flocs formed at the optimal concentration. Excessive polymers are reabsorbed from the solution because new adsorption sites are reformed after the flocs are damaged. The reason why flocs cannot form effectively is that there is a repulsion force in the excessively adsorbed polymer. Therefore, the optimal concentration needs to be achieved under certain stirring conditions.

(3) Particle size

For fine silica particles, the optimal polymer dosage/solid particle number ratio is related to the specific surface of the particles, but generally speaking, the smaller the particle size, the higher the flocculant dosage.

(4) Slurry density

Slurry density affects the optimal concentration because it affects the collision between particles. Although the optimal concentration decreases at low slurry density, the change of optimal concentration from 4 to 50% has little effect on the flocculation effect (using silica). The formation rate of the aggregate is directly proportional to the square of the particle concentration, and the coagulation time (such as the time when the number of particles is halved) is inversely proportional to the initial concentration. The stability of flocs during stirring increases with the increase in slurry density. After the flocs are damaged, the time for the flocs to reform depends on the collision between particles. At low slurry density, this time is relatively long, and polymer molecules may adsorb on the same particles, reducing the bridging effect of the system. For high-density slurry, the possibility of particle collision will be higher.

(5) Molecular weight

When using low molecular weight polymers, each polymer molecule may adsorb on the same particle. The degree of flocculation decreases with the addition of polymer. For the same type of polymer with a higher molecular weight, it is easier to be adsorbed and utilized by flocs. The optimal concentration and precipitation rate increase with the increase of molecular weight. In the treatment of coal washing water, the ideal flocculant is high molecular weight anionic, because it is very important to achieve a fast precipitation rate. Of course, low molecular weight products, as mentioned before, are more suitable for filtration.

Molecular weight is not the only measure of flocculation effect, as two polymers with the same apparent molecular weight may have different molecular weight distributions, as shown in Fig. 4.7. Polymers A and B have the same average molecular weight, but they exhibit different flocculation effects. Polymer A has a broad molecular weight distribution, including a large number of high molecular weight and low molecular weight parts. The difference in actual application depends on the nature of the slurry and the nature of the polymer.

(6) pH Value

The pH value of the slurry determines the surface charge of the particles:

$$\text{High pH value:} \ -\mathrm{Si}-\mathrm{OH} == -\mathrm{Si}-\mathrm{O}- + \mathrm{H}^{+}$$
$$\text{Low pH value:} \ -\mathrm{Si}-\mathrm{OH} + \mathrm{H}^{+} == -\mathrm{Si}- + \mathrm{OH}_2$$

The isoelectric point of silicon dioxide is around a pH value of 3.

Under conditions of excessively low or high pH values, flocculation is difficult to carry out; although the optimal concentration varies with the pH value. When the

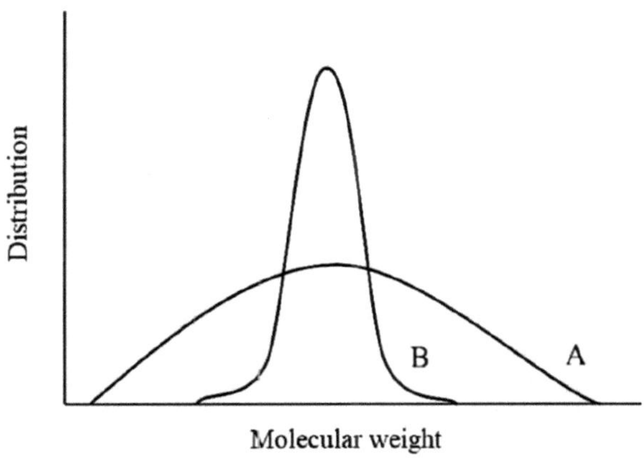

Fig. 4.7 Polymers with the same apparent molecular weight have different molecular weight distributions

particle surface has a high charge that separates the particles, the bridging effect is significantly weakened as more polymers adsorb to individual particles. The pH value also controls the degree of ionization of the polymer, the number of charges on the polymer chain, and the extension of the polymer molecule. These factors naturally affect bridging and flocculation, and the suitable pH value for different flocculants is determined by their type.

(7) Temperature

It is generally believed that as the temperature increases, the flocculation effect improves, but this is not always the case, because temperature changes have different impacts on different systems. The diffusion of flocculants and the collision rate between particles increase with temperature, but the adsorption process is exothermic, and an increase in temperature is not conducive to adsorption. Depending on the interaction between the solvent and the solute, the linear extension of the polymer molecule also changes with temperature, so it is difficult to accurately judge the impact of temperature on a given system.

4.3.2 *Application of Flocculants*

For an effective synthetic flocculant, it must transition from a transport state to a freely dissolved and fully extended single molecule. Considering factors of transportation and price, contemporary high-efficiency polymers are generally provided in solid form, and solid products have a particle size distribution around the average molecular weight. Each particle is randomly wrapped by long polymer chains, resembling a skein. For the dissolution of long chains, they first need to adsorb water molecules. The particle size determines the degree of water molecule adsorption, and dissolution occurs through hydrolysis and activation of repulsive ion groups. Unfortunately, many gels form on the wet polymer surface, preventing free water from wetting the polymer at the particle center. This is the main problem with polymer dissolution. When the surface is wetted and the polymer is not fully separated, the outer surface of the particle cluster is wetted, forming a gel layer, thereby preventing water molecules from penetrating. This is particularly disadvantageous for large polymer particles, which generally take several hours to dissolve and wet. Many diffusion devices provided by flocculant manufacturers have effectively solved this problem. Once the initial adsorption is complete, each particle forms a gel cluster, the polymer molecules continue to unfold, and the flocculation effect weakens with the formation of gel on the outer surface of the particle. The formation of the gel is related to the size of the particle. Figure 4.8 shows the dissolution rate of large and small particles of polymer flocculants. Small particles generally take 15–30 min to dissolve, while large particles take 2 h. As can be seen, as the particle size increases, the dissolution time increases, so it is best to provide flocculants in small particle sizes. Of course, small particle sizes also have two drawbacks. Firstly, it is challenging to completely separate fine particle powder after diffusion. Additionally, fine particles can pose safety risks and are difficult to manage. They easily generate aerosols, creating a hazardous environment around the reactor. A new invention tackles these issues by reducing the size of the polymer, eliminating unnecessary fine particles, and shaping it into bead form. This manufacturing process results in a particle size that is much smaller than traditional flocculants and effectively removes fine particles. Bead flocculants have good solubility and, due to their strong fluidity, increase the convenience of storage and operation. To accelerate the solubility of gel particles, a solution of higher concentration than actual application needs to be prepared. High-concentration solutions can promote the destruction of gel clusters by stirring, increasing the surface area of particles, and improve solubility. Another advantage of preparing high-concentration solutions is that it can reduce mixing, each mixing can last a long time, and a large amount of flocculant is stored for emergency use. The role of particle size mentioned above only applies to difficult-to-dissolve polymers (such as extremely high molecular weight or extremely low molecular weight), where particle size plays an important role. The dissolution time of low molecular weight or high molecular weight ionic products is shorter than the predicted dissolution time in Fig. 4.8.

During the dissolution process, stirring should follow these factors:

1. ① Maintain the suspension of flocculant particles to avoid sedimentation to the bottom of the reactor to form a coagulation layer.

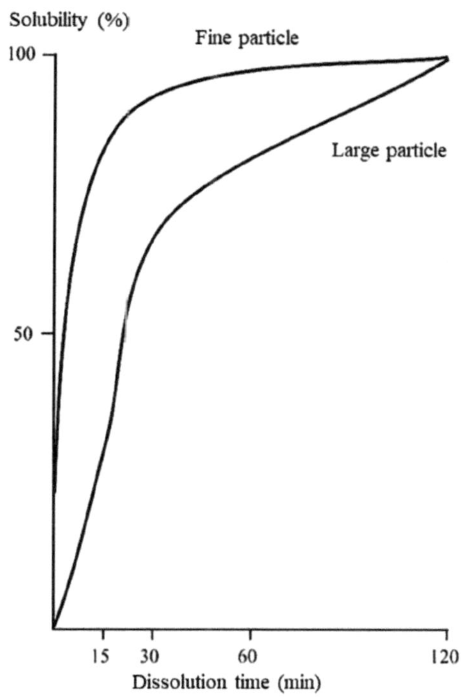

Fig. 4.8 Dissolution rate of large and small particles of polymer flocculants

2. ② Release the free polymer chains in the expanded colloidal particles.
3. ③ Reduce the size of the gel particles and increase the release of free long chains.

Different types of stirring need to provide enough power to meet the above requirements, but they cannot provide high central shear conditions that would cause the destruction of flocculant molecules. The ideal stirring condition is to use large paddle for slow stirring or air stirring.

Once the flocculant solution is prepared, it should be transmitted to the dosing point. Generally, the flocculant is first sent to the pulp tank to prepare a fresh solution. After that, the dosing amount is controlled by a valve. If the dosage amount is exceeded, the valve is closed. If it is less than the dosing amount, the valve is opened. However, this can easily cause an uneven concentration of flocculant in the dissolution pool due to changes in water head. A commonly used solution is to pump the solution to a small tank with a continuous water head. The required solution is controlled by a valve according to the dosing amount. If there is too much solution, it is returned to the pulp tank. For the transmission of flocculants, pumps with high shear force should not be used.

Figure 4.9 describes the effect of a centrifugal pump on the sedimentation rate of a 0.1% polymer solution. Figure 4.9 show that as the molecular weight decreases, the sedimentation rate decreases. First, the solution is damaged when it passes through the pump, and second, the reflux reduces the throughput of the pump. This destructive effect can be reflected by the decrease in solution viscosity.

The rational use of flocculant solution determines the effectiveness of the treatment of suspended solutions. Polymers cannot be evenly distributed in suspended particles due to their own viscosity, which is sometimes a problem. Flocculants are easily adsorbed on the surface of particles, generally irreversible adsorption. Uneven distribution of flocculants leads to waste at the flocculant enrichment site, and the treatment efficiency for high concentration suspended matter is not high. In general, low concentrations of flocculant result in low solid/liquid separation efficiency. The concentration of flocculant should be increased until the effect is good, at which point only higher costs are required.

Polymer solution stirring requires an appropriate speed to ensure enough power for the flocculant to be evenly distributed in the suspension, but the power should not be too high, causing the destruction of the floc. Figure 4.10 shows the change in floc size with increasing mixing intensity. It can be seen that the floc effect increases with the increase in stirring intensity, but as the floc is destroyed, the flocculation effect gradually decreases.

Good stirring of the flocculant solution can be improved by adding a dispersant. Figure 4.11 shows the same concentration of flocculant used in solutions of different concentrations. Of course, the better the dispersion, the more it conflicts with the process of previously preparing high concentration solutions. Linear diffusion along the high concentration line can overcome the above shortcomings.

Another important method to improve processing efficiency is to divide the reagent into two or more water streams and add them to different addition points of the

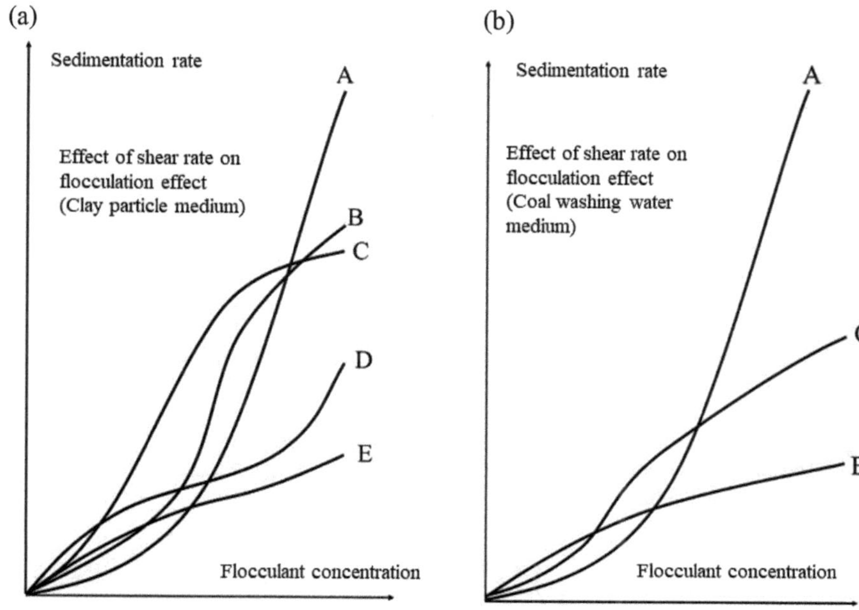

Fig. 4.9 The effect of a centrifugal pump on the sedimentation rate of a 0.1% polymer solution, **a** Clay particle medium, **b** Coal washing water medium (Solution A—before being pumped, viscosity 600 cps; Solution B—once through the pump, viscosity 540 cps; C—after the fourth full flow, viscosity 420 cps; Solution D—after 10% transport capacity, viscosity 330 cps; Solution E—after 3% transport capacity, viscosity 240 cps)

Fig. 4.10 Change in floc size with increasing mixing intensity

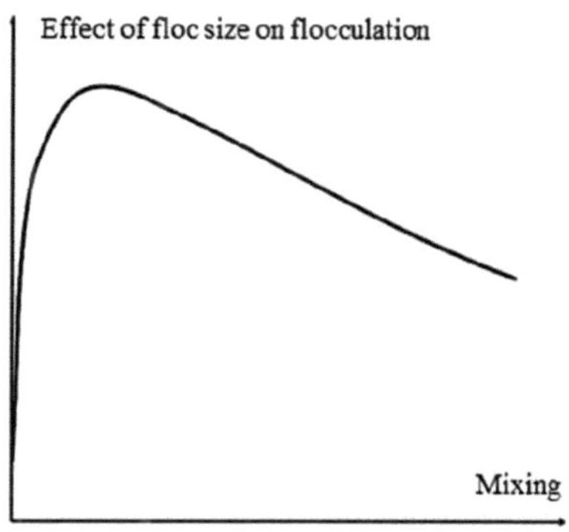

processing solution. The two-stage addition points of the flocculation process are reflected in Fig. 4.12.

Figure 4.10 has simply described the impact of floc size on flocculation effect, the impact of mixing strength on solution clarification (capturing fine particles), and shows that additional mixing can enhance the floc's capture of dispersed particles. As shown in Fig. 4.12, in the two-stage addition process, the maximum amount

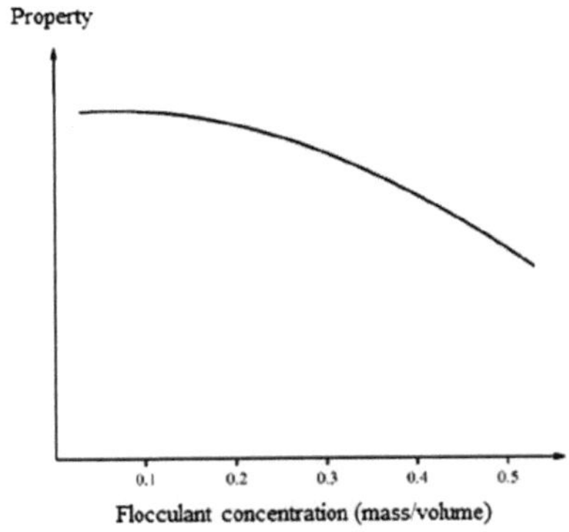

Fig. 4.11 Flocculation performance of the same concentration of flocculant used in solutions of different concentrations

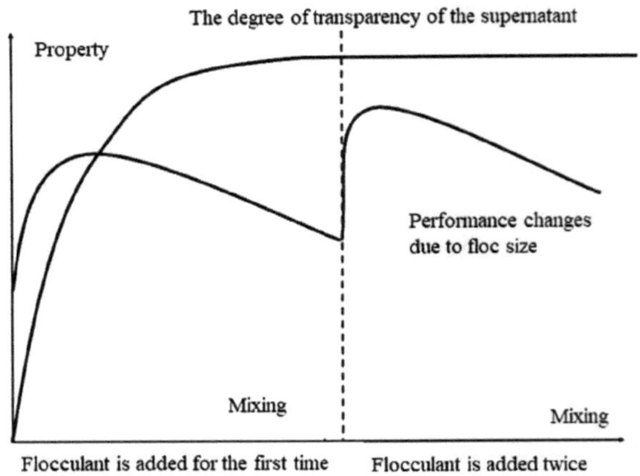

Fig. 4.12 Establishment of two-stage addition points of reagent during flocculation process

of net capture effect is obtained after the floc formed in the first stage is broken; secondary addition can re-form optimized floc under controlled mixing conditions. Compared with the same amount of addition, the two-stage addition system can provide the maximum amount of capture, the largest floc size, and small flocs can form large flocs. Finally, it is very important to fully consider flocculation theory and choose different flocculants according to different reaction systems; in addition, the maximum flocculation treatment efficiency can be obtained through effective dissolution and dispersion.

4.4 Application Fields and Selection of Polyacrylamide

4.4.1 Application Fields of Polyacrylamide

Polyacrylamide (PAM) and its derivatives are multifunctional polymers widely used in papermaking, petroleum, chemical, metallurgy, geology, coal, textile, water treatment, and other fields. It is also one of the most important varieties in water-soluble synthetic high polymer electrolytes. Internationally, polyacrylamide was applied to the above fields earlier, with the largest application field being water treatment agents (Table 4.1 shows the application fields of polyacrylamide). China began to use and produce polyacrylamide in large quantities in the late 1980s. According to statistics, the global PAM product exceeded 490,000 tons in 2000, with a production of 60,000 tons in the United States, about 20,000 tons in Japan, and about 100,000 tons in China. By 2020, China's PAM production was 1.09 million tons, accounting for about half of the global production, and has become the world's largest polyacrylamide production and sales market. With the increasing demand for PAM in various industries, new varieties and new synthesis processes are constantly emerging.

Acrylamide can be polymerized through free radical polymerization to obtain polyacrylamide with different molecular weights, as shown in Fig. 4.13. The value of n can range from a few to several hundred thousand. Radiation, ultraviolet light,

Table 4.1 Application fields of polyacrylamide

Fields of application	Fields purpose	Polymer type
Municipal sewage	Treatment of biological sludge	Cation
Municipal sewage	Physico-chemical treatment	Anion
Steelmaking	Circulating water	Anion
Aluminum sulfate	Removal of impurities during production	Anionic or non-ionic
Brewing industry	Wastewater	Positive or anionic
Clay & Ceramic	Concentrate	Anion
Coal washing	Coal slurry sedimentation, tailings sedimentation, coal filtration	Anion
Dairy processing	Biological treatment of waste liquids	Cation
	Sludge treatment	
Iron ore	Sedimentation	Anion
	Fine particle filtration	Non-ionic
Iron and steel industry	Blast furnace gas scrubbing	Anion
	Purification of pickling liquids	
Drinking water	Improve processing efficiency	Anion
	Reduce the amount of aluminum and iron salt coagulants	Anion

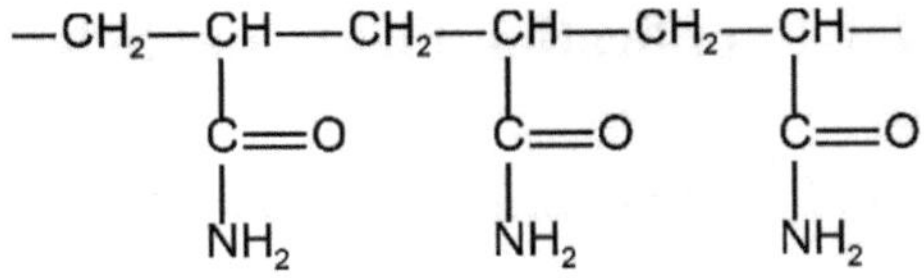

Fig. 4.13 Molecular structure of polyacrylamide

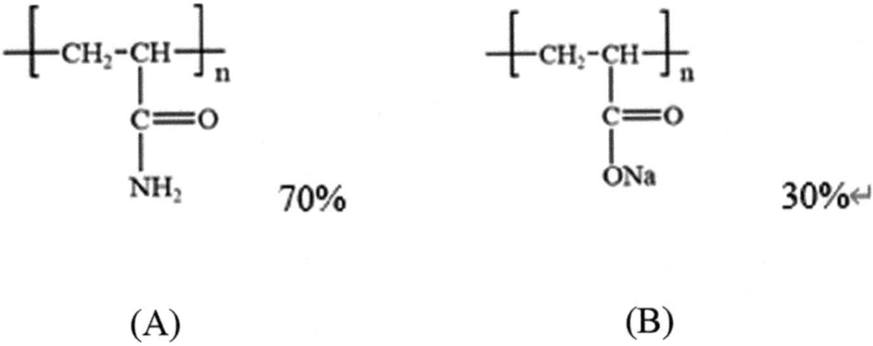

Fig. 4.14 **a** 70% of the acrylamide chain ends, **b** Sodium acrylate chain ends occupy 30%

ultrasound, and easily decomposable peroxides or azo compounds can all initiate the polymerization reaction. The initiators commonly used in industry are organic peroxides, azo compounds, and inorganic peroxides. Common organic peroxides include benzoyl peroxide, dodecane peroxide, and isopropyl peroxide; there are many types of azo initiators.

4.4.2 The Issue of Reasonable Selection of Polyacrylamide

In order to improve the dewatering performance of sludge in water treatment plants, a suitable amount of organic high molecular polymer polyacrylamide (PAM) is generally added evenly to the concentrated sludge before mechanical dewatering to reduce the specific resistance of the sludge, making it easier to dewater. Polyacrylamide comes in three types: anionic, cationic, and nonionic. They should be comprehensively evaluated from both technical and economic perspectives, and the most suitable PAM type and brand for each water treatment plant should be selected through testing.

The sludge in wastewater treatment plants mainly consists of hydrophilic colloidal particles with organic components. The colloidal particles have a strong negative Zeta potential. When the sludge is mechanically dewatered, a suitable amount of cationic PAM is usually added first to neutralize the charge of the colloidal particles and bridge the particles, making the sludge easier to dewater. The sludge in water treatment plants mainly consists of colloidal particles with inorganic components such as silt, and aluminum or iron salt coagulants have already been added during the water purification process in the plant, significantly reducing the negative Zeta potential of the colloidal particles. Therefore, a suitable amount of PAM is mainly added to the concentrated sludge in water treatment plants before dewatering to promote bridging flocculation between the mud particles and reduce the specific resistance of the sludge. Both laboratory tests and production tests at water treatment plants have confirmed that both anionic PAM and cationic PAM can achieve similar ideal effects of reducing the specific resistance of sludge and achieving a good effect of more than 35% solid content in centrifugally dewatered sludge at similar dosages (the cationic dosage is generally slightly higher than the anionic dosage). The turbidity of the separated liquid after sludge dewatering is lower when cationic PAM is used. Since the unit weight price of cationic PAM is about twice that of anionic PAM, it is advisable to use high molecular weight anionic PAM with an acrylamide monomer content of less than 0.05% as the PAM conditioner for dewatering sludge in water treatment plants. Nonionic PAM is generally not used for sludge conditioning due to its slow dissolution rate.

4.4.3 The Molecular Weight Issue of Polyacrylamide

The "Water Treatment Agent Polyacrylamide" GB 17514-2008 specifies the polymer content of water treatment agent polyacrylamide as shown in Table 4.2:

Table 4.2 Technical requirements for polyacrylamide (water treatment agent polyacrylamide GB 17514-2008)

Project	Index	
	Class I	Class II
Solids content (solid), $w/\% \geq$	90.0	88.0
Acrylamide monomer content (Dry basis), $w/\% \leq$	0.025	0.05
Dissolution time (Anionic type) /min $\leq$	60	90
Dissolution time (Non-ionic) /min $\leq$	90	120
Sieve residues (1.00 mm screens), $w/\% \leq$	5	10
Sieve residues (180 μm Screens), $w/\% \geq$	85	80
Insoluble matter (Anionic type), $w/\% \leq$	0.3	2.0
Insoluble matter (Non-ionic), $w/\% \leq$	0.3	2.5

Class I: For drinking water treatment; Class II: For industrial and wastewater treatment

4.4.4 *Quality and Salt Content Issues of Polyacrylamide*

From the polymer content of polyacrylamide above, its effective content should exceed 88%. That is, the content of acrylamide and sodium acrylate polymerized at the chain end should exceed 88%. With a hydrolysis rate of 30% for polyacrylamide, that is, the chain end of sodium acrylate occupies 30%:

The molecular weight of the chain end of acrylamide is 71, and the molecular weight of sodium acrylate is 95. If the molar ratio of ionicity is 30%, the actual mass ratio is $95 \times 30\%/(71 \times 70\% + 95 \times 30\%) = 36.4\%$, the mass of the acrylamide chain end is 100%-36.4% = 63.6%. Based on commercial habits, the anionic degree is generally expressed by the mass ratio method. The industry standard requires that the effective polymer content exceeds 89%, with the remainder consisting of added dispersants, moisture, and other substances. Calculated based on an effective content of 88%, the mass of the acrylamide chain segment contained therein should be $[63.6\% + 36.4 \times (71/95)] \times 88\% = 90.8 \times 88\% = 79.9\%$. That is to say, the actual polymer acrylamide chain segment content in the finished product should not be less than 79.9%. Further, the cost of the acrylamide monomer accounts for 79.9% or 80% of the polymer raw material cost.

If calculated based on the acrylamide monomer discount price of 12,000 yuan/ton (most companies purchase acrylamide solution to produce polymers), then the cost of the monomer is $12{,}000 \times 80\% = 9600$ yuan/ton (price of June 2023), which is also the largest cost of acrylamide. Before polyacrylamide polymerizes, it needs to purify the monomer solution with ion exchange resin, add initiator to polymerize, cut into pieces, dry, crush, sieve, package, etc. The entire polymerization and drying crushing cost is estimated to be 1000 yuan/ton, so the direct cost of the polymer is $9600 + 1000 = 10{,}060$ yuan. This does not include energy consumption, labor, management fees, taxes, and profits. If a 15% profit and 13% tax are added, it is close to 13,000 yuan/ton. If the product price is much lower than the monomer price, many companies have added different salts, such as sodium chloride, sodium sulfate, etc. Since there is no method to detect the effective content of organic polymers in the standard, the thermogravimetric method can quickly determine the percentage content of organic matter. Above 300 °C, organic polymers will lose weight, while salts will not volatilize between 300 °C and 400 °C.

During the coagulation process, to enhance sedimentation, it is common to add less than 1 mg/L of polyacrylamide. This addition significantly improves the flocculation effect, particularly in situations involving high-turbidity water or in low-temperature, low-turbidity conditions during winter. Many designs for high-density sedimentation tanks typically incorporate the use of polyacrylamide.

4.5 Polymerization Method of Acrylamide

The polymerization methods of acrylamide mainly include bulk polymerization, solution polymerization, suspension polymerization, emulsion polymerization, and dispersion polymerization, as shown in Table 4.3.

Table 4.3 Polymerization methods of acrylamide

Polymerization method	Ontology polymerization	Solution polymerization	Suspension polymerization	Emulsion polymerization	Decentralized polymerization
Formulation	Monomer	Monomer	Monomer	Monomer	Monomer
Main ingredients:	Initiators	Initiators, solvents	Water, oil-soluble initiators, dispersants	Water, water-soluble initiators, water-soluble emulsifiers	Water, water-soluble initiators, salts
Aggregation sites	In the main body	In solution	Inside the droplet	Within micelles and latex particles	Disperse the surface of the particles
Polymerization mechanism	The reaction speed is fast and the molecular weight is low	At low concentrations, the molecular weight is low; At high concentrations, the molecular weight is high	Same as bulk aggregation		
Production characteristics	It is not easy to dissipate heat and is generally not used	Low concentration polymerization heat dissipation solution, high concentration is difficult	Easy heat dissipation	Easy heat dissipation	Easy heat dissipation
Product characteristics	The molecular weight distribution is wide	Low concentration products are easy to dissolve and use directly; The highly concentrated product is further processed into a dry powder		A reversed-phase agent is added during use, or further processed into a dry powder	Use directly for dilution

(1) Bulk polymerization

Bulk polymerization is the polymerization by directly adding an initiator to the acrylamide monomer. This method is generally not used in the actual production process.

(2) Solution polymerization

Solution polymerization is the polymerization of the monomer and initiator dissolved in a solvent.

(3) Suspension polymerization

Suspension polymerization is the polymerization of the monomer suspended in water in droplet form, with the polymerization system mainly composed of the monomer, water, oil-soluble initiator, dispersant, etc.

(4) Emulsion polymerization

Emulsion polymerization is the polymerization form where the monomer is dispersed in water, and the formed polymer enters the emulsion. This method is the main method for preparing water-in-oil polyacrylamide emulsion.

(5) Dispersion polymerization

Dispersion polymerization is a new polymerization method that has emerged in recent years, where the monomer and initiator are dispersed in a high-salt-content aqueous solution, and the tiny polymer particles formed by polymerization are dispersed in the aqueous solution in the form of a suspended emulsion due to salting-out.

A common problem encountered in the use of water-soluble polymers is how to quickly and completely dissolve the polymer. In the early days, polymer dilute solutions were used. With the continuous advancement of technology, the molecular weight of the polymer keeps increasing, and the transportation and storage of the polymer become very difficult, even if the concentration of the polymer is only a few percent. Later, manufacturers began to produce solid products, and various mechanical devices were used to dissolve them during use. Although this solves the transportation problem, the polymer partially degrades due to shear forces during the dissolution process, and sometimes it cannot be completely dissolved, resulting in a "fish-eye" phenomenon. This not only causes waste of the polymer, but also brings great inconvenience to use. Incomplete dissolution of the polymer in papermaking often reduces the quality of the paper. In the 1970s, oil-in-water emulsion was successfully developed. Using oil-in-water technology, fast-dissolving polymers can be produced, and this technology has been widely recognized by the industry. However, the biggest drawback of oil-in-water emulsion is that it contains a large number of hydrocarbons. Kyoritsu Yuki Co. Ltd. was the first to apply for a method of producing cationic polymers by dispersion. This method uses hydrophobic monomers to prepare high molecular weight polymers in a salt solution, and the obtained polymers can be quickly dissolved in water. U.S. Pat. No. U.S. Patent

5,605,970 proposes a method for preparing a water-soluble anionic polymer dispersion. This method details the preparation of acrylic acid-(ethyl) acrylate hexyl ester. (Ethyl) acrylate hexyl ester is a hydrophobic monomer, which gives the polymer a certain degree of hydrophobicity.

During the dispersion polymerization process, both the monomer and the initiator should be soluble in the dispersion medium, but the medium should be a poor solvent for the polymer. At the start of the reaction, the system should be homogeneous, with polymerization initiated in the homogeneous system.

Since the dispersion medium is a poor solvent for oligomers, large free radicals, and large molecules, phase separation will occur shortly after the reaction begins. The precipitated polymer forms nuclei, which, due to the adsorption of stabilizers, become stable colloids. The formed nuclei contain a certain amount of dispersion medium and monomer swelling. The diameter of the general particles is between 0.1 and 10.0 μm.

Factors to control during the dispersion polymerization process include the concentration of stabilizers, monomers, and initiators, the solubility of the dispersion medium, reaction temperature, etc. These conditions have a significant impact on particle diameter, final polymer molecular weight, and polymerization kinetics.

If no stabilizer is present, polymer particles can easily agglomerate. Adding a few percent of stabilizer can effectively stabilize the obtained polymer particles. Generally, the stabilizer is a polymer or oligomer with moderate affinity for the polymer and has relatively low solubility in the polymer system. An increase in stabilizer concentration leads to a decrease in particle size. This means that as the stabilizer concentration increases, the number of formed nuclei increases. The flocculation nucleation theory can explain the dependence of particle size on concentration very well. The more stabilizer adsorbed, the slower the agglomeration speed, resulting in more particles becoming mature particles, thus producing smaller particle sizes.

There are already many patents using polymer dispersions to treat various water bodies. A Japanese patent introduces a method for preparing a cationic polymer containing amine and urea groups, which can be used for deinking treatment of water bodies in the paper industry. There are also reports of copolymers of ethylene amine with other monomers used for coal washing, paper decolorization, and coal slurry concentration.

4.6 Synthesis Methods of Acrylamide Cationic Polymers

With the increasing efforts of China in water pollution control, the application range of various water treatment agents is becoming wider and wider, and the consumption is rapidly increasing. Among the types of organic flocculants, polyacrylamide drugs are used the most. Among them, anionic and nonionic PAM account for a large proportion, while cationic and various modified PAM only account for a small proportion, and there are fewer types of cationic polyacrylamide. Cationic flocculants have excellent treatment effects on negatively charged colloids, and are widely used for the removal of turbidity, color, and organic colloids. They are favored in dyeing, papermaking, food fermentation, and other industrial wastewater and sludge dewatering. The development and application of cationic flocculants are increasingly attracting people's attention. There are several methods for the synthesis of acrylamide cationic polymers.

4.6.1 Copolymer of Acrylamide and Diallyl Dimethylammonium Chloride

Diallyl dimethylammonium chloride, abbreviated as DADMAC, is a quaternary ammonium salt cationic monomer. After polymerization, it can form a water-soluble cationic polymer, namely poly-DADMAC. This polymer has a relatively low molecular weight, generally less than 1 million, and is widely used abroad as a coagulant. It also has good algaecidal properties, but the application cost of this coagulant is relatively high. In order to improve its flocculation performance, acrylamide, acrylic acid, and other monomers are now commonly used to increase the molecular weight and enhance its flocculation performance. For example, the copolymers of DADMAC with acrylic acid, hydroxypropyl acrylate (molar ratio 87:10:3), or DADMAC with acrylic acid, are excellent amphoteric polymers, which have good effects on removing ink and oil from pulp wastewater.

The synthesis method of the copolymer of acrylamide and diallyl dimethylammonium chloride is as follows: 2.89 kg of 60% DADMAC monomer solution, 2.11 kg of 50% AM solution, 3.31 kg of deionized water, 27 kg of 40% diethylenetriaminepentaacetic acid sodium, 2 g of 2,2'-Azo(2-methylpropylamidine)dihydrochloride, 5 g of azobis(-2-iminopropane) hydrochloride, 27 g of sodium persulfate, are added to a 30-L reactor. The reactor is equipped with a stirring device, a heating jacket, a vacuum interface, a reflux condenser, a feeding hole, and an intermediate feeding during the reaction process. The system is degassed by vacuum and aeration, and the vacuum pumping and aeration are repeated three times to completely deoxygenate the system. First, the reactants are heated, and polymerization is initiated during the stirring process, while the pressure inside the reactor is controlled at 50–55 MPa. This reaction is exothermic. Once a temperature rise is detected, the remaining 11.62 kg of 50% AM solution and 3.0 g of 2,2'-azobis (N,N'-2-methylisobutyramide) hydrochloride are added. The feeding speed should be properly controlled. After all the acrylamide is added, the temperature is maintained at 35 °C, and stirring is continued for 40 min. The reaction pressure is controlled at atmospheric pressure by introducing nitrogen, and then 0.5 kg of 10% sodium hypophosphite solution is added to the gel product within 30 min and thoroughly mixed. During this process, the temperature will gradually rise to 90 °C, and it is maintained at this temperature for 2 h to reduce the residual monomer content. The colloidal product is crushed and dried in a fluidized bed at 90 °C for 50 min to obtain a granular polymer with a uniform monomer distribution. The intrinsic viscosity of the polymer in a 4% sodium chloride solution can reach 16dL/g.

Emulsion polymerization can also be used, with the potassium bisulfite and ammonium persulfate redox system initiating the reaction to prepare acrylamide and diallyldimethylammonium chloride copolymers. Impurities in the monomer have a significant impact on polymerization. To prepare high molecular weight copolymers, the content of impurities such as metal ions in acrylamide, trimethylamine, methoxy quinone, and allyl dimethylamine, dimethylamine, allyl alcohol, etc., in diallyldimethylammonium chloride should be strictly controlled.

4.6.2 Copolymer of Acrylamide and Dimethyl Aminoethyl Methacrylate Chloride

Currently, this type of polymer is the largest in production and the most widely used cationic flocculant. Its synthesis process is to adjust the molar ratio of acrylamide and dimethyl aminoethyl acrylate chloride according to the cationic degree of the final product, with a total monomer concentration of 20–30% aqueous solution. The reaction generally uses inorganic–organic composite initiators, such as ammonium persulfate-ferrous sulfate, persulfate-sulfite and organic initiator systems, such as azoalkyl imidazolium chloride, dibenzoyl peroxide, etc. The reaction is generally carried out at 20–70 °C for several hours to produce a colloidal copolymer. After crushing, drying, and pulverizing, the final powdered polymer is obtained. In order to accelerate the dissolution rate of the solid polymer, a small amount of surfactant is generally added.

4.6.3 Acrylamide-Dimethyl Aminoethyl Acrylate Chloride-Methoxyethyl Acrylate Copolymer

In a flask equipped with a nitrogen inlet, nitrogen outlet, and initiator addition port, add 16.6 g of 50% acrylamide solution, 30.3 g of 75% dimethyl aminoethyl acrylate chloride solution and 1.6 g of methoxyethyl acrylate, 250 g of deionized water and potassium salt of ethylenediaminetetraacetic acid as a chelating agent. The pH of the monomer solution is adjusted to 3.5 with a 1 mol/L HCl solution, and then N_2 is passed through to remove oxygen for 1 h. Add 1 g of 5.0% 2,2'-Azo(2-methylpropylamidine) dihydrochloride solution, and polymerize for 24 h. High-performance liquid chromatography analysis found that the monomer conversion rate can reach 99.9%, and the relative molecular weight can reach 6.7 million.

4.6.4 Preparation Method of N-Vinyl Formamide Homopolymer Dispersion Polymerization

In a reactor equipped with a reflux condenser, a nitrogen diffuser, a star stirrer, and a thermocouple, add 75 g of N-vinyl formamide monomer, 40 g of 10% polyvinyl alcohol solution, 250 g of water, 6 g of glycerol, and 120 g of sodium nitrate. Heat the mixture to 45 °C under stirring, and add 0.2 g of V-50 initiator (dissolved in 20 mL of water). After the reaction starts, the viscosity gradually increases. During the reaction, 125 g of NaCl is gradually added. The reaction time is 3–3.5 h. Finally, a milky white dispersion is obtained. The above polymer dispersion is made into a 2% aqueous solution and an equivalent amount of sodium hydroxide is added, heated to 80–90 °C, and reacted for 3 h, 90% of the amide group is hydrolyzed to an amino group. Alternatively, acid can be used as a hydrolyzing agent. If an equivalent mole of acid is added, the hydrolysis forms a copolymer containing amine and formamide, with a ratio of 70/30. Hydrolysis can be performed by passing hydrogen chloride gas or ammonia gas into the polymer aqueous solution.

4.6.5 Preparation Method of N-Vinyl Formamide-Ethyl Hexyl Acrylate Copolymer Dispersion

In a reactor equipped with a reflux condenser, nitrogen gas distribution pipe, star-shaped stirrer, and thermocouple, add 54 g of N-vinyl formamide monomer, 3 g of ethyl hexyl acrylate, 30 g of 10% polyvinyl alcohol solution, 150 g of water, 45 g of glycerol, 40 g of sodium nitrate, and 2 g of sodium chloride. Heat the mixture to 45 °C under stirring, then add 0.3 g of V-50 initiator (dissolved in 20 mL of water). After the reaction starts, the viscosity gradually increases. Gradually add 100 g of ammonium sulfate during the reaction. The reaction time is 3–3.5 h. Finally, a milky dispersion is obtained. The ratio of N-vinyl formamide to ethyl hexyl acrylate can be adjusted arbitrarily.

4.6.6 Acrylamide-Acrylic Acid N, N-Dimethyl Ethyl Ester Chloromethylamine Copolymer Acrylamide-N, N-dimethylaminoethyl Ester Ammonium Chloride Copolymer

The addition amounts of various monomers, initiators, and water are calculated based on the cationic degree, solid content index, and synthesis product volume. Add the measured 150 kg of acrylamide (AM) to the dissolution pool, then add 230 kg of water, filter after dissolution, transfer to the reaction tank, add the measured dimethyl

Table 4.4 Cost/yuan when solid acrylamide is used as raw material

Cation degree/%	12.5	25	33	50
Raw material costs	18,200	22,550	26,200	29,350
Production costs	2000	2000	2000	2000
Total cost	20,200	24,550	28,200	31,350

Table 4.5 Cost per yuan when using liquid acrylamide as raw material

Cation degree/%	12.5	25	33	50
Raw material costs	15,500	20,160	24,100	27,530
Production costs	2000	2000	2000	2000
Total cost	17,500	22,160	26,100	29,530

acryloyloxyethyl trimethyl ammonium chloride (DMC) 72L, azodiisobutyronitrile 0.7L, 5% sodium bisulfite solution 72 mL, 0.15% N, N-methylenebisacrylamide 4–8 mL, close the flask, start stirring and exhaust valve, and start introducing nitrogen. About 0.5 h later, add the measured 5% ammonium persulfate solution 315.6 mL, continue to introduce nitrogen, reduce the gas flow rate, then let it react, stop stirring after the reaction starts, and raise the stirrer. After 3–4 h, the reaction reaches the maximum temperature, continue to keep warm for 5 h, after the end, take out the colloidal product, then cut it preliminary, add it to the extruder, and you can get the granular colloidal product. The granular product is dried at 80–95 °C, crushed to obtain a powdery solid product. The relative molecular weight is determined to be 3–5 million by the viscosity one-point method.

Tables 4.4 and 4.5 are the economic benefit prediction analyses of various cationic polyacrylamides produced from solid and liquid acrylamide as raw materials, respectively.

4.7 Divinyl Propyl Dimethyl Ammonium Chloride Class Polymers

Research on DADMAC monomers and polymers commenced in China during the 1990s, and they were later applied in the treatment of high-turbidity water. High molecular weight DADMAC polymers or copolymers with AM can be used as filter aids, and pigment dewatering agents, and can also be used as flocculants in wastewater treatment and are particularly effective for high turbidity water. The copolymer of DADMAC and AM is the most widely used flocculant in water and wastewater treatment. It is effective for treating pulp waste liquids, printing and dyeing wastewater, refinery wastewater, and oilfield wastewater.

Usually, the monomers used to synthesize this type of polymer are DADMAC, although some literature now mentions that other halides can be used to replace DADMAC as raw materials, but due to cost reasons, most manufacturers currently use DADMAC as raw materials to synthesize polymers.

Below is an introduction to the synthesis methods and processes of DADMAC monomers and polymers.

4.7.1 Raw Material Requirements

The raw materials are all industrial first-class or superior products.

(1) 3-Chloropropene, superior product
(2) Sodium hydroxide (ion membrane alkali), superior product
(3) Dimethylamine aqueous solution, superior product
(4) Ammonium persulfate, analytical grade reagent
(5) Reducing agent, analytical grade reagent.

4.7.2 Brief Introduction to Synthesis Method

The introduction of DADMAC synthesis methods at home and abroad can be simply divided into two types: one-step method and two-step method. The two-step method is to synthesize allylpropyl dimethylamine first, then distill and separate it and react with allyl chloride; the one-step method is to react allylpropyl dimethylamine directly with allylpropyl chloride without separation. The one-step method is superior to the two-step method, with a higher yield, and our research is based on the one-step method.

(1) Add 600 parts of dimethylamine solution (40%) to the jacketed reaction kettle, and lower the temperature in the reaction kettle to 0 °C.
(2) Add 400 parts of chloropropene and 200 parts of sodium hydroxide solution (45%) alternately in 10 h, ensuring that the pH of the reaction system is between 12 and 13.
(3) After the reaction is completed, add 420 parts of chloropropene, stop cooling, continue stirring for 1 h, then gradually heat to 30–40 °C, and reflux for 8 h.
(4) If there is no oil–water stratification phenomenon and the pH is between 7 and 8. Then heat to 40–50 °C, evaporate the unreacted chloropropene. Gradually heat to 80–90 °C, add a small amount of sodium hydroxide, vacuum extract unreacted impurities, after 2 h, react.
(5) Finally, add an appropriate amount of water, control the monomer pH to 7, the refractive index to 1.464 (concentration is 68%), then filter, separate the sodium chloride in the solution, the filtrate is the finished monomer solution. If the production concentration is 40%, dilute with water.

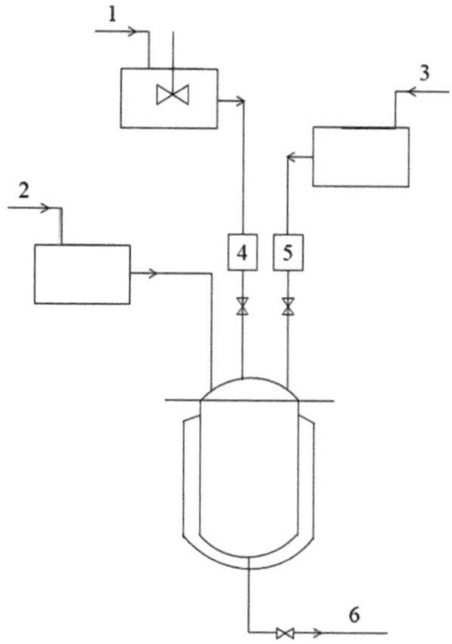

Fig. 4.15 DADMAC Synthesis Process Flow Diagram (1 Sodium hydroxide solution (45%) metering tank, 2 Dimethylamine solution (40%) metering tank, 3 Chloropropene, 4 Sodium hydroxide metering tank, 5 Chloropropene metering tank, 6 DADMAC monomer)

Table 4.6 Raw material consumption table

Raw material	Unit consumption (tons)
Chloropropylene	0.5
Dimethylamine	0.37
Acrylamide	0.03
Caustic soda	0.16
Barrels	20
Water	20
Electricity	400

The production process flow is shown in Fig. 4.15:

According to the above ratio, the consumption of raw materials is as shown in Table 4.6:

4.7.3 Polymer Production Process

In 1981, W. Jaeger and others determined the residual double bonds through radiological chemistry and calculated that PDADMAC (Poly dimethyl diallyl ammonium chloride) mainly contains a pyridine ring, which is connected at the 3rd and 4th positions with $-CH_2-CH_2-$. Theoretically, the synthesis of PDADMAC is carried out as follows:

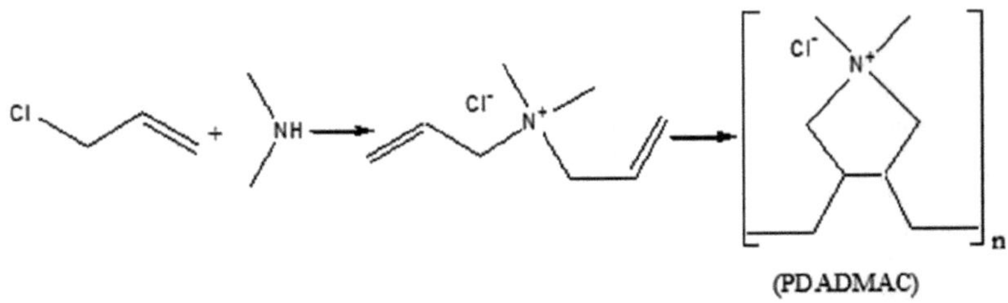

(4.1)

(1) Preparation of DADMAC-acrylic acid copolymer aqueous solution: 426.9 parts of DADMAC monomer, 78.9 parts of deionized water, 14.0 parts of acrylic acid, and 0.2 parts of EDTA are added to the reaction kettle, nitrogen is introduced, the oxygen in the kettle is washed, and the temperature is raised to 65 °C. After 30 min, 17.1 parts of 25% ammonium persulfate are added, the viscosity of the solution increases, and the temperature of the mixture can reach up to 80 °C.
(2) DADMAC-acrylamide dispersion emulsion polymerization: 25.667 parts of 49.0% acrylamide solution and 161.29 parts of 62.0% DADMAC solution are added to the reactor, along with 200 parts of ammonium sulfate and 40 parts of sodium sulfate, 303.85 parts of deionized water, 0.38 parts of sodium formate, 45 parts of 20% acrylic acid (trimethylamine) ethyl ester quaternary ammonium salt chloride, and 0.2 parts of EDTA. The temperature is raised to 48 °C. 2.50 parts of 4% 2,2-azobis(2-amidinopropane)dihydrochloride are added. High-purity nitrogen is introduced into the reactor for 15 min to replace oxygen. About 15 min later, the viscosity of the solution increases, and after 4 h, the temperature of the solution reaches 50 °C. Then, a small amount of acrylamide and EDTA mixture is added with a syringe, and the final Brinell viscosity of the solution can reach 4200 cps.

Because DADMAC homopolymers and acrylamide copolymers have good coagulation and flocculation properties in water treatment, water-soluble cationic polymers have a wide range of uses, and DMDAAC polymers are among the best. Not only because it is highly efficient and non-toxic, but more importantly, it is low cost. From the current international market, the price of PDMDAAC is 1–3 times lower than other cationic polymers, thus showing a broad application prospect.

In the process of sludge dewatering, one should not blindly pursue higher molecular weight of cationic polymers. The cationic functional groups of cationic polymers are mainly adsorbed at the negatively charged sites of the sludge. When the molecular weight of the cationic polymer is very large, it is not conducive to adjusting its own conformation and cannot exert a greater adsorption and bridging effect. Therefore, the molecular weight of the cationic polymer should be a good combination of high, medium, and low, so that it can match the dual effects of bridging and neutralization, achieving maximum dewatering and optimal water permeability. For example, in winter, the organic matter in the sludge of a certain wastewater company in Shanghai exceeded 70%. Professor Li Fengting's team used a low molecular weight cationic polymer, which worked very well. With low-temperature pressure filtration, the moisture content of the sludge can be reduced to below 40%, and the lowest can reach 25%.

Chapter 5
Microbial Flocculants

Traditional chemical flocculants include inorganic metal salt flocculants represented by aluminum and iron polymers, as well as organic polymer flocculants represented by polyacrylamide, chitosan, starch, and cellulose. Although the production process of commonly used chemical flocculants is very mature, their use is costly and can easily cause secondary pollution. For example, if the residual aluminum content in drinking water is high, human intake can increase the probability of developing Alzheimer's disease. The residual monomer of polyacrylamide has neurotoxicity and a risk of cancer. Therefore, the search for widely sourced, biodegradable, and environmentally friendly new generation flocculant products has received much attention.

As a typical representative of new flocculants, microbial flocculants (MBF) are high molecular weight polymers produced by microbes with flocculating activity. The addition of MBF in water treatment processes can promote the formation of flocs, accelerate the aggregation and sedimentation of suspended particles and colloidal particles, achieve solid–liquid separation, and have great application potential.

The history of human discovery of microbial flocculation can be traced back to the end of the nineteenth century. In 1876, French scholar Louis first discovered that yeast and its metabolites can flocculate solid particles. In 1899, Bordet's research showed that bacteria also have similar flocculating effects. In 1935, Butterfield screened out strains that could secrete flocculating active ingredients from activated sludge. In 1976, Nakamura obtained 19 types of microbes with flocculating ability from 214 strains of molds, bacteria, actinomycetes, yeasts, etc. Since then, the research and application of MBF have gradually received attention and development. In 1985, Takagi verified the good flocculating effect of PF101 produced by Pseudomonas. In 1986, Kurane used Rhodotorula to make NOC-1, treating yeast and mud wastewater, with significant flocculation and decolorization effects. Reports on microbial flocculation technology in China began in the 1980s and 1990s. Although it started relatively late, it has made positive progress. Entering the twenty-first century, research

L. Fengting, *Water Treatment Chemicals: Coagulants and Flocculants*,
https://doi.org/10.1007/978-981-96-8221-8_5

on MBF has become increasingly in-depth. Over the years, nearly 2000 related documents have been published in the China Knowledge Network and Web of Science databases. According to the visualization metric analysis of microbial flocculant-related literature (China Knowledge Network and Web of Science), the research hotspots at home and abroad mainly focus on the screening and fermentation culture of flocculating bacteria, MBF component analysis and function enhancement, MBF action mode and mechanism, MBF application range and influencing factors, which will also be the main content of this chapter.

5.1 Classification of Microbial Flocculants

At present, many types of microbes have been discovered at home and abroad for the production of MBF, including bacteria, fungi, and algae. According to the type of microorganism, MBF can be subdivided into bacteria, actinomycetes, molds, yeasts, and algae. According to different sources, MBF is mainly divided into four types:

(1) The microbial cells themselves, which can be isolated, screened, and produced from various soils, activated sludge, sediments, and even waste. The use of *Aspergillus niger* and Saccharomyces cerevisiae cell suspensions as flocculants resulted in a flocculation rate of over 50% for kaolin suspensions.
(2) Active ingredients in microbial cell walls, such as chitin hydrolysis products extracted from filamentous fungal cell walls, glucans inside yeast cell walls, and alginic acid in brown algal cell walls, etc.
(3) Metabolic products secreted by microorganisms into the extracellular environment, also known as extracellular polymeric substances (EPS), such as capsules and mucous substances. The fermentation supernatant of Penicillium and *Streptomyces aureofaciens* was used to treat kaolin suspension, and the flocculation rate exceeded 80%.
(4) Products of highly efficient flocculating strains constructed using gene cloning technology. For example, U9-EPS was produced after overexpression of the UDP-glucose dehydrogenase gene of oleaginous yeast V19, and MBF-FB-5 generated by low-temperature UV-induced mutation of *Bacillus cereus.*

From the perspective of the proportion of functional components, the most common are polysaccharide type, protein type, and MBF where both play a flocculating role. Reports on MBF components such as lipids and nucleic acids are less common. Taking EPS as an example, it explains the flocculation activity and characteristics of MBF. EPS is a high molecular weight polymer secreted outside the body during the metabolic process of microorganisms and surrounds the bacteria. Its composition varies with the producer, but the most abundant are proteins and polysaccharides, along with small amounts of nucleic acids, humic substances, and lipids. The polar and non-polar structures of functional components directly affect the hydrophilicity and hydrophobicity of EPS, thereby affecting the coagulation action and flocculation effect of target pollutants. Functional groups such as phosphate, hydroxyl, thiol, carbonyl, and peptide bonds can react with heavy metals. At the same time, these negatively charged groups make the isoelectric point of EPS higher, so EPS is usually negatively charged. For EPS, based on different solubility and separation difficulty, it can be divided into soluble EPS (S-EPS), loosely bound EPS (LB-EPS), and tightly bound EPS (TB-EPS), with the extraction difficulty increasing in order. Table 5.1 lists some reported categories, sources, and functional components of MBF-producing bacteria.

5.2 Flocculation Mechanism and Influencing Factors of Microbial Flocculants

Unlike traditional flocculants, MBF is a microbial preparation, and its flocculation mechanism and influencing factors are more complex. Variations in sources and uses of MBF can lead to significant differences in the functional components of the flocculant, the flocculation formation process, and the overall flocculation effect. Therefore, studying the flocculation mechanism of MBF and the factors that influence it is crucial for optimizing selection and ensuring the efficient application of MBF.

5.2.1 *Flocculation Mechanism of MBF*

The widely recognized mechanism of flocculation in MBF involves several processes working together, including electro-neutralization, bridging, sweeping, and chemical reactions. Additionally, various hypotheses have been proposed in the study of MBF to help explain the flocculation phenomenon.

(1) Electro-neutralization

The surface of pollutant colloidal particles is generally negatively charged, and the positively charged MBF can neutralize the negative charge on the surface of the colloidal particles, reduce the repulsion between the colloidal particles, help compress the double electric layer, lower the Zeta potential, and promote the aggregation of colloidal particles. In actual operation, metal cations can be added, or the pH value can be adjusted to change the charge state of MBF. For example, the carboxyl group commonly found in MBF can chelate with Fe^{3+}, reducing the degree of ionization of the carboxyl group, promoting the adhesion and co-precipitation of MBF, Fe^{3+}, and pollutants, and achieving better flocculation effects. Flocculant substance I, which contains proteins, polysaccharides, and humic substances, can be produced by fermentation of flocculating bacteria XD-3 extracted from activated sludge. It contains numerous protonated amino groups that can neutralize the negative charge of the system. Therefore, after adding this substance, the Zeta potential of the flocculated supernatant becomes about 0 mV, showing a significant electro-neutralization effect. Flocculant substance I is also rich in hydroxyl and carboxyl groups, which can

Table 5.1 Common flocculating bacteria categories, sources, and active components of flocculation

Category of flocculents	Chinese name of flocculent	The foreign name of the flocculent	Source of flocculents	MBF active ingredient
Bacteria	红平红球菌	*Rhodococcus erythropolis*	China culture collection center	Polysaccharides, proteins
	肠杆菌	*Enterobacter* sp.	Activated sludge	Polysaccharide
	芽孢杆菌	*Bacillus* sp.	Soil	Polysaccharide
	胶质芽孢杆菌	*Bacillus mucilaginosu*	Activated sludge	Polysaccharides
	厚壁芽孢杆菌	*Bacillus firmus*	Soil	Polysaccharides
	克雷伯氏菌	*Klebsiella* sp.	Activated sludge	Polysaccharides
	克雷伯氏肺炎菌	*Klebsiella pneumoniae*	Activated sludge	Polysaccharide
	农杆菌属	*Agrobacterium* sp.	Activated sludge	Polysaccharides
	棒状杆菌	Corynebacterium	Soil	Polysaccharides
	产碱假单胞菌	*Pseudomonas alcaligenes*	Strain Collection Center of Nanjing Agricultural University	Polysaccharides, nucleic acids
	荧光假单胞菌	*Pseudomonas fuorescens*	Soil	Acidic proteins, acidic polysaccharides
	铜绿假单胞菌	*Pseudomonas aeruginosa*	Soil	Acidic proteins, acidic polysaccharides
	动胶菌属	*Zoogloea* sp.	Soil	Polysaccharide
	侏囊菌属	*Nannocystis* sp.	Saline soil samples	Polysaccharides and proteins
	纤维素堆囊菌	*Sorangium cellulosum*	Soil	Polysaccharide
	金黄杆菌	Chryseobacterium	Activated sludge	Proteins, polysaccharides and nucleic acids
	无花果沙雷氏菌	*Serratia ficaria*	Activated sludge	Neutral sugar
	普氏沙雷菌	*Serratia plumuthica*	Bay mudflats	Polysaccharide
	氧化亚铁硫杆菌	*Acidithiobacillus ferrooxidans*	Soil	Protein

(continued)

Table 5.1 (continued)

Category of flocculents	Chinese name of flocculent	The foreign name of the flocculent	Source of flocculents	MBF active ingredient
	罗斯氏菌	*Rothia* sp.	Bay floor sediments	Polysaccharide
	枝芽胞杆菌	*Virgibacillus* sp.	Bay floor sediments	Polysaccharide
	类芽孢杆菌	*Paenibacillus*	Soil	Polysaccharide
	嗜麦芽窄食单胞菌	*Stenotrophomonas maltophilia*	Soil	Polysaccharides, proteins
	玫瑰色考克氏菌	*Kocuria rosea*	Bay floor sediments	Polysaccharides, proteins
	嗜盐单胞菌	*Halomonas* sp.	Bay floor sediments	Polysaccharide
Fungi	寄生曲霉	*Aspergillus parasiticus*	China Culture Collection Center	Polysaccharides, proteins
	黄曲霉	*Aspergillus flavus*	Microbial culture collection room, Universiti Putra, Malaysia	Polysaccharide
	白地霉	*Geotrichum candidum*	Activated sludge	Polysaccharide
	裂褶菌	*Schizophyllum commune*	Soil	Polysaccharide
	黑曲霉	*Aspergillus niger*	Soil	Polysaccharide
	黑酵母	Black yeast	Soil	Polysaccharide
	酿酒酵母	*Saccharomyces cerevisiae*	Guangxi Normal University Culture Collection Center	Polysaccharide
Actinomyces	黄三素链霉菌	*Streptomyces flavotricini*	Soil	Polysaccharide
	囊腹链霉菌	*Streptomyces cyslabdanicus*	Hubei Mushroom Collection Center	Polysaccharide
	诺卡氏菌	*Nocardia* sp.	Soil	Acidic proteins, acidic polysaccharides

(continued)

Table 5.1 (continued)

Category of flocculents	Chinese name of flocculent	The foreign name of the flocculent	Source of flocculents	MBF active ingredient
Algae	微拟球藻	*Nannochloropsis oceanica*	Dalian University of Technology Microbial culture collection chamber	Neutral sugar
	小球藻	Chlorella	Laboratory at the University of Göttingen, Germany	Polysaccharides, proteins
	斜栅藻	*Scenedesmus obliquus*	Laboratory at the University of Göttingen, Germany	Polysaccharides, proteins
	新氯藻	*Neochloris oleoabundans*	University of Texas Laboratory in United States	Polysaccharides, proteins
	黑曲霉	*Aspergillus niger*	Soil	Polysaccharide

provide many cation binding sites, and enhance flocculation ability in conjunction with Ca^{2+}.

(2) Bridging

With the help of electrostatic attraction, van der Waals forces, hydrogen bonds, etc., MBF serves as a bridge between colloidal particles, allowing pollutant colloidal particles to flocculate into groups without direct contact, and precipitate in a network. MBFS1, derived from coal chemical wastewater, has a glycoprotein with multiple binding sites, which realizes the flocculation of colloidal particles through bridging. Pharmaceuticals and personal care products (PPCPs) are a new type of water pollutant that is difficult to degrade and easily bioaccumulates, such as carbamazepine. Compared with CPAM, the microbial flocculant MFX has a higher removal effect on carbamazepine. Although the charge state of carbamazepine is the same as that of MFX, both are negatively charged, which affects the performance of the flocculant's electro-neutralization effect, but MFX is a high polymer compound with more active flocculation groups, mainly through adsorption and bridging, achieving the removal of PPCPs.

(3) Chemical Reaction

The active groups of MBF cause colloidal particles to aggregate through chemical reactions, forming obvious particle precipitation. Usually, by changing the active composition, functional groups, and operating environment of MBF, and comparing the flocculation effect before and after the change, it can be used to judge whether the flocculation is based on chemical reactions. Chen Ting screened a polysaccharide-type MBF, established thermodynamic and kinetic models for the removal process

Fig. 5.1 Comparison of microscope images of kaolin suspension. **a** Without adding flocculant, **b** Adding $CaCl_2$ solution, **c** Adding $CaCl_2$ solution and PCF-1767

of Pb^{2+}, Cd^{2+}, and Ni^{2+} in water, and combined with the characterization of the composition, functional groups, and heavy metal elements before and after, verified that the flocculation mechanism of MBF treating heavy metal wastewater is mainly chemical reaction.

(4) Sweeping

After the addition of microbial flocculants, a net-like floc is formed in the water, which mechanically sweeps and nets colloidal particles by gravity, gradually gathering to form large particles and settling. The yellow spore original hairy flat leather fungus BKMF-1767 can produce extracellular polysaccharide PCF-1767, and its flocculation activity is determined by kaolin suspension. Through the observation of the state of kaolin suspension before and after flocculation under a microscope (Fig. 5.1), combined with the analysis of Zeta potential changes, it shows that the added $CaCl_2$ only weakens the electronegativity of the system, and kaolin and PCF-1767 still carry a charge, "electrical neutralization" does not work; in fact, kaolin particles adhere to the long chain of polysaccharides, continuously compressing and sweeping during the settling process, forming large and dense three-dimensional flocs, completing the flocculation.

(5) Other theories

In addition to the above mechanisms, many scholars have proposed several hypotheses. Table 5.2 lists the formation process and content of some hypotheses, their focus is different, usually can only explain part of the flocculation phenomenon.

5.2.2 Factors Influencing MBF Flocculation

The flocculation effect of MBF is not only closely related to the properties of the flocculant itself, but also involves the pH value of the environment, temperature, water pollution characteristics, and hydraulic conditions, etc.

Table 5.2 Hypotheses of common MBF flocculation mechanisms

Hypothesis name	Hypothesis formation and content
PHB esterification hypothesis	Poly β-hydroxybutyrate (PHB) is a lipid polymer that many microbial cells can produce. Crabtree's research found that after removing the PHB accumulated outside the branched motion gum cells, the flocculation effect was greatly reduced. Therefore, it is believed that PHB is the direct cause of microbial flocculation
Mucous hypothesis	In the early development of MBF, Butterfield separated microbes from activated sludge that could secrete extracellular mucus, using this sticky substance to adhere particles or pollutants, forming flocs
"Pseudo-exogenous flocculation protein" hypothesis	Special surface proteins on the cell wall of flocculating bacteria can form specific binding with the mannose residues of other cells, thereby causing flocculation. This hypothesis was proposed by Miki, who observed that the fusion of flocculating yeast cells and non-flocculating yeast cells caused cell wall deformation, and the interaction between non-flocculating yeast cells was weak and difficult to contact flocculation. Surface-treated yeast cells, after the structure of their flocculation donor and receptor changes, this specific binding is destroyed, ultimately causing irreversible loss of flocculation activity
Virus hypothesis	Strantford believes that exogenous flocculation proteins are produced by viruses infecting yeast, not by yeast itself. For example, yeast flocculation may be the result of the expression of Kill-L virus coat protein, because Kill-L virus is accompanied by FLO phenotype, and LdsRNA is consistent with the assumed flocculation structure gene. This hypothesis is mainly based on indirect evidence and needs to be proven by experiments
Extracellular cellulose of bacteria	For cellulose-like MBF, Friedman pointed out that the filament is part of the bacterial cell rather than a cell secretion, and some bacterial bodies form floc precipitation by connecting and aggregating with colloidal particles through external filaments
Hydrophobic hypothesis	The functional components and surface structure of MBF affect the hydrophilicity and hydrophobicity of flocculants, and good hydrophobicity is crucial for the adhesion of colloidal particles, which in turn affects the flocculation process and flocculation effect

5.2.2.1 Properties of the Flocculant Itself

MBF is generally a biological macromolecule compound such as protein, polysaccharide, cellulose, and DNA. Its molecular structure and molecular weight directly affect the flocculation effect: usually, linear molecules have better flocculation effects, while branched or cross-linked structures have poorer flocculation performance;

MBF molecular weights are mostly in the tens of thousands to hundreds of thousands, and can even reach millions, the larger the molecular weight, the more active sites, and the higher the flocculation activity.

Hydrocarbon groups, ester groups, and other self-carrying groups can enhance the hydrophobicity of the MBF surface, increase adsorption sites, maintain the microscopic conformation of the flocculant, and greatly affect the activity of the flocculant. Polysaccharide MBF has a wide range of applications through metal ion-mediated bridging and electrical neutralization effects. A large number of studies have shown that different strains produce different amounts and activities of MBF at different growth stages. Most MBF-producing bacteria increase the hydrophobicity of the cell surface and produce higher activity flocculants after the stable period and entering the endogenous respiration period.

Standard kaolin suspension is used to measure the flocculation activity of MBF. The typical flocculation activity detection process can be described as: take 2 mL of MBF fermentation broth, 4 mL of $CaCl_2$ (1 wt%) solution and add them separately to 94 mL of kaolin suspension (0.4 wt%). Adjust the required pH value with NaOH or hydrochloric acid solution, stir quickly at 200 r/min for 1 min, and then stir slowly at 100 r/min for 3 min. After standing for 10 min, take the supernatant and measure the absorbance of the sample at 550 nm using spectrophotometry. At the same time, use un-inoculated culture medium instead of fermentation broth for control experiments. According to formula (5.5–5.1), calculate the flocculation rate to obtain the flocculation activity results.

$$E(\%) = (B - A)/B \times 100\% \tag{5.1}$$

where E is the flocculation rate;

A is the absorbance value of the control group;

B is the absorbance value of the sample.

5.2.2.2 Water Pollution Characteristics

The composition of actual wastewater is complex. In different waters to be treated, the types, components, properties, and concentrations of pollutants are different, and the presence of interfering substances inevitably affects the flocculation effect of MBF. In the flocculation process, it is first necessary to identify which impurities are interfering substances, and then consider how to weaken their adverse effects on flocculation. In chromium-containing leather dyeing wastewater, there are many impurities such as industrial salt, leveling agent, fat liquoring agent, anionic dye, complex tanning agent, etc. Prepare simulated chromium-containing wastewater and add each component of leather wastewater one by one. Use straw fermentation to produce MBF for treatment and compare the flocculation effect. Among them, fatliquoring agents, acidic dyes, and polyacrylic acid complex tanning agents containing anionic groups significantly weaken the chromium removal effect of MBF and require appropriate pretreatment to remove impurities.

5.2.2.3 Flocculation Process Parameters

The flocculation process parameters include the amount of flocculant added, hydraulic conditions, and flocculation sedimentation time. To achieve a more ideal flocculation effect, parameter optimization is often carried out through flocculation experiments. Based on the consideration of flocculation effect and medication cost, the addition of MBF should be the best flocculation effect and the least amount used.

Hydraulic conditions are often achieved in the form of mechanical or hydraulic stirring, and the stirring intensity must ensure the smooth progress of rapid mixing and slow flocculation. Stirring promotes MBF to quickly and evenly diffuse into the water, flocculate with colloidal particles, and promote the formation, growth, and sedimentation of flocs. This factor is generally adjusted in conjunction with the flocculation sedimentation time to ensure the flocculation effect and complete solid–liquid separation. Flocculant WK2-19 is produced from the gas-producing single bacterial strain separated from the sludge at the sewage outlet of the paper mill. When the flocculation time is 5 min, the sedimentation time is 20 min, and the stirring speed is increased from 100 r/min to 250 r/min and 350 r/min, the flocculation rate of kaolin suspension increased from 64.78 to 78.50% and then decreased to 67.56%. If the stirring speed is too small, the flocculant cannot fully contact the target; if the stirring speed is too large, the formed flocs are easily destroyed, which also affects the flocculation effect. Zhang Wei discussed the flocculation time and sedimentation time in turn, and the flocculation effect of SNUX-MBF derived from activated sludge. The stirring speed is fixed at 160 r/min, the flocculation time and sedimentation time are set to 1 ~ 9 min and 5 ~ 25 min respectively, and the optimal value of this experiment is obtained through the comparison of the flocculation rate of kaolin suspension (flocculation 7 min, sedimentation 15 min).

5.2.2.4 Flocculation Environmental Conditions

(1) Temperature

MBF is mostly bioactive components, and its flocculation performance is closely related to temperature, especially proteins, nucleic acids and other components are extremely sensitive to temperature, high temperature can easily cause denaturation, causing the flocculation activity of MBF to decrease or even inactivate. Phanerochaete chrysosporium (BKMF-1767) is a white rot fungus. Its produced PCF-1767 shows a high flocculation rate for kaolin suspension with an initial concentration of 3 ~ 6 g/L in the range of 30 ~ 50 °C, but when the temperature is > 50 °C, the flocculation rate begins to drop sharply, indicating that high temperature will inhibit its flocculation activity.

(2) pH value

Different flocculants have different sensitivities to pH value, and even the same flocculant has different optimal pH value requirements for different wastewaters. The pH value of the solution will affect the surface electricity, charge state and electric neutralization ability of MBF and pollutant colloidal particles, causing changes in the properties of the aqueous solution, thereby affecting the actual flocculation effect. Within the appropriate pH range, the repulsion between MBF and colloidal particles weakens, the bridging action between the two strengthens, and flocculation sedimentation can be carried out. Using a combination of ultraviolet mutation and chemical mutation, the strain screened from activated sludge was artificially selected to obtain the mutant strain HZU-1. Adjust the initial pH value of the kaolin suspension to 4 ~ 10. As the pH increases, the flocculation rate first increases and then decreases, reaching a maximum at pH = 7. Compared with PAC, the dosage is less, the sludge production is low, and the flocculation effect is better.

(3) Metal cations

Added Ca^{2+}, Mg^{2+}, Fe^{3+} and Metal ions such as Al^{3+} can enhance the flocculation effect of MBF. An appropriate amount of metal cations can significantly strengthen the electro-neutralization and bridging action of MBF, but their concentration should not be too high, otherwise it will increase the repulsion, alienate the interaction between the flocculant and colloidal particles, and inhibit flocculation. Ca^{2+} is the most commonly used coagulant. M-C11 from activated sludge is used for sludge conditioning. Compared with the control group without $CaCl_2$, the sludge specific resistance to filtration (SRF) and moisture content show a more obvious decreasing trend, but excessive dosage will lead to worse sludge dewatering performance. Under the optimal addition combination of pH 6, 3 mL M-C11, and 4 mL $CaCl_2$ (1%), after conditioning 25 mL sludge, the SRF and moisture content decrease from 11.64 × 1012 m/kg and 98.86% to 4.66 × 1012 m/kg and 83.74% respectively. The dewatering effect of M-C11 is significantly better than that of inorganic flocculants such as aluminum sulfate and PAC. A type of polyglutinous Bacillus L2019-1 was separated from the sludge of a sewage treatment plant. The MBF it produces is mainly acidic polysaccharides. The presence of appropriate amounts of Ca^{2+}, Fe^{3+}, and Al^{3+} in the flocculation system can promote the flocculation activity of MBF. It was found that the flocculation effect of low-dose Al^{3+} (0.2 mmol/L) is good.

5.3 Preparation and Production of Microbial Flocculants

Generally speaking, the preparation and production of MBF follow the process shown in Fig. 5.2, including the isolation and screening of MBF-producing bacteria, strain identification, fermentation, MBF extraction and purification, often combined with the determination of flocculation rate to judge the flocculation activity of MBF.

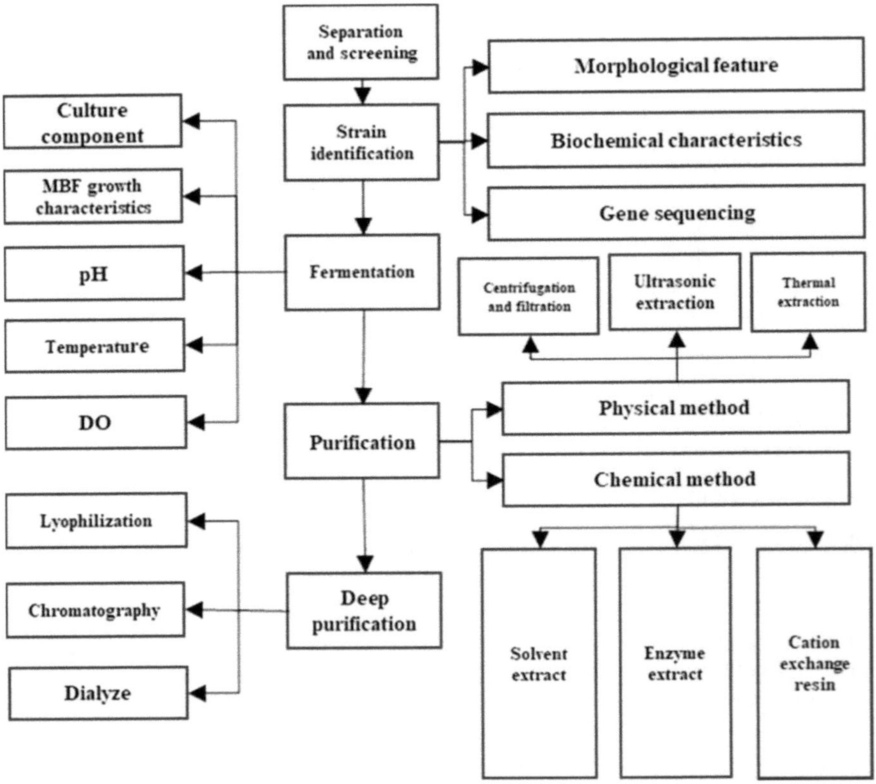

Fig. 5.2 MBF preparation and production process

5.3.1 Isolation and Screening of MBF-Producing Bacteria

The sources of MBF-producing bacteria are wide, and microbial samples can be collected from sewage, sludge, soil, water bodies, animals and plants, diluted at different concentration gradients, spread on solid culture medium (such as agar, beef extract peptone, etc.) plates, incubated at constant temperature, and observed regularly. Different morphological single colonies from the spread plate are taken, and the streak plate method is used for multiple inoculations and separations until the colony morphology is single and stable and then transferred to the slant culture medium for standby. For example, soil samples are taken from the bottom mud of hot springs to screen α-amylase-producing bacteria. Firstly, the enrichment liquid is diluted and spread on the separation medium, and incubated upside down at 40 °C for 24 h; then the single colony is picked with a sterile toothpick and streaked on the separation medium plate, an iodine solution is added, the diameters of the transparent circle and the colony are measured respectively, and the ratio of the two is calculated. The colony with a larger ratio is inoculated on the slant tube.

Select well-separated strains and transfer them into a liquid fermentation medium. The bacteria are then cultured under constant temperature and shaking to produce a bacterial solution. Using kaolin suspension as a simulated wastewater, MBF-producing bacteria are screened through flocculation activity comparison. This work can also be subdivided into two steps: preliminary screening and re-screening. The strains with higher flocculation activity obtained through preliminary screening can be transferred to a new medium for further enrichment and cultivation, resulting in strains with stable flocculation effects and genetic performance. For example, samples of pig farm facultative pond mud, experimental fish pond mud, and experimental field soil are collected, added to physiological saline, sealed and shaken, and the supernatant bacterial solution is taken after standing and layering. 5 mL of bacterial solution is placed into 100 mL of liquid medium and cultured under sealed shaking at 30 °C and 160 r/min for more than 72 h, using the kaolin flocculation method for preliminary screening. The fermentation broth of the preliminary screening samples is diluted with physiological saline to different multiples, spread on a solid medium plate, and cultured at a constant temperature for more than 24 h until larger colonies grow. These larger colonies are transferred to a new liquid medium and re-screened using the same method. The screening samples are repeatedly diluted and plated for separation, and several single bacterial strains are obtained by picking colonies with an inoculation loop and streaking multiple times.

5.3.2 Strain Identification

Traditional MBF strain identification combines morphological observation of the strain and physiological and biochemical property testing to determine the type of bacteria. In recent years, with the rapid development of biological detection technology, gene sequence determination based on PCR amplification has become an efficient means of MBF strain identification.

(1) Morphological Characteristics

The selected strains are inoculated on a solid medium, and after a certain period of constant temperature culture, the culture is taken for Gram staining and lipid staining. The colony texture, bacterial shape, color, size, and other characteristics are then observed with the naked eye and microscope. The morphological characteristics of the bacteria are recorded with photos, tables, and text descriptions. For example, the Z11 colony from activated sludge appears round, pale yellow, and opaque on LB solid medium. The colony is sticky, the surface is not smooth, has wrinkles, and is easy to pick (Fig. 5.3a Colony morphology). Under the microscope, the bacteria are rod-shaped, some are curved, and there is not much difference in thickness at both ends (Fig. 5.3b Bacterial microscope photo).

Fig. 5.3 Z11 colony on LB solid medium

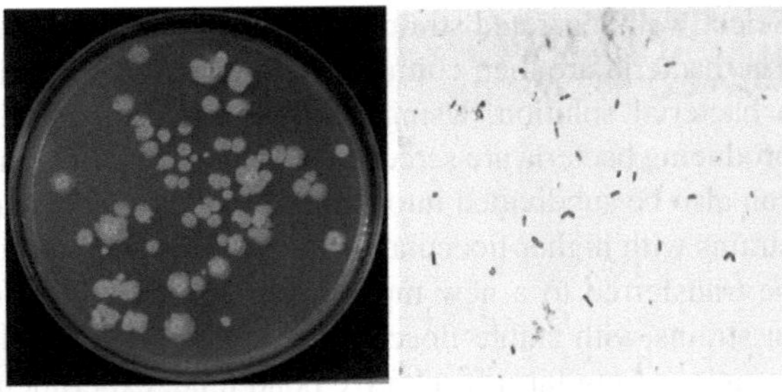

(a) Colony morphology (b) Bacterial microscope photo

(2) Physiological and Biochemical Characteristics

Different strains have different physicochemical properties. After the strain is expanded and cultured in a liquid medium, tests such as nutritional type, carbon and nitrogen source utilization ability, various metabolisms, and enzyme-catalyzed reactions are carried out. By referring to the "Common Bacterial System Identification Manual", "Bergey's Manual of Bacterial Identification", and "Common Identification Methods for General Bacteria", the strain's affiliation is clarified. Common physiological and biochemical detection items are listed in Table 5.3.

(3) Gene sequence determination

PCR amplification is a molecular biology technique used to amplify specific DNA fragments. By adding designed primers and going through denaturation, annealing,

Table 5.3 Physiological and biochemical characteristic detection items

Test items	Result options	Test items	Result options
Nitrate reduction	+/–	Oxidative fermentation of lactose	+/–
Hydrolyzed starch	+/–	Oxidative fermentation of sucrose	+/–
Citrate	+/–	Oxidative fermentation of glucose	+/–
Methyl red	+/–	Oxidative fermentation of mannitol	+/–
Exposure to enzymes	+/–	Oxidative fermentation of maltose	+/–
Urate	+/–	Phenylalanine deoxygenase	+/–
V-P	+/–	Indigo matrix	+/–
H_2S production	+/–	hydrogen sulfide	+/–
Catalase	+/–	L-aspartic acid	+/–
Hydrolyzed cellulose	+/–	Oxidase	+/–
Indole-producing	+/–	Ethanol fermentation	+/–
Liquefied gelatin	+/–	Oxidative fermentation of glycerol	+/–

Note "+" in the table indicates positive, "–" indicates negative

and extension steps, the target gene is replicated in vitro. The PCR amplification product of MBF can be viewed and photographed in a DNA electrophoresis image analyzer after gel electrophoresis. The target fragment is sequenced using a gene analyzer, and the obtained gene sequence is entered into a database (such as GenBank) for multiple comparisons to identify homology and construct a phylogenetic tree of the strain, identifying the microbial spectrum. Combined with the results of physiological and biochemical tests of the strain, the species of MBF is determined.

Figure 5.4 shows the identification process of the 16S rDNA sequence of strain Y1 from the secondary screening fermentation of distillery wastewater. As detected by agarose gel electrophoresis, the sequence length of the PCR amplification product of Y1 is about 1437 bp (Fig. 5.4a). Comparing this sequence with the nucleic acid data in GenBank for homology, a phylogenetic tree is constructed using MEGA5.0 software. It has a high similarity of 99 ~ 100% with strains of the genus Bacillus (Fig. 5.4b), identifying strain Y1 as *Bacillus subtilis*.

5.3.3 Optimization of the Fermentation Process

Through fermentation, flocculating bacteria are cultured on a large scale, promoting biochemical reactions, thereby producing and accumulating a large amount of MBF with flocculating activity, which is convenient for further research and application. At different growth stages and under different fermentation conditions, the metabolic products of the bacterial cells are different, and the flocculating activity of MBF is also different. Therefore, the growth characteristics of flocculating bacteria, the components of the culture medium, and the parameters of the fermentation environment are crucial to the success or failure of fermentation and MBF culture.

(1) Growth characteristics of the strain

The production and efficacy of MBF are closely related to the growth characteristics of the strain. Usually, the target strain is inoculated into a liquid medium, samples are taken at regular intervals, the concentration of bacterial cells in the fermentation broth is determined by the turbidity method, and the flocculating activity of the bacterial liquid is detected at the same time. The fermentation growth curve is drawn, and the relationship between the fermentation stage (or culture time) and the corresponding flocculating activity of the fermentation broth is established. Combined with the fitting of the growth kinetic model of the strain, the growth pattern and flocculating characteristics of the strain are clarified.

The growth of bacterial strains often goes through four stages: adjustment period, logarithmic period, stable period, and decline period. The flocculation activity of MBF changes with the growth stage. Generally, after the bacterial strain enters the logarithmic period, it begins to produce flocculating substances, but at this time the bacterial concentration is low and the flocculation rate is low. As the fermentation time extends, the bacteria continue to multiply, and the flocculation active substances

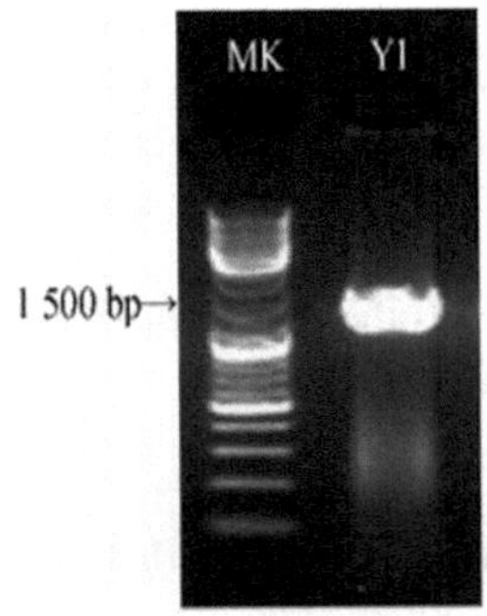

(a) Electrophoresis diagram of PCR amplification product for gene sequence identification of strain Y1

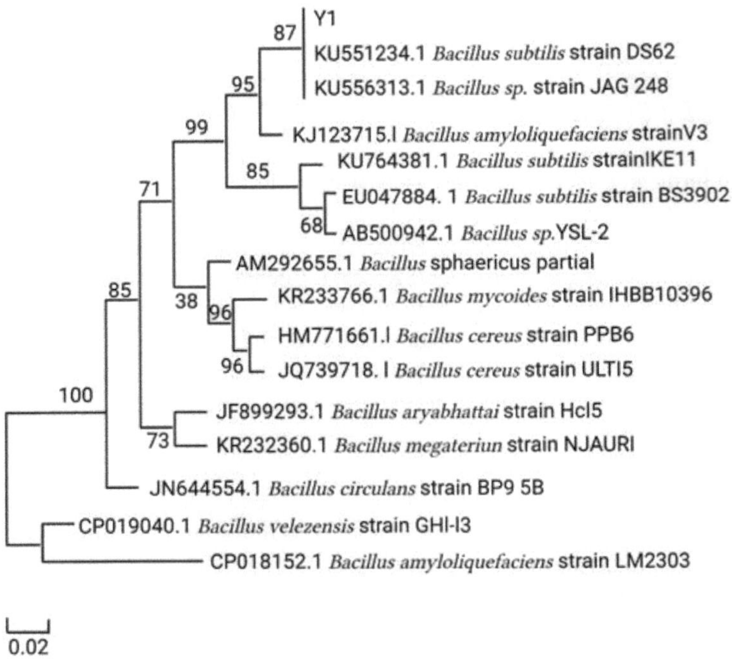

(b) Y1 phylogenetic tree diagram

Fig. 5.4 Identification process of the 16S rDNA sequence of strain Y1 from the secondary screening fermentation of distillery wastewater

also increase significantly. The flocculation rate reaches its maximum value in the later stage of the stable period, and then drops sharply due to entering the decline period.

(2) Components of Fermentation Medium

Nutrients such as carbon, nitrogen, phosphorus, sulfur, trace elements, and vitamins are important foundations for the growth and reproduction of MBF bacteria. Their functions and common sources are shown in Table 5.4. The source, concentration, and ratio of the substrate directly affect the growth of the bacteria, which in turn leads to differences in flocculation activity. For example, excessive heavy metal ion content can inhibit microbial growth, thereby affecting the yield of MBF. To investigate the fermentation effect of different media, the bacteria can be separated by centrifugation and dried for comparison by weighing.

To further reduce the cost of fermentation substrates, MBF researchers and manufacturers have been committed to seeking economical, readily available, and high-quality alternatives to culture media. Dairy wastewater, starch wastewater, molasses wastewater, aquaculture wastewater, and liquor yellow water, all rich in various nutrients, can be used for MBF fermentation and production. However, it should be noted that necessary dilution and pretreatment of the wastewater should be carried out, controlling and balancing the concentration of nutrient substrates, and avoiding the interference of high turbidity and high color of wastewater with absorbance testing.

(3) Fermentation pH Value

Based on different pH tolerance levels, MBF-producing bacteria can be divided into acidophilic bacteria, neutrophilic bacteria, and alkaliphilic bacteria. During the

Table 5.4 Functions and common sources of fermentation medium components

Nutrient	Function	Common source
Carbon	Most important substance for cell growth	Glucose, sucrose, starch, maltose, lactose, potassium acetate, methanol, fructose, glycerol, etc.
Nitrogen	Plays an important role in the production of enzymes, proteins, and nucleic acids	Ammonium chloride, peptone, urea, ammonium nitrate, yeast extract, ammonium nitrate, etc.
Phosphorus	Essential for the synthesis of proteins and coenzymes	Sodium dihydrogen phosphate, potassium dihydrogen phosphate, etc.
Sulfur		Magnesium sulfate, calcium sulfate, sulfur-containing amino acids, etc.
Trace elements	Assist enzyme reactions, promote cell growth	Copper, iron, zinc, cobalt, sodium, potassium, magnesium, and manganese salts, etc.
Vitamins	Maintain cell growth, regulate cell metabolism, constitute active components of enzymes, etc.	Two major categories, water-soluble and fat-soluble, including vitamins A, B, C, D, E, K, pantothenic acid, folic acid, niacinamide, pyridoxal, thiamine, riboflavin, choline, inositol, and biotin, etc.

fermentation cultivation process, changes in pH value will alter the redox potential of the system, the surface charge of bacterial cells, and the ionization of organic nutrients, thereby affecting the utilization of the culture medium by microorganisms. At the same time, the bio-enzyme that catalyzes the synthesis of flocculating substances can only fully exert its activity within the optimal pH range. Therefore, the fermentation pH value has a significant impact on the yield and flocculation efficiency of MBF. The Klebsiella strain M1 from food factory sediment sludge, with an optimal fermentation pH of 4.5, and supplemented with optimized cultivation conditions such as rotation speed, temperature, and time, has a flocculation rate of up to 82.03%, making it the preferred strain for wheat starch alcohol wastewater. The flocculation rate of MBF obtained from the fermentation of *Streptomyces putida* varies significantly with different pH values of the culture medium. The MBF produced by the neutral culture medium (pH = 7) can have a flocculation rate of nearly 90%, which is far superior to the corresponding products of weakly acidic culture medium and alkaline culture medium, and saves the amount of acid and alkali used to adjust the pH value.

(4) Fermentation Temperature

The optimal fermentation temperature range for most MBF-producing bacteria is between 25 and 40 °C. If the temperature is too low, microbial growth metabolism slows down, bio-enzyme activity decreases, and MBF yield is low. If the temperature is too high, it is not conducive to normal bacterial growth, and the structure and function of enzyme proteins are easily damaged, leading to a decrease or even inactivation of enzyme activity, and a significant decrease in flocculation rate. The C2-5 strain was obtained from paddy soil and activated sludge through ultraviolet mutation. Within the fermentation temperature range of 20 ~ 45 °C, the flocculation rate of this strain to kaolin suspension shows a trend of first increasing and then decreasing, with the highest flocculation rate (84.83%) at 30 °C, producing the highest flocculation activity. The LL6 strain of *Yersinia pestis*, sourced from the sludge of a sewage treatment plant's biochemical pool, has an optimal fermentation temperature of 20 °C, with a corresponding flocculation rate of up to 88.55%.

(5) Fermentation DO

Most MBF-producing bacteria are aerobic or facultative bacteria, which use DO to decompose nutrients and carry out metabolic activities. To ensure the DO environment of the fermentation liquid, stirring speed adjustment is often used to control the ventilation volume. An appropriate rotation speed can meet the oxygen demand and prevent bacterial adhesion, promoting bacterial growth and reproduction and the production of flocculating substances. If the rotation speed is too low, the lack of DO will inhibit bacterial growth. If the rotation speed is too high, the increase in DO can easily lead to rapid microbial proliferation, huge consumption of nutrient components, causing a relative decrease in the concentration of the culture medium, affecting the duration of microorganisms in the stable period, and slowing down the secretion of flocculating active substances. Parasitic Aspergillus produced more EPS in stirred culture than in static culture. Using pig farm wastewater for fermentation cultivation of Rhodotorula, the DO dropped rapidly from 4.5 to 0.38 mg/mL within 0 ~ 12 h, and after 12 h, the DO in the system remained at 0.55 ~ 0.75 mg/mL. The large ventilation volume in the early stage of fermentation is beneficial for the rapid proliferation of flocculating bacteria, while avoiding bacterial aggregation, affecting the metabolism of internal bacteria and the transfer of internal and external substances. The P33 flocculating bacterial strain used for petrochemical refining wastewater is an aerobic bacterium, and the shaking bed rotation speed should be around 160 r/min. If the rotation speed is too high, it affects the secretion of EPS, and the flocculation effect is not good; if the rotation speed is too low, the growth of the flocculant is restricted.

The fermentation effect of MBF-producing bacteria is influenced by various factors. In actual operation, single-factor experiments, orthogonal experiments, and uniform experimental designs are often used. The flocculation rate is used as an evaluation index to compare the response values of different combinations of indicators such as culture medium, fermentation time, fermentation temperature, pH value, and rotation speed, in order to seek suitable fermentation parameters and achieve process optimization.

5.3.4 Extraction of MBF

The methods of MBF extraction are mainly divided into two categories, physical methods and chemical methods. Centrifugation, filtration, ultrasonics, and heating belong to physical extraction, while solvent extraction, enzyme-assisted methods, and cation exchange resin methods are classified as chemical extraction. Successful extraction of MBF often requires the combined use of multiple methods. For flocculating active substances dispersed in the fermentation broth, they can be easily obtained through centrifugation and filtration, but MBF adhering to the bacterial body needs to be processed using extraction, heating, and enzyme-assisted methods. Centrifugation and filtration are well-known separation methods, easy to operate, and often used as pre-treatment and post-treatment means for extraction. The following will focus on other extraction techniques.

(1) Heat Extraction

Heat extraction is achieved by heating the medium, transferring heat to the extraction container, evaporating the fermentation broth, and obtaining MBF. Hydrothermal method and microwave-assisted method are the two main heat extraction methods currently. Hydrothermal extraction is widely used, it usually uses a water bath for heating, requires higher temperatures and longer extraction times, and consumes more energy. The rough basket fungus CC-1 can produce polysaccharide EPS. Comparing the extraction effects of 5 °C water bath heat extraction, centrifugal precipitation, cation exchange resin, and NaOH-ultrasonic method, the results show that the efficiency of heat extraction is the highest (EPS extraction amount is 940 mg/L), and it does not cause damage to the flocculating active substances.

The microwave-assisted method has a small equipment size, short extraction time, and is simpler than traditional heating methods, but high microwave power and extraction temperature can easily cause thermal degradation of effective components, so the requirements for heating and temperature control are strict. In addition, the choice of solvent is extremely important, high dielectric or polar solvents are good microwave heating media, such as water and organic solvents.

(2) Ultrasonic Assisted Extraction

With the cavitation effect, ultrasonic technology is widely used in the environmental field. During the propagation in liquid, ultrasound acts on the tiny bubbles in it and gathers the sound field energy. When the energy reaches the cavitation threshold, the cavitation bubbles collapse and instantly form a high-temperature and high-pressure microenvironment, which can improve the extraction effect and shorten the extraction time. To promote the resource utilization of residual sludge and reduce the chemical sludge production of oilfield-produced water treatment, MBF is extracted from the biochemical sludge of the sewage treatment plant using ultrasonic waves. At a power density of 1.5 W/mL and ultrasound for 20 min, the flocculation rate of the extracted MBF is as high as 97.25%.

The ultrasonic-assisted method is especially suitable for the pre-treatment of MBF extraction, improving the effectiveness of extraction. To further reduce cavitation hindrance, solvents with low viscosity and small surface tension, such as acidic or organic solvents, are usually chosen, while temperature control is maintained to prevent the inactivation of heat-sensitive components of MBF and to avoid the evaporation of organic solvents as much as possible.

(3) Solvent Extraction

Substances often have different solubilities or distribution coefficients in immiscible (or slightly soluble) solvents. Using this principle to transfer substances from one solvent to another is called solvent extraction (liquid–liquid extraction). Through repeated extractions, substance separation is achieved. MBF extraction typically uses a combination of one or more solvents, such as water, acid–base reagents, salt solutions, and organic solvents, and the extraction effect is greatly influenced by temperature, time, and material-to-liquid ratio. Appropriate extraction parameters help maintain MBF activity, reduce solvent loss, and increase MBF yield.

Water extraction is one of the most commonly used methods for MBF extraction and can also be used in conjunction with hydrothermal methods. Heating can reduce the viscosity of the aqueous solution, enhance the mutual permeation between the solvent and the bacterial cell, and promote the rapid mass transfer of water-soluble MBF to the extraction phase. The extraction temperature may vary depending on the source of the MBF, and the water extraction process may also vary.

Extracting MBF with a certain concentration of salt solution prepared with water is more effective than water extraction, but the concentration of the salt solution has a significant impact on the performance of the extracted MBF. Generally speaking, higher concentrations of salt solutions such as NaCl, KCl, and cetylpyridinium chloride (CPC) significantly improve the extraction effect. Acid–base reagents adjust the pH of the fermentation broth, combined with stirring, centrifugation, heat treatment, and filtration steps, can also achieve MBF extraction, but the choice of acid–base concentration must be cautious, too high concentration can cause denaturation of MBF itself. The product BF-1 (80 °C, 60 min, pH = 12) was obtained by treating sewage sludge from a sewage treatment plant with the alkali dissolution method. The flocculation rate and energy consumption ratio selection results show that the extraction effect of the alkali dissolution method is better than that of the ultrasonic method and the freeze–thaw method. BF-1 can remove more than 90% of the suspended solids and silica from the oilfield oil removal settling tank effluent.

Ethanol can effectively separate water-soluble and fat-soluble components, is easily compatible with acetone, methanol, etc., and is one of the most commonly used organic extraction solvents. Compared with water extraction, ethanol extract has a longer service life and better extraction effect, but it is mostly used in cold soaking steps and in large quantities. In addition, ethanol extraction usually requires two centrifugation processes, including the solid–liquid separation of MBF and fermentation broth, and the re-collection of crude MBF. Using the ethanol precipitation method, the crude flocculant produced by Klebsiella strain M1 was extracted. After centrifuging 50 mL of M1 fermentation broth for 15 min (1500 r/min), the supernatant was taken and 1.54 times (volume ratio) of ethanol extractant was added. The mixture was placed in the refrigerator (4 °C) for 12 h, then centrifuged at 4000 r/min for 20 min, the precipitate was freeze-dried at low temperature for 24 h, and the average yield of the flocculant could reach 3.998 g/L.

(4) Enzyme-assisted extraction

Enzyme-assisted extraction is a modern extraction process based on bio-enzymatic technology. It assists in the rapid release of flocculating active substances by enzymatically breaking down cell walls. Due to the specificity, mild conditions, and easy control of enzymatic hydrolysis, enzyme-assisted methods are particularly suitable for the extraction of heat-sensitive products, significantly enhancing extraction efficiency and the quality of the extract. It's worth noting that the effects of pH and temperature on enzymes cannot be ignored. The optimal pH for most enzymes is near their isoelectric point, and the optimal temperature is between 30 and 60 °C. Additionally, the type and number of enzymes, reaction time, etc., also play a significant role in the yield of MBF. Using response surface analysis, with extraction rate as the evaluation index, the optimal enzyme ratio, temperature, pH, and reaction time are optimized to obtain the best enzyme-assisted control parameters to ensure the yield and cost-effectiveness of MBF. The EPS secreted by *Bacillus subtilis* B69 has a flocculating effect. After crude extraction with ethanol, it still contains impurities such as proteins and inorganic salts. The crude product can be dissolved in deionized water and treated with papain for 3 h (pH = 6.4, 60 °C), and the supernatant is collected by centrifugation.

(5) Cation exchange resin extraction

According to current research reports, cation exchange resins are mainly used to extract MBF from activated sludge. Microorganisms in the sludge secrete EPS, which promotes the formation of activated sludge flocs. Using the ion exchange function of the resin, the cations in the sludge flocs are replaced, destroying the three-dimensional structure of the flocs, transforming the active flocculating substances in the sludge into nearly soluble substances, and then separating and extracting them with ultrasonic or centrifugal means.

In the fermentation broth of Rough Stalked Basket Fungus CC-1, domestic sodium type 732 cation exchange resin is added, shaken at a speed of 250 r/min for 90 min, the mixture is filtered through a 0.45 μm filter membrane, and then through ethanol crude extraction, dialysis, and freeze-drying, EPS flocculant is obtained. Compared with the heat extraction method, the total amount of EPS obtained by the resin method is slightly less, but the nucleic acid content is higher (227.5 mg/L). The disadvantage is that the cation exchange process may cause cell destruction and intracellular DNA contamination of the EPS composition. Using the resin-ultrasonic method, MBF is directly extracted from the residual sludge, and its yield and flocculation activity are better than the situation of using ultrasound or resin alone, and the total amount of MBF increases with the increase in sludge concentration. Cation exchange resin (60 g/g VSS pretreatment) is added to the residual sludge, stirred continuously at a speed of 650 r/min for 3 h, and then continuously sonicated for 2 min (20 kHz, 270 W, pulse 4 s), and then centrifuged for collection. The cation exchange resin weakens the bridging effect between the flocculating active substance and the sludge floc, and the shear effect and cavitation effect of the subsequent ultrasound further weaken the connection between the two, promoting the separation and extraction of the flocculating active substance.

5.3.5 Purification of MBF

To remove impurities from the crude MBF and further improve its purity while maintaining its flocculating activity for better flocculation performance, freeze-drying, dialysis, and chromatography are often used for purification. These three methods can be used alone or in combination.

(1) Freeze-drying

Freeze-drying is a short form of vacuum freeze sublimation drying, also known as lyophilization. This method uses low temperature and low pressure to directly sublime water, simultaneously promoting the porous structure of the material, increasing the specific surface area, and facilitating the mass transfer process of flocculation. The product has strong rehydration and can also preserve the flocculating active components and functions well, especially suitable for the purification treatment of heat-sensitive MBF.

(2) Dialysis

Dialysis is a highly efficient membrane separation method for removing small molecule impurities. Typically, the MBF to be purified is sealed in a dialysis bag made of semi-permeable membrane, and then the dialysis bag is placed in a dialysis solution. By utilizing the difference in osmotic pressure and the specifications of the dialysis membrane pores, the separation of small molecules and biomacromolecules is achieved. The most commonly used dialysis solution is water. Sometimes, to better maintain the flocculation activity of MBF, a buffer solution with an appropriate pH value and ionic strength, or even a concentrated solution of inert polymers, is chosen. In addition, to improve the dialysis rate and accelerate the diffusion of small molecules to the dialysis solution, the dialysis process often requires continuous stirring, increasing the frequency or usage of dialysis solution replacement, and artificially enhancing the osmotic pressure difference. The polysaccharide polymer EPS-160 secreted by Raoultella ornithinolytica 160-1 has good flocculation activity. To detect the polysaccharide structure of EPS-160 and quantify its flocculation rate for kaolin suspension, a purification process was carried out. Using a dialysis bag, impurity ions and residual culture medium in the ethanol crude extract of EPS were removed, freeze-dried to constant weight, and 2.76 g/L of pure EPS-160 was obtained. It was found to contain protein (5.7%) and polysaccharides (80.3%), with a molar ratio of glucose, rhamnose, and mannose of 3.1:1:2.4. When the dosages of Al^{3+} and EPS-160 were 0.08 mg/mL and 43.2 mg/L respectively, the flocculation rate reached as high as 97.94%.

(3) Chromatography

Chromatography refers to the specific zones formed by different colored or colorless components arranged on a filled column carrier. Therefore, chromatography is also known as chromatographic analysis or layering method. Using a mixture containing MBF as the mobile phase, the flocculation active components are selectively separated based on the differences in adsorption, affinity, ion exchange, and distribution with the stationary phase (filled carrier). This method requires the selection of suitable mobile and stationary phases based on the active components of MBF, and the adjustment of flow rate to maximize separation efficiency and minimize loss of functional components. Using a DEAE-52 cellulose ion exchange column and Sephadex-200 dextran gel column, the extract of *Bacillus cereus* GIM1.16 fermentation broth was stepwise chromatographed. During the first chromatography, distilled water, 0.3 and 0.5 mol/L NaCl were used to elute at a flow rate of 0.6 mL/min, resulting in three elution peaks at 490 nm, which were concentrated, freeze-dried, and made into three samples. The three samples were dissolved in distilled water and subjected to a second chromatographic purification at the same flow rate. The effluent was then dialyzed and freeze-dried to form three pure polysaccharide flocculants PSB1-1, PSB2-1, and PSB3-1, with corresponding polysaccharide contents of 96.8%, 95.7%, and 97.8% respectively. It can be seen that different types and concentrations of mobile phase can lead to differences in product purity and composition. Acidic polysaccharides carry a negative charge and have good removal performance for heavy metals. The crude extract of the fermentation broth of Bacillus ps-5, after two-step chromatography, dialysis, and freeze-drying, was made into purified EPS. It contains a large number of –OH, –C=O, –COOH, and –C–O–C– functional groups. Under optimized conditions of pH, dosage, and reaction time, its adsorption capacities for Pb^{2+}, Cu^{2+}, and Cr^{6+} were 225.40 mg/g, 183.67 mg/g, and 86.97 mg/g respectively.

5.4 Application of Microbial Flocculants in Water Environment

Coagulation is an important step in water purification, and traditional inorganic and organic coagulants play a key role in water treatment. However, chemical coagulants can cause secondary pollution, posing certain environmental hazards and health risks. As a green water treatment agent for the future, MBF can overcome the drawbacks of commercial coagulants, meeting the country's environmental protection needs. It has also shown advantages in many areas such as raw water purification, wastewater treatment, sludge conditioning, and microalgae harvesting.

5.4.1 Raw Water Purification

MBF has a high removal efficiency for turbidity, natural organic matter, and pathogens in raw water, with less dosage, good floc settling properties, and higher safety. It is one of the most important development directions for coagulant products in water supply and drinking water treatment (the application of MBF in raw water purification is shown in Table 5.5).

5.4.2 Wastewater Treatment

MBF is suitable for the pretreatment and enhanced treatment of wastewater, with obvious advantages in decolorization, oil removal, SS reduction, heavy metal removal, and treatment of high-concentration refractory wastewater. It is a good substitute for traditional coagulants (the application of MBF in wastewater treatment is shown in Table 5.6).

5.4.3 Sludge Conditioning

A large amount of sludge produced during the water treatment process, if not properly disposed of, can cause a huge economic and environmental impact. Usually, the coagulation method is used to condition the sludge to improve the settling performance of the sludge, eliminate sludge expansion, and enhance the dewatering effect

Table 5.5 Application of MBF in raw water purification

MBF name and source genus	Active substance category	Treatment object	Treatment conditions and effects
MBF1 Klebsiella	Protein	Yangtze River Raw Water	Under conditions of pH 1 ~ 10 and temperature 20 ~ 45 °C, with the addition of 40 mg/L of MBF1, the turbidity removal rate is over 95%. When the pH is 7.59 and the temperature is 20 °C, the raw water turbidity and COD decreases from 56.2 NTU and 42.67 mg/L to 0.6 NTU and 13.47 mg/L respectively, with turbidity and COD removal rates of 99.0% and 68.4% respectively, which are better than the flocculation effect of PAC under the same conditions and at a lower cost
Immobilized beer yeast ball yeast	Protein, nucleic acid	Taiyuan Qing Xu high hardness, high mineralization groundwater	The mineralization and hardness of the raw water are as high as 3109 mg/L and 2856 mg/L, respectively, which is classified as Class V water. Using sodium alginate embedding method, yeast immobilization (10% bacterial body embedding ratio) is achieved. The surface of the formed ball is uneven and covered with a large amount of protein, nucleic acid, and other bacterial metabolic products. After adding 1 g/L of immobilized yeast and flocculating for 60 min, the mineralization and hardness of the groundwater decrease to 940 mg/L and 658 mg/L, respectively, with flocculation rates of 76.97% and 69.77%, respectively, reaching or approaching the Class IV water standard of "Groundwater Quality Standard" (GB/T14848-93)

(continued)

Table 5.5 (continued)

MBF name and source genus	Active substance category	Treatment object	Treatment conditions and effects
MBF-B-16 Klebsiella	Polysaccharide, protein	Low turbidity Yellow River Water	The raw water quality indicators are: pH = 8.23, turbidity 4 ~ 6 NTU, COD 3 ~ 4 mg/L, total number of algae 35.5 × 104/L. The optimal addition range of MBF-B-16 is between 0.05 ~ 0.08 mg/L. When the treatment effect is similar, the addition amount of MBF-B-16 is only 20% ~ 30% of PAM. If MBF-B-16 (0.06 mg/L) is combined with PAFC (3 mg/L), the removal rates of turbidity and COD can reach 85.5% and 41%, respectively, and the capture rate of algae is 60.6%
CF compound flocculant	Polysaccharide, protein	High turbidity drinking water	CF is a compound of PAFC and MBF-B-16 at a mass ratio of 2.39:1. At CF dose of 80 mg/L and pH 8.44, the turbidity decreases from 3000 NTU to 4.9 NTU. The compound flocculant reduces the amount of PAFC by 56.2%, reducing the secondary pollution of chemical flocculation and saving operating costs
MBF-ZS-7 Lichen Bacillus	Glycoprotein	Low temperature drinking water	The optimal pH range for MBF-ZS-7 is 6.5 ~ 9. When the dosage is 5 mg/L, the turbidity and COD values of the effluent can reach 1.4 NTU and 3.5 mg/L (pH = 8, 37 °C), respectively, which meets the "Sanitary Standards for Drinking Water" (GB5749-2006). In comparison, when the dosages of PAC and PAM are 30 mg/L and 6 mg/L, respectively, the turbidity of the effluent reaches 3.3 NTU and 3.1 NTU, while the COD of the effluent reaches 4.5 mg/L and 4.2 mg/L, which are not as good as the flocculation effect of MBF

Table 5.6 Application of MBF in wastewater treatment

MBF name and source genus	Active substance category	Treatment object	Treatment conditions and effects
LF-Tou2 Bead Bacillus	Protein, polysaccharide	University wastewater treatment station wastewater initial water quality	Indicators: COD 126.3 mg/L, turbidity 7.3 mg/L, SS 21 mg/L, pH = 7.5. When the dosage is 19 mL/L, temperature 27 °C, and the stirring time is 3 min, the removal rates of COD and SS reach 93.6% and 100%, respectively, and the removal rates of Fe^{3+}, Cu^{2+}, Mn^{2+}, Zn^{2+} and Hg^{2+} in the water all exceed 90%
XM09 Soy Sauce Aspergillus	Polysaccharide	City wastewater treatment plant primary sedimentation tank water	When XM09 is added alone at 30 mg/L, the flocculation rate is 87.74%, which is close to the treatment effect of PAC added alone at 133 mg/L (89.96%). When XM09 (15 mg/L) and PAC (50 mg/L) are added together, the flocculation rate significantly increases to 91.57%; the residual aluminum in the effluent can be as low as 0.8 μg/L, which is 99.57% lower than when PAC is used alone
MBF-P71 Phanerochaete Chrysosporium	–	Pig farm wastewater	Initial Water Quality Indicators: pH = 7.04, COD 334 mg/L, NH_3-N 135 mg/L, TN 142 mg/L. When 10 mL/L of MBF-P71 and 5 mL/L of 10%$CaCl_2$ are added, the flocculation rate can reach 93.63%, and the removal rates of COD, NH3-N, and TN are 88.32%, 40.63%, and 38.50%, respectively

(continued)

Table 5.6 (continued)

MBF name and source genus	Active substance category	Treatment object	Treatment conditions and effects
YH39 Burkholderia	Polysaccharide	Dyeing wastewater	YH39 has high flocculation activity in alkaline environments (pH > 8), and the form of direct addition of fermentation broth can save the extraction and purification steps of MBF. When the dosages of YH39 fermentation broth and 0.2 mol/L $CaCl_2$ are 10 mL/L and 20 mL/L respectively, the COD decreases from 418.7 to 115.7 mg/L, and the decolorization rate is comparable to the effect of traditional chemical flocculants
MBFA18 Black Aspergillus	Polysaccharide	Potato starch wastewater	Initial Water Quality Indicators: COD 10130 mg/L, TN 604.6 mg/L, pH = 6.3, turbidity 467 NTU, color 128 times. When the dosages of MBFA18 and 0.5 mol/L $CaCl_2$ are 2 mL/L and 10 mL/L respectively, under the conditions of flocculation and settling time of 10 min and 20 min, the removal rates of COD and turbidity reaches 61.01% and 92.97%, respectively
G13	–	Pharmaceutical wastewater	Initial Water Quality Indicators: COD 10,800 mg/L, color 80 times, turbidity 1.574 NTU, odor 5. The optimized flocculation conditions of orthogonal experiment are: the volume fraction of G13 bacterial fluid is 50%, the flocculation time is 6 h, the temperature is 30 °C , and the pH value is 6.0. After mixing the 18 h cultured G13 bacterial fluid with pharmaceutical wastewater at a 1:1 ratio and settling for 4 h, the removal rates of COD, color, and turbidity are 35.19%, 75%, and 59% respectively, and the odor decreases from 5 to 3

(continued)

Table 5.6 (continued)

MBF name and source genus	Active substance category	Treatment object	Treatment conditions and effects
LBF	–	Papermaking wastewater	Initial water quality indicators: SS 194 mg/L, COD 1157 mg/L, color 600 times. When the dosages of LBF and 1% $CaCl_2$ are 5 mL/L and 120 mL/L respectively, the removal rates of SS, color, and COD are 87%, 71%, and 39% respectively, which are better than the treatment effects of PAC's 64%, 12%, and 30% (2 g/L, pH = 12)
Flocculating Bacteria F2 Rhizobium	Polysacch-aride	Wastewater containing Pb^{2+}, Cd^{2+}, Ni^{2+}	The initial concentrations of Pb^{2+}, Cd^{2+} and Ni^{2+} are 30 mg/L, 20 mg/L, and 15 mg/L respectively. When the dosages of MBF-F2 are 200 mg/L, 700 mg/L, and 800 mg/L respectively, the removal rates of Pb^{2+}, Cd^{2+}, and Ni^{2+} are 99.13%, 94.44%, and 84.55% respectively (pH = 6)
BS-04 Secretes EPS Bacillus	Polysaccharide	Compost leachate	Using EPS flocculation to treat compost leachate after biochemical treatment. By adding 0.5 g/L of EPS, the removal rates of COD, phosphate, and nitrate can reach 29%, 66%, and 85.2% respectively (pH = 9); when the pH value is 8, the removal rate of ammonia nitrogen can reach as high as 98%, while the removal rates of K, Na, Mg, Al, and Ca up to 88%, 91%, 87%, 88%, and 82% respectively

(continued)

Table 5.6 (continued)

MBF name and source genus	Active substance category	Treatment object	Treatment conditions and effects
MEMA composite bacterial	Community waxy Bacillus, *Staphylococcus aureus*, Amylolytic Bacillus, Lactococcus, Casein Lactobacillus, etc.	Thermal power plant circulating cooling water	Industrial trial of a 25,000 m^3 circulating cooling water system. Using MEMA treatment, the average PSI of the circulating water is 6.25, the system has no scaling trend, and the corrosion rates of the system's carbon steel and copper alloy are reduced by 22.5% and 26.8% respectively. The total number of heterotrophic bacteria and the amount of biological sludge are better than the "Industrial Circulating Cooling Water Treatment Design Specification" GB/T 50050-2017. Compared with chemical treatment, the average SS of the system's concentrated discharge water decreased from 39.3 to 9.1 mg/L, the average TP decreased from 2.4 to 0.5 mg/L, and the average COD decreased from 86.3 to 43.1 mg/L
MBFS1	Polysaccharide, protein	Coal chemical wastewater	Wastewater MBFS1 and $CaCl_2$ are used together for the pretreatment of coal chemical wastewater (COD 7500–8500 mg/L). By adding 2 mL of MBFS1 and 5 mL of 1% $CaCl_2$ to 30 mL of wastewater, the flocculation rate reaches 99.13%, while the COD removal rate up to 72.56%
ZW5 golden rod bacteria	Polysaccharide, protein	Lead–zinc mine ore washing wastewater	Initial water quality indicators: Pb^{2+} 30 mg/L, Zn^{2+} 10 mg/L, yellow medicine 40 mg/L. ZW5 is better than PAC in removing lead and zinc, but PAC is better than ZW5 in removing yellow medicine. Under optimized flocculation conditions, the removal rates of ZW5 fermentation broth for Pb^{2+}, Zn^{2+}, and yellow medicine are 97.92%, 90.72%, and 60.84% respectively

of the sludge. The cost of chemical coagulants is high, and subsequent treatment is difficult. In contrast, the appropriate addition of MBF can promote the formation of biological flocs, reduce the shear action and dispersion degree between sludge particles, helping to significantly reduce the sludge volume index, thereby restoring the settling ability of the sludge, alleviating sludge expansion, and promoting sludge dewatering and volume reduction (the application of MBF in sludge conditioning is shown in Table 5.7). Although many studies have shown the superiority of MBF in conditioning sludge, further exploration and evaluation are needed in terms of sludge types and applicable conditions (cation concentration, pH value, and organic matter content, etc.) to tap the potential of MBF in treating sludge.

5.4.4 *Microalgae Harvesting*

As a type of autotrophic aquatic plant widely distributed in land and sea, microalgae are rich in proteins, algal polysaccharides, β-carotene, lipids, and various vitamins, and have been applied in various fields such as pharmaceuticals, food, feed, environmental purification, and renewable energy. Using economical and effective methods to collect and capture microalgae from the water environment is a necessary condition for the resource utilization and industrial development of microalgae. At present, the main methods suitable for microalgae harvesting include sedimentation, centrifugation, filtration, flotation, and coagulation, especially in the production of biofuels, coagulation has received much attention. Physical means such as ultrasound, electricity, and magnetism are often needed to promote microalgae aggregation and separation, which require special equipment and high energy consumption. In contrast, the harvesting of conventional inorganic and organic coagulants is more efficient, but the cost of drug input is high and it is easy to pollute the microalgae and the harvesting environment. Using MBF to harvest microalgae can avoid the shortcomings of physical and chemical coagulation methods and promote the green and clean production of microalgae.

Microbial flocculation of microalgae harvesting mainly includes bacterial-induced flocculation, fungal-mediated flocculation, and algal self-flocculation. In practical applications, the specificity of MBF and algae species should be fully considered, and the efficient flocculation of bacteria or their secretions with microalgae should be promoted through the addition of flocculants or co-cultivation. Many studies have shown the potential of MBF for harvesting microalgae (as shown in Table 5.8), but future focus should be on the competition for nutrients in co-cultivation, harvesting conditions, environmental safety of MBF harvesting, and expansion and innovation of suitable bacteria for harvesting.

Table 5.7 Application of MBF in sludge conditioning

Source and characteristics of sludge	Sludge conditioning methods and results
Anaerobic digestion sludge from wastewater treatment plant Sludge characteristics (1) pH is neutral: 6.79 ± 0.04 (2) Oxidation–reduction potential (ORP): -213 ± 8.74 mV (3) Conductivity: 14.25 ± 0.04 dS/m (4) Dry sludge content (DS): 2.4 ± 0.056% (5) Capillary cuction time (CST): 38.7 ± 0.07 s (6) SRF: 3.29 × 1013 ± 5.26 × 1012 m/kg (7) Calorific value is 3492 cal/g (8) Poor dehydration	0.2% (based on DS) PAM added After treatment, CST = 16.25 s, SRF = 1.08 × 1013 m/kg, a reduction of 67.17% After treatment, the organic matter content was 66.8 ± 1.2%, while the calorific value was 3815 ± 9 cal/g
	10% of ferrous sulfate rod bacteria solution (10^8 cell/mL) was added After treatment, the moisture content of the sludge cake was reduced to 70.3 ± 0.8%, TSS was 208 ± 4 mg/L, TDS at 12,524 ± 344 mg/L and COD at 224 ± 8 mg/L, which was better than the pollution reduction effect of PAM After treatment, CST = 10.1 s, SRF = 0.36 × 10^{13} m/kg, a reduction of 89.05%, significantly improved the dewatering performance of the sludge After treatment, the organic matter content was 74.5 ± 1.4%, and the calorific value was 4013 ± 4 cal/g. The sludge cake recovered after treatment with the biological flocculant had higher organic matter and calorific value, which is beneficial for resource utilization
Wastewater treatment plant biochemical section filamentous bacteria expanded sludge Sludge properties (1) MLSS: 2500 mg/L (2) SVI: 235 mL/g (3) Sludge settling ratio (SV30): 55 (4) SV5: 92 (5) Severe expansion	Set up R1, R2, R3, and R4 SBR reactors, and add different substances to improve the expansion performance of the sludge R1: Add 7.5% mycelial balls (made from *Aspergillus niger* 557) according to the volume ratio of the SBR reaction container; R2: Add 7.5% mycelial balls and 0.5 g/L activated carbon; R3: Add 5% mycelial balls, 0.5 g/L activated carbon, and 5 mL/L biological flocculant F2; R4: Add 1 g/L activated carbon Adding mycelial balls effectively improves the settling performance of filamentous bacteria expanded sludge. After 2 days of operation, the SVI and SV30 of R2 were reduced to 107 mL/g and 27 respectively, and the SVI and SV30 of R4 were reduced to 73 mL/g and 21 respectively. R2 and R3 can achieve similar SVI treatment effects to R4 Unlike physical conditioning, adding mycelial balls can increase the protein content of EPS in the activated sludge system and enhance the hydrophobicity of the sludge. At the same time, the mycelial balls can self-dissolve and break, which helps to form new bacterial clumps. Not only will it not increase the total amount of sludge, but it can also promote the maturation and stability of the sludge system. From an economic point of view, the cost of activated carbon treatment is high. Using the biological regulation of mycelial balls can not only alleviate sludge expansion but also help to reduce treatment costs

(continued)

Table 5.7 (continued)

Source and characteristics of sludge	Sludge conditioning methods and results
Secondary sedimentation tank sludge in wastewater treatment plant Sludge properties (1) Dry sludge content (DS): 12.6% (2) SRF: 11.3 × 1012 m/kg (3) pH value: 6.4	The polysaccharide MBF (neutral sugar 45.2%, uronic acid 5.9%, amino sugar 4.1%, molecular weight 3.79×10^5 Da) is extracted from *Rhodotorula glutinis* fermented with slaughterhouse wastewater When MBF is added alone at a pH of 7.5 and an addition amount of 12 g/kg DS, the DS increases to 19.8%, and the SRF decreases to 4.6×10^{12} m/kg (stirred at 200 r/min for 10 min, left to stand for 30 min)
	When a chemical flocculant is added alone, the stirring conditions and flocculation sedimentation time are the same. The dewatering effect of PAC and PAM is slightly better than that of MBF, but the dewatering effect of $FeCl_3$ and $Al_2(SO_4)_3$ is significantly worse than that of MBF $FeCl_3$: When the pH is 6.5 and the addition amount is 30 g/kg DS, the DS increases to 17.6%, and the SRF decreases to 6.8×10^{-2} m/kg; $Al_2(SO_4)_3$: When the pH is 6.5 and the addition amount is 30 g/kg DS, the DS increases to 17.8%, and the SRF decreases to 6.7×10^{12} m/kg; PAC: When the pH is 7.5 and the addition amount is 15 g/kg DS, the DS increases to 20.1%, and the SRF decreases to 4.5×10^{12} m/kg; PAM: When the pH is 7.5 and the addition amount is 6 g/kg DS, the DS increases to 21.5%, and the SRF decreases to 4.1×10^{12} m/kg
	When MBF is combined with PAC, the stirring conditions and flocculation sedimentation time are the same. The dosage of the drug is determined through response surface optimization. The combined conditioning method of the two can improve the structure of the sludge floc, facilitate the release of internal water wrapped in the sludge, and significantly improve the dewatering effect of the sludge compared to a single drug When the additional amounts of MBF and PAC are 10.5 g/kgDS and 12.4 g/kgDS respectively, the DS increases to 23.6%, and the SRF decreases to 3.2×10^{12} m/kg (without adjusting the original sludge pH); if the sludge pH is adjusted to 7.5, the DS increases to 24.1%, and the SRF decreases to 3.0×10^{12} m/kg

5.5 Development Trend of Microbial Flocculants

MBF has a wide range of sources, is easily degradable, and has high flocculation activity. However, some problems such as high production and usage costs, difficulty in strain preparation, weak applicability, and an incomplete evaluation system limit the large-scale production and application of MBF. With the continuous improvement of environmental protection requirements and new technologies, research on

Table 5.8 Application of MBF in the process of microalgae harvesting

Target microalgae	MBF	Flocculation form	Flocculation mechanism	Flocculation influencing factors and effects
Chlorella	hsn06 Actinomycetes Streptomyces	Bacterial induction	Under the joint action of Ca^{2+}, hsn06 neutralizes the negative charge on the surface of algal cells, promoting the mutual attraction of MBF bacteria and algal cells, forming heavy flocs, and harvesting algae	hsn06 has good thermal stability and flocculation activity, with the optimal pH value being 9 ~ 11. A moderate increase in the concentration of hsn06 can improve flocculation activity. When the addition amount is 10 mg/L, the flocculation rate reaches 68.7%. In actual applications, the addition amount of hsn06 is generally 20 mg/L, and the appropriate addition of Ca^{2+} can further improve flocculation activity
High-temperature Chlorella F51	EPS *Bacillus beveridgei*	Bacterial induction	The EPS molecules produced by the bacteria contain functional groups such as carbonyl, hydroxyl, and amino groups. On the one hand, they disrupt the charge balance on the microbial surface, causing the microalgae cells to destabilize. On the other hand, the high molecular weight of EPS is conducive to the full extension of large flocculant molecules, forming a net-like floc with microalgae cells through adsorption bridging	EPS maintains a flocculation efficiency of over 60% at pH = 4.0 ~ 11.0. When high-temperature Chlorella grows to a period (pH = 8.0), adding 2 mL of EPS gives the best effect after 15 min of flocculation, reaching 82.1%

(continued)

Table 5.8 (continued)

Target microalgae	MBF	Flocculation form	Flocculation mechanism	Flocculation influencing factors and effects
Oil-producing Chlorella	QM9414 *Trichoderma reesei*	Fungal mediation	After co-cultivation of *Trichoderma reesei* QM9414 and Chlorella, on one hand, the fungus forms hyphae and intertwines into granules, encapsulating Chlorella. On the other hand, the hyphae of QM9414 contain positively charged polysaccharides, while the fungal body and Chlorella surface carry negative charges, allowing Chlorella cells to adhere to the fungal cell wall through hyphal traction. Therefore, Chlorella cells are not only trapped in fungal granules but also fixed on the viscous extracellular polysaccharides secreted by the fungus	The harvest rate is closely related to the input ratio. When the input ratio of *Trichoderma reesei* QM9414 to Chlorella is 1:10, the harvest rate is 69.3% after 4 h, and the total biomass and lipid yield are 2.17 ± 0.35 g/L and 0.60 ± 0.09 g/L, respectively. When the input ratio increases to 1:5, the harvest rate quickly stabilizes to 83.6% within 1.5 h, and the total biomass and lipid yield increase to 3.78 ± 0.21 g/L and 0.75 ± 0.19 g/L, respectively. When the input ratio is 1:2, the harvest rate reaches 94.5% within 10 min, nearly 100% within 30 min, and the total biomass and lipid yield are 6.64 ± 0.65 g/L and 1.70 ± 0.09 g/L, respectively
Chlorella	hsn26 *Aspergillus niger*	Fungal mediation	Under the light microscope, Chlorella cells are distributed and aggregated along the surface of the hyphae, interweaving to form a networked fungal-algae complex. The granules on the hyphae contain low molecular weight surface proteins, and added metal ions such as Ca and Mg bridge between the hyphae granules and Chlorella cells, promoting the surface binding of hyphae and Chlorella, and enhancing the hydrophobic action of the floc	In an alkaline environment ($pH = 8 \sim 9$), the flocculation effect of hyphae granules on Chlorella cells is the best. After 20 h of flocculation, the color of the hyphae granules changes from white to green, and the harvest rate reaches 86.9%. The hyphae granules show good flocculation activity below 40 °C, but they are easily inactivated above 60 °C

(continued)

Table 5.8 (continued)

Target microalgae	MBF	Flocculation form	Flocculation mechanism	Flocculation influencing factors and effects
Euglena gracilis	EPS *Euglena gracilis*	Algae-algae self-flocculation	*Euglena gracilis* cells usually carry a negative charge (– 25 mV), and its EPS also shows negative electricity (– 10 ± 0.6 mV), the electrostatic repulsion in the system affects the aggregation of microalgae. At a pH of 6, the added Al^{3+} forms various positively charged aluminum flocs, and the EPS and *Euglena gracilis* cells are electrically neutralized, and then through hydrogen bonding, electrostatic attraction, and bridging, the flocculation process of the two is formed and improved	The efficiency of EPS alone for the harvest of *Euglena gracilis* is as low as 8.3 ± 0.9%, and does not increase with the increase in EPS dosage. After adding Al^{3+} (4.5 mg/g), the harvest efficiency significantly increases with the increase in EPS dosage. Compared with the Al^{3+} dosage of 77.6 mg/g, the combination of EPS (3.2 mg/g) and Al^{3+} (4.5 mg/g) reduces the harvest cost from 282 to 71 dollars/t

key directions such as efficient strain development, cheap culture medium substitution, preparation production improvement, flocculation strengthening, application strategy and safety assessment will strongly promote the future development of MBF, highlighting its economic, environmental and social benefits.

5.5.1 Developing Highly Applicable MBF Producing Bacteria

Isolation and screening are the first steps to obtain MBF bacteria. Numerous bacteria, fungi, actinomycetes, and algae are sources of MBF, but the process of isolating and screening MBF strains is tedious and time-consuming. Currently, the efficacy of most MBFs is judged by kaolin water distribution or flocculation tests with a few specific types of actual wastewater, with few microbial preparations universally applicable to various types of wastewaters, especially lacking MBF products for extreme environments such as low temperature, poor nutrition, and high salinity. Traditional strain mutation and emerging genetic engineering technologies are crucial for future efficient strain screening and modification.

Using physical or chemical means to induce mutation is a simple and effective classic method to enhance the productivity and adaptability of flocculating strains. Through ultraviolet mutation (30 s) and low-temperature (5 °C) stress culture, a certain wax-like Bacillus produced FM-5 mutant strain. After optimizing the flocculation conditions, its removal rates of BOD, COD, turbidity, and color from domestic sewage all exceeded 97%, and the effluent quality reached the Level A standard of "Urban Sewage Treatment Plant Pollutant Discharge Standard" (GB 18918–2002). Chemical inducers can enhance MBF output. In the culture medium of *Agrobacterium tumefaciens* F2 strain, adding 0.45 μmol/L of N-caproyl-L-homoserine lactone increased the MBF yield and flocculation activity by 175% and 10%, respectively.

Strain mutation cannot guarantee genetic stability, but this defect can be effectively compensated by gene recombination. In the process of strain breeding, the exogenous target gene is introduced into the recipient cell, causing it to replicate, transcribe, and express in large quantities, constructing genetically engineered bacteria, and endowing the strain with special functions to increase MBF yield and adapt to the water environment. It should be noted that MBFs are diverse, large in molecular weight, and complex in structure. Searching and determining key functional genes is not easy, which to some extent limits the genetic modification of strains. In the future, integrating molecular biology and genetic technology, identifying functional genes through comparative genomics, and reasonably regulating gene deletion and overexpression will cultivate high-yield MBF strains, reduce production costs or increase treatment functions, and expand the application range of MBF. For example, using whole-genome sequencing technology, it was determined that the sinR gene is a key inhibitor of polysaccharide synthesis. Deleting this gene helps to increase the yield of polysaccharide MBFs. The whole-genome spectrum analysis of Lichen

Bacillus CGMCC2876 showed that the overexpression of the EpsB gene increased its flocculation performance and yield by 224% and 36.62%, respectively, which is conducive to promoting the industrial production of this bacterial agent.

5.5.2 Developing Green and Low-Cost Alternative Culture Media

Traditional culture media are important carriers for the screening, preparation, and production of MBF-producing bacteria, providing essential nutrients such as carbon, nitrogen, sulfur, and phosphorus for strain fermentation. However, their high cost is not suitable for the industrial development of MBF. Agricultural and industrial waste and wastewater are widely available and cheap. Their sugars, proteins, and amino acids can be used as nutrient sources for MBF strains, reducing production costs and achieving waste resource utilization. Because it contains biodegradable organic components such as proteins, fats, and cellulose, brewery wastewater has become one of the typical representatives of low-cost culture media. By adding glucose (8.94 g/L) to brewery wastewater and culturing Escherichia coli P3, the yield of BW-P3 can reach 1.274 g/L. When 1 g/L of BW-P3 is added under conditions of 50 °C and pH = 7, the color and suspended solids removal rates of backwash wastewater reach 85% and 52%, respectively.

Depending on the different characteristics of alternative culture media, it is often necessary to add a pretreatment step to ensure the effectiveness of the culture. Activated sludge, after heat treatment or acid–base treatment, is suitable for low-purity MBF culture. The hot acid hydrolysate of fibrous waste can be used as a fermentation carbon source for MBF production, but it needs to be neutralized with alkali in the later stage. Moreover, this type of hydrolysate often contains phenolic and furan toxic by-products, which inhibit fermentation and even remain in the MBF product, causing environmental pollution. Therefore, the use of waste fermentation to produce multifunctional strains with both enzymatic and flocculating activities is key to expanding the range of green and cheap culture media. Pseudomonas G22 can secrete xylanase and produce MBF-G22 while decomposing agave, corn straw, elephant grass, wood chips, and wheat bran. At a temperature of 30 °C and a pH of 10, using 10 g/L of wood chips as the only carbon source, the yield of MBF-G22 after 96 h of culture is as high as 3.75 mgMBF/g dry biomass. In the same temperature and pH environment, adding 80 mg/L of MBF-G22, the flocculation rate of Scenedesmus can reach 95.7%. In addition, most reports focus on the research of alternative carbon sources, and the exploration of cheap nitrogen sources, phosphorus sources, etc., needs to be strengthened in the future. Agar Bacillus C9 can produce a variety of degrading enzymes such as amylase, protease, lipase, cellulase, xylanase, and pectinase, which can directly convert kitchen waste into molecular-level nutrients for efficient use by the strain. The resulting MBF is as high as 6.92 g/L and can be used for mining wastewater treatment.

5.5.3 Intelligent Control and Method Improvement of MBF Production

At present, the production of MBF is mostly at the laboratory level, with very few entering the pilot stage, lacking data and experience in large-scale fermentation production. The most suitable fermentation environment conditions determined by experimental research often deviate when scaled up. Using γ-PGA produced by lichen Bacillus, algae are harvested. When the pilot volume is 5 and 50 L, the recovery rate is 99.2%, but when further scaled up to 200 L, the corresponding recovery rate drops to 98%. Therefore, it is necessary to vigorously develop intelligent detection instruments and devices, track the changes in flocculation conditions during process amplification in real time, clarify the changing rules of substrates, strains, and product concentrations, study the metabolic pathways of efficient strains, combine fermentation kinetics, microbial production kinetics, and product synthesis kinetics analysis, accurately control fermentation parameters such as substrate composition and concentration, pH, DO, and temperature, and optimize MBF production.

The cost of MBF extraction and storage accounts for more than 30% of the total cost, and the product is prone to lose some flocculation activity during this process. The fresh MBF-C9 produced by Agar Bacillus C9 has a flocculation rate of 95.29% (pH 6.53, 29 °C) for kaolin suspension, which drops to 92.8% after half a month of storage and 89.1% after six months. Therefore, exploring efficient and cheap extraction and storage methods to ensure the yield and activity of MBF is also an urgent issue to be addressed in the future. For example, under the condition of affordable transportation costs, the direct use of strain fermentation broth as MBF addition can save extraction costs and reduce inactivation. Lichen Bacillus lacking β-glucosidase cannot decompose its own polysaccharide flocculant, which can greatly increase the yield and storage stability of MBF. In the future, attention should be paid to the design of the extraction process, analyze its impact on the structural components and flocculation activity of MBF, avoid the introduction of pollutants and impurities as much as possible, cultivate and select highly stable strains, and promote the industrial application of MBF.

5.5.4 Strengthening of MBF Flocculation Through Compounding, Combination, and Co-use

Compounding, combination, and co-use have always been important methods to strengthen flocculation. Numerous research reports have proven that these three methods can realize the complementary advantages of flocculation components or combined processes, enhance flocculation effects, increase the adaptability of MBF, reduce the dosage of reagents, lower usage costs, simplify the treatment process, and effectively weaken or avoid the secondary pollution of traditional coagulants.

Due to the presence of a large number of hydroxyls, carboxyl, and other negatively charged functional groups, most MBFs remain negatively charged even under neutral pH conditions. To promote the interaction of MBF with target objects carrying negative surface charges, metal cations or inorganic coagulants are often added to assist flocculation. By using MBF and coagulants in combination, treatment effects are ensured. The CBF (0.26 g/L) produced by the combined cultivation of Rhizobium radiobacter F2 and *Bacillus sphaericus* F6, when used in combination with $AlCl_3$ (0.18 g/L) and $CaCl_2$ (0.12 g/L) under room temperature and neutral pH conditions, achieved a harvesting rate of 96.77% for Chlorella, which is significantly better than the flocculation effects of using CBF (48%), $AlCl_3$ (51%), and $CaCl_2$ (34%) separately. The MBF (10.5 g/kg) produced by *Rhodococcus erythropolis* and PAC (19.4 g/kg) combined for sludge dewatering, reduced the SRF value from 11.3×10^{12} m/kg to 3×10^{12} m/kg. To achieve similar treatment effects, the individual dosages of $FeCl_3$, $Al_2(SO_4)_3$, and PAC would need to be at least 60 g/kg, 60 g/kg, and 30 g/kg respectively. Therefore, it is crucial to determine the appropriate compounding ratio of different flocculants for different water treatment environments. In the future, according to the flocculation characteristics of MBF, it is necessary to carry out in-depth experimental research on complex combination, combine the formation and growth law of floc, analyze the mechanism of complex action, optimize flocculation conditions, and guide the flocculation process practice.

Composite MBF originally came from the mixed fermentation of multiple bacterial strains. Compared to a single bacterial group, a composite group has higher biodiversity, more stable ecological structure, more complete functions, and is easier to produce and use. Different strains have different ecological niche characteristics, and their metabolic and flocculation adaptation ranges vary. Strains with homologous DNA but significant specificity are more conducive to the formation and functional coupling of composite groups. Using a mixture of four different types of yeast, 11 types of composite MBF (MBF-HJ1 ~ MBF-HJ11) and 4 types of single MBF (MBF-J1 ~ MBF-J4) were produced. The single type MBF-J2 had a flocculation rate of up to 83.5% (pH = 4, 30 °C) for kaolin suspension; in comparison, five types of composite MBF had flocculation rates exceeding 83.5%, with the highest being 94.7% (pH = 4.5, 30 °C), demonstrating the superiority of composite MBF. In the future, it is necessary to fully utilize modern microbial identification technology to select the best combination of bacterial groups. Based on the functional matching of the bacterial group, it is vital to optimize the combination of nutritional components and cultivation parameters, create a more resilient and adaptable symbiotic fermentation environment, and achieve increased production and efficiency of composite MBF. In recent years, the direct production of composite MBF from environmental materials rich in multiple bacterial species and their metabolic products has also been a focus of this type of flocculant, such as the extract from activated sludge showing good flocculation activity. In addition, polymers formed by MBF and other active components through chemical reactions or modification treatments can also be considered as composite MBF for application. The key technical issue of this method is to form a homogeneous system that is mutually soluble and stable, while ensuring the positive synergistic effect of each component. The integrated system of MBF-FeO(OH) formed by the active components extracted from waste yeast fermentation broth and hydroxyl iron oxide, had a decolorization rate of up to 95.5% for Direct Yellow DY4, which is 82.6 and 33.6% higher than the decolorization rates of using MBF and FeO(OH) separately.

The combined use of MBF usually has two implications. The first is the combined use of MBF and traditional flocculants. In practical applications, it is often synonymous with "compounding", which will not be elaborated here. The second is the combination of MBF with adsorption, oxidation, filtration, etc., to enhance the flexibility and reliability of the water treatment system. The combined use of MBF and adsorption materials such as activated carbon, fly ash to treat dye wastewater and heavy metal wastewater can not only play the role of adsorbents in decolorization and heavy metal removal, but also promote the flocculation and sedimentation effect of MBF, which is beneficial for wastewater to meet the standard discharge. Polyglutamic acid (PGA) is a high molecular weight water-soluble amino acid produced by microbial fermentation, and NaClO has a decolorizing and oxidizing coagulation effect. The use of PGA, NaClO, and filtration combined coagulation/sand filtration/sodium hypochlorite/sand filtration system to treat tequila distillers' wastewater, the SS removal rate increased from 34% of PGA alone to 70%. Although there are not many reports on the combined use of MBF with other physical and biochemical methods, there is no doubt that this will be one of the new directions for the promotion and application of MBF.

5.5.5 *Strengthen MBF Safety Assessment and Application Research*

In a broad sense, the safety evaluation of MBF involves three aspects: the toxicological judgment of the flocculant itself, the product technical protection of the manufacturer, and the quality assurance period determination of the commodity flocculation effect. Given that the application field of MBF is closely related to human health and the ecological environment, the commonly mentioned safety evaluation is often in a narrow sense—the toxicity test and safety evaluation of MBF. At present, the majority of MBF research and development assumes that it is safe and non-toxic, and no systematic toxicological research has been conducted, which will become a potential environmental risk for the application of MBF. Therefore, in the future, it is important to strive to build a diversified MBF gene resource library, build an intelligent resource management platform, improve MBF product standards and toxicity evaluation system, and provide the institutional and technical basis for the safety evaluation of genetic toxicity, reproductive toxicity and general toxicity of flocculating bacteria and their MBF products.

Compared with chemical flocculants, the flocculation activity and water treatment performance tests of MBF are mostly based on beaker tests, and the data is difficult to explain the actual flocculation application situation. In addition, the evaluation index of MBF flocculation activity is single, and the flocculation rate index of kaolin suspension alone cannot fully measure the treatment effect of various types of wastewater with different properties. The use of MBF lacks specificity and has not formed a mapping relationship between the types of flocculants, active ingredients, and types of wastewater. Therefore, it is necessary to actively advocate the application research of MBF. According to the biological flocculation characteristics of MBF, it is vital to develop automated, integrated flocculation equipment, regulate flocculation conditions, monitor the dynamic changes of flocculation effect, summarize the kinetics and morphology rules of flocculation, explore the flocculation mechanism of MBF in different application occasions, improve the actual water treatment efficiency, and further promote the industrial development of MBF.

Chapter 6
Natural Polymer Flocculants

Natural polymers refer to high molecular compounds that exist in nature and in organisms. As a sustainable resource, natural polymers are receiving more and more attention due to their abundant sources, low prices, and wide uses. Natural polymers used in water and wastewater treatment systems include starch, galactomannan, cellulose, chitosan, lignin, tannin, microbial polysaccharides, animal glue, and gelatin. Most of these natural polymers are water-soluble polymers. Their structure, molecular weight, biodegradability, solubility, and synthesis temperature vary. This chapter focuses on the first five natural polymers, and only briefly introduces tannin, microbial polysaccharides, animal glue, and gelatin.

Organic synthetic high polymer coagulants have many advantages over inorganic drugs in many aspects, such as being less affected by pH, forming large and strong flocs, and producing less sludge. However, there are still some unavoidable problems with their use. For example, organic synthetic high polymers have a great selectivity for the colloids they coagulate, and they are not easily biodegradable, which affects subsequent treatment. Moreover, it is not yet clear whether they will have adverse effects on human health, such as toxicity, carcinogenicity, and mutagenicity. Therefore, from the perspective of sustainable environmental development, it is very important and has broad application prospects to develop natural high polymer coagulants that can be biodegraded and are harmless to human health as alternatives to organic synthetic high polymer coagulants.

L. Fengting, *Water Treatment Chemicals: Coagulants and Flocculants*,
https://doi.org/10.1007/978-981-96-8221-8_6

6.1 Starch

6.1.1 Overview

Starch is a type of sugar substance found in the natural world within plant bodies, mainly deposited in the seeds, tubers or roots, fruits, and leaves of plants, and is spread throughout the plant world. Starch mainly comes from potatoes, corn, cassava, taro, peas, broad beans, wheat, and tobacco, among which the varieties used for industry are mainly potatoes, corn, cassava, and wheat. The world's starch production broke through 20 million tons in the early 1990s, and it has now exceeded 36 million tons. In the United States, over 95% of starch comes from corn; while in Europe, especially in the Netherlands, France, Poland, Germany, Denmark, Sweden, and Russia, the production of potato starch is high. In recent years, the use of wheat starch has been increasing. Cassava starch is mainly produced in Thailand and Brazil. In recent years, the development in the southern regions of China has also been rapid. Currently, corn starch accounts for about 80% of China's total starch production, cassava starch accounts for 14%, and other tubers, grains, and wild starches account for 6%.

6.1.2 Structure and Properties of Starch

Starch is a polysaccharide carbohydrate composed of glucose, with the chemical structure formula $(C_6H_{10}O_5)_n$, where $C_6H_{10}O_5$ is the dehydrated glucose unit, and n is the degree of polymerization, i.e., the number of dehydrated glucose units that make up the starch polymer.

The properties of starch from different plants vary greatly, with differences in particle size, shape, moisture content, gelatinization temperature, swelling rate in solvents, and the ratio of amylose to amylopectin.

Starch appears as a white powder, and under a microscope, it is seen as transparent particles of various shapes and sizes. The shapes of starch granules are generally round, oval (or egg-shaped), and polygonal. The shape and size of the granules of the same type of starch are also inconsistent. Table 6.1 compares the granular properties of different types of natural starch.

Starch has a high moisture content, but because the hydroxyl groups in starch molecules form hydrogen bonds with water molecules, the starch appears as a dry powder. Different types of starch have different moisture contents, which is due to the different degrees of self-binding of the hydroxyl groups in the starch molecules and their binding with water molecules. For example, the self-binding degree of hydroxyl groups in corn starch molecules is higher than that in potato starch, so the number of free hydroxyl groups that can bind with water molecules through hydrogen bonds is relatively reduced, hence the moisture content of the starch is lower.

Pouring starch into cold water and stirring can form a milky suspension, i.e., starch milk. If stirring is stopped, the starch will slowly settle. If starch is poured

Table 6.1 Granular properties of natural starch

Granular properties	Corn starch	Potato starch	Wheat starch	Cassava starch	Waxy corn starch
Type of starch	Grain seed	Tuber	Grain seed	Root	Grain seed
Granule shape	Round, polygonal	Oval, spherical	Round, lentil-shaped	Round, truncated	Round, polygonal
Diameter range, μm	2 ~ 30	5 ~ 100	0.5 ~ 45	4 ~ 35	2 ~ 30
Average diameter, μm	10	23	8	15	10
Specific surface area, m^2/kg	300	110	500	200	300
Density, g/cm^3	1.5	1.5	1.5	1.5	1.5
Number of starch granules per gram, $\times 10^6$	1300	100	2600	500	1300

into hot water, the starch granules start to expand when heated, and continue to expand as the temperature continues to rise. When heated to a certain temperature, the starch becomes a viscous, semi-transparent gel or colloidal solution, known as starch paste. Even if stirring is stopped, the starch will not settle. This phenomenon is called gelatinization or gelling (gel). The temperature at which gelatinization occurs is called the gelatinization temperature or gelling temperature. The essence of starch gelatinization is that the hydrogen bonds between starch molecules in the starch microcrystalline bundles are broken to form a colloidal solution. The process of starch gelatinization can be divided into three stages:

(1) Reversible water absorption stage. Water enters the amorphous part of the starch granule and the volume expands slightly, at which point the granule can be restored, and the birefringence phenomenon remains unchanged.
(2) Irreversible water absorption stage. As the temperature rises, water enters the gaps between the starch microcrystals, irreversibly absorbing a large amount of water, the birefringence phenomenon gradually blurs, and the starch granules expand to 50 ~ 100 times their original volume.
(3) Complete disintegration of starch. Starch molecules enter the solution, and the birefringence phenomenon completely disappears.

Different starches have different temperatures, viscosities, solubilities, expansion forces, critical concentrations, etc., when gelatinization occurs. Table 6.2 lists the gelatinization characteristics of different plant starches. The data in the table shows that potato starch has the greatest expansion ability.

Ordinary starch contains two structures, linear and branched. Natural starch exists as white solid granules, with the outer layer being branched starch, i.e., the cortex of the starch, and the inner layer being linear starch, i.e., the granular substance of the starch. When treated with hot water, linear starch can dissolve, while branched

Table 6.2 Gelatinization characteristics of starch

Gelatinization characteristics	Corn starch	Potato starch	Wheat starch	Cassava starch	Waxy corn starch
Gelatinization temperature, °C	52 ~ 72	56 ~ 66	58 ~ 64	59 ~ 69	62 ~ 72
Viscosity (5% concentration) peak range, Bu units	300 ~ 1000	1000 ~ 5000	200 ~ 500	500 ~ 1500	600 ~ 1000
Average viscosity, Bu units	600	3000	300	1000	800
Expansion force (95 °C)	24	1153	21	71	64
Solubility (95 °C), %	25	82	41	48	23
Critical concentration, g	4.4	0.1	5.0	1.4	1.6

starch does not dissolve. Figure 6.1 shows the composition and structure of linear and branched starch. The side chains of both linear and branched molecules are linear, tending to align parallelly, and adjacent hydroxyls are bonded into a scattered crystalline "bundle" structure by hydrogen bonds. Linear and branched starches are often distinguished by the way glucose units that make up large molecular chains are connected and the shape of the chain. In linear starch, the dehydrated glucose structural units are mainly connected by α-D-(1,4) glycosidic bonds, and in recent years, a small amount of branching has been found in some linear starches. In addition to α-1,4 glycosidic bonds, branched starch molecules also connect glucose units with α-1,6 glycosidic bonds, so branched starch molecules are branched, with about 1 branch for every 20 ~ 30 glucose units, and their molecular weight is much larger than that of linear starch molecules. The content and degree of polymerization (DP) of linear and branched starch in starch vary with the variety. Generally speaking, the proportion of linear starch in plant starch is less, usually 20 ~ 25%. The content of linear starch in beans is higher, some up to 70% or more. Table 6.3 lists the properties of linear and branched starch in natural starch.

Retrogradation is an important property of starch paste. When thin starch paste is stored for a long time, it gradually becomes turbid, with white sediment sinking and water being released, destroying the colloidal structure. This is due to the re-condensation and precipitation of the dissolved starch, a phenomenon known as retrogradation, also known as aging or reduction. The dissolved amylose tends to align parallelly, forming a crystalline structure through hydrogen bonding, which cannot dissolve in water. When it grows to a certain extent, it becomes white sediment that falls, destroying the colloidal structure of the paste, and water is released.

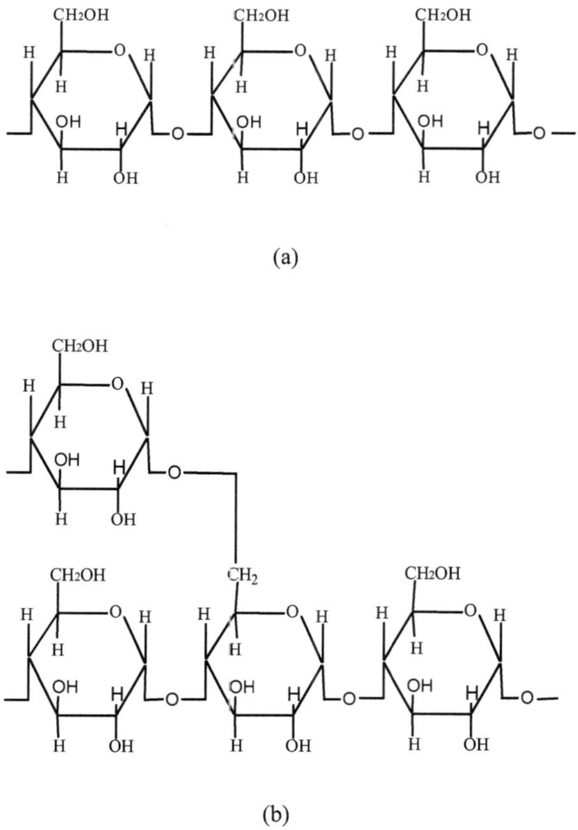

Fig. 6.1 Structure of linear starch (**a**) and branched starch (**b**)

The crystalline structure of these condensed precipitates has been confirmed by X-ray diffraction. To redissolve, it needs to be heated at 100 ~ 160 °C. At the same concentration, retrogradation is less likely to occur within the range of about 63 ~ 66 °C. Both < 63 °C and > 66 °C are sensitive temperature zones for retrogradation. Different pH ranges also affect the speed of retrogradation, at pH = 5 ~ 7, the retrogradation speed is fast, and at higher or lower pH values, the retrogradation speed is slow. When cooled at a higher concentration, it quickly condenses into a semi-solid gel, also due to retrogradation.

Retrogradation is mainly due to the binding between starch molecules. Branched starch molecules are not easy to retrograde due to their branched structure, and they also inhibit the retrogradation of linear starch, weakening the retrogradation. However, under high concentration or cold low temperature conditions, the side chains of branched starch molecules will also bind and retrograde. The length of linear starch molecules is related to the strength of retrogradation, and the retrogradation is strongest and fastest for molecules with a degree of polymerization between 100 and 200. Corn starch contains 27% linear starch, with a degree of polymerization of 200 ~ 1200, strong retrogradation, and also contains 0.6% lipid compounds, which

Table 6.3 Comparison of linear and branched starch

Item	Linear starch	Branched starch
Molecular shape	Linear molecule	Branched Molecule
Degree of polymerization	100 ~ 6000	1000 ~ 3,000,000
Terminal base	One end of the molecule is a non-reducing terminal, the other end is a reducing terminal base	The molecule has one reducing terminal base and many non-reducing terminal bases
Iodine coloring reaction	Deep blue	Reddish purple
Iodine adsorption/%	19 ~ 20	< 1
Sedimentation properties	Solution is unstable, strong sedimentation	Easily soluble in water, solution is stable, very weak sedimentation
Complex Structure	Can form complexes with polar organic substances and iodine	Cannot form complex structures with polar organic substances and iodine
X-ray diffraction analysis	Highly crystalline	Amorphous
Acetyl derivatives	Can be made into high-strength fibers such as films	The films made are very brittle

promote retrogradation. Potato starch contains 20% linear starch, with a degree of polymerization of about 1000 ~ 6000, weak retrogradation.

Most modified starches have greatly reduced retrogradation due to the introduction of ionic groups into the starch molecules. The repulsion between the same charges of the ionic groups makes it difficult for hydrogen bonds to form between linear starch molecules, thus greatly reducing retrogradation.

6.1.3 Applications

Natural starch is an important raw material for food and chemical industries. It is a water-soluble polymer that, in addition to having adhesive properties, can also form gels, making it widely used in papermaking, fiber processing, food, medicine, oil fields, casting, agriculture, and more. The known uses of starch currently exceed 2000. Natural starch is just the primary product of various starch plants, and its properties such as decay, aging, and thickening after heating limit its industrial applications. To further improve economic benefits, or to apply these natural polymers more widely, starch is modified to improve some of its functions, allowing for more reasonable use of resources. Currently, the main areas of starch application are fivefold: utilizing the polymer characteristics of starch; utilizing the decomposition products of starch (such as glucose, maltose, etc.); using starch as a fermentation raw material; comprehensive utilization of starch by-products; and the use of modified starch and starch derivatives.

In recent years, the research and development of modified starch flocculants have attracted much attention. Each unit of the large starch molecule contains three free hydroxyl groups. Through esterification, etherification, oxidation, and cross-linking reactions of these hydroxyl groups, the properties of starch can be changed to produce starch derivatives. The starch molecule chain is attached to artificially synthesized polymer chains, giving the copolymer the properties of both natural polymers and artificially synthesized polymers, paving the way for the preparation of new products. China's research on starch derivatives as water treatment flocculants has achieved good results, especially cationic graft starch has a great development momentum. China has rich plant resources and its own characteristics; the raw material cost is low, non-toxic, and the process route is relatively simple, it is possible to obtain lower-cost products. From the perspective of China's rescurce conditions, the research on modified flocculants using starch as raw material has strong practical significance. In the foreign water treatment agent market, there are many modified starch flocculants, such as Aerofloc from AmercianCynamid Co., Budond from Buckman, Zfloc-Aid and Starches 613-45 from National Starch and Chemical Corp, and Wisproloc from ZyorkShiree, Dyeware.

Cationic starch is an important type of starch derivative prepared by the reaction of starch with cationic reagents. The preparation of cationic starch usually uses the etherification method, that is, introducing an ammonia atom group into the starch molecule. The cationic starch sold on the market is generally divided into tertiary amine (tertiary amine) alkyl starch ether and quaternary amine (quaternary amine) alkyl starch ether. The preparation method of quaternary amine alkyl starch ether is usually as follows.

① Firstly, prepare the cationic reagent. The most common and important cationic reagents are N-(2,3-epoxypropyl) trimethylammonium salt and 3-chloro-2-hydroxypropyl trimethyl ammonium chloride, the reaction formula is:

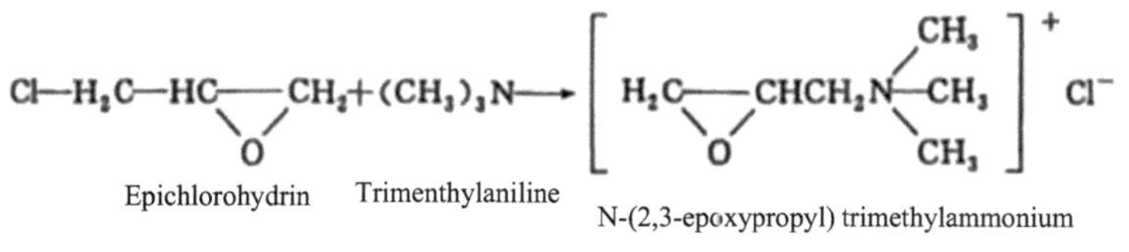

(6.1)

3-chloro-2-hydroxypropyl trimethyl quaternary ammonium salt can undergo reversible reactions at different pH values:

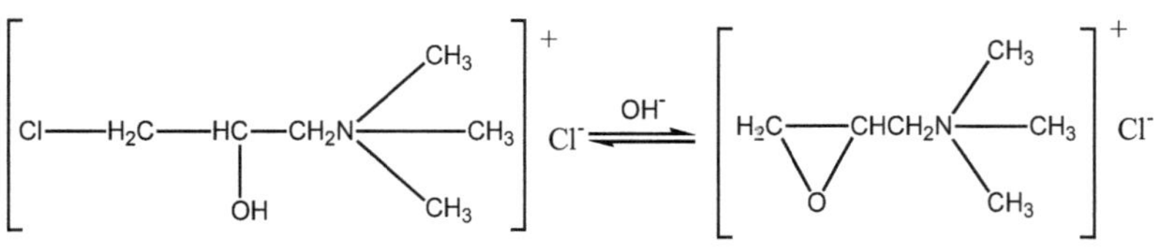
(6.2)

Then carry out the cationization reaction of starch.

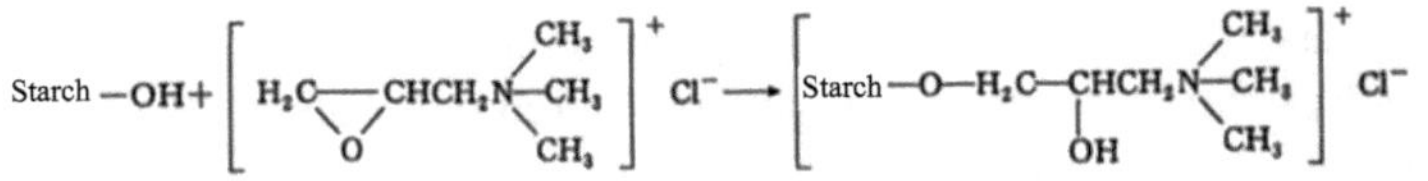

(6.3)

The key to the practicality of cationic starch is that cationic starch can ionize positive charges in water dispersions and has affinity for negatively charged substances. After introducing the cationic reagent into the starch molecule structure, the colloidal properties and chemical properties of the starch will change significantly, thereby affecting the use of the product. The gelatinization temperature of cationic starch is lower than that of natural starch, and the viscosity of cationic starch paste is slightly higher than that of natural starch. The Zeta potential of cationic starch is positive, it is the most important wet-end chemical aid, and it is widely used in the paper industry. The positive charge carried by cationic starch itself is easy to adsorb with cellulose in the pulp, thereby reducing the Zeta potential of the pulp and producing a micro-flocculation structure, encapsulating the fine fibers and various paper fillers in the pulp suspension in this structure, achieving a high retention rate and other good usage effects. In the paper industry, the combined use of colloidal borosilicate and cationic starch can improve paper products. Cationic starch is an excellent flocculant for negatively charged inorganic or organic suspensions. Kaolin, inorganic ore, ore mud, silicon, coal, carbon, anionic starch, cellulose, sewage sludge, and slurry suspensions can alflocculated with cationic starch. L. Järnström et al. used potato starch as a raw material, oxidized it with hypochlorite, stabilized it, and then cationized it to produce a modified cationic starch. The flocculation performance of this starch on kaolin suspension was studied using fiber optic sensing technology. The study found that this quaternary ammonium starch ether is an effective flocculant even at very low concentrations, within a temperature range of 23 ~ 50 °C and NaCl concentrations of 1 ~ 100 mM. However, if the kaolin suspension is pretreated with sodium polyacrylate, the flocculation performance of the modified starch will deteriorate, and flocculation cannot be observed when the NaCl concentration is 1 mM.

The highly substituted cationic starch produced by reacting epichlorohydrin with trimethylamine and then with starch under certain conditions is used to treat papermaking white water with good results. The reaction process is as follows:

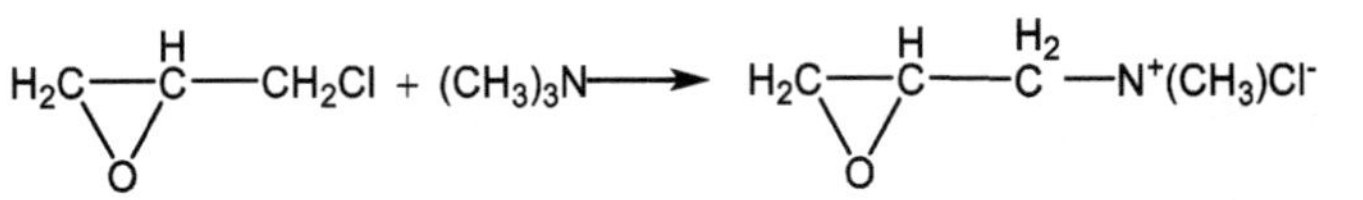

(6.4)

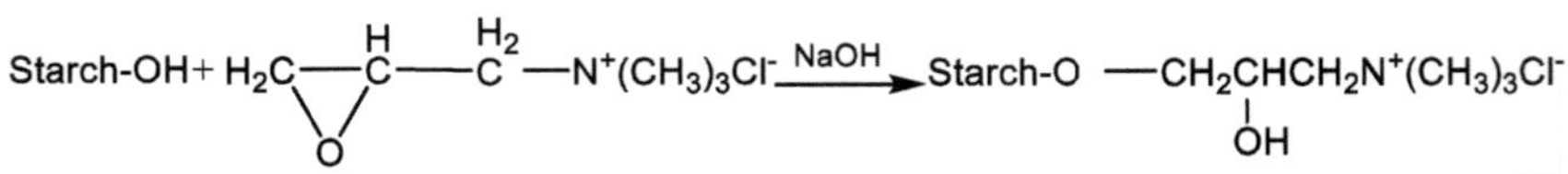

(6.5)

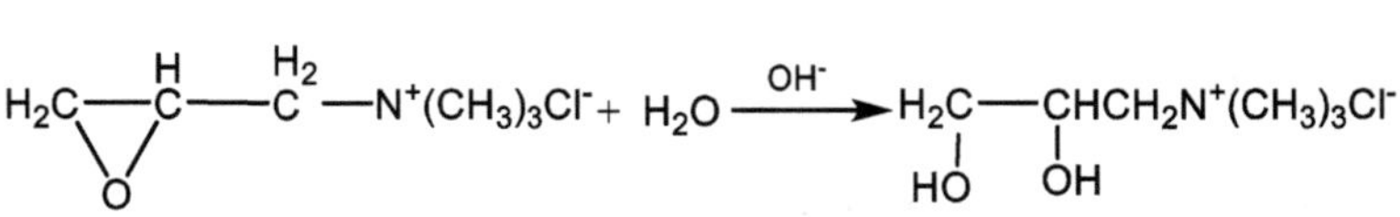

(6.6)

$$\text{Starch-O—}\underset{\text{OH}}{\text{CH}_2\text{CHCH}_2}\text{N}^+(\text{CH}_3)_3\text{Cl}^- + \text{H}_2\text{O} \xrightarrow{\text{OH}^-} \text{Starch-OH} + \underset{\text{HO}}{\text{H}_2\text{C}}\text{—}\underset{\text{OH}}{\text{CH}}\text{CH}_2\text{N}^+(\text{CH}_3)_3\text{Cl}^- \quad (6.7)$$

Using a similar reaction principle, a quaternary ammonium cationic starch flocculant is produced from natural starch using 3-chloro-2-hydroxypropyl trimethyl ammonium chloride as the cationic etherifying agent and the microwave drying method. When treating high-turbidity water, the effect is comparable to polyacrylamide at the same dosage. This cationic starch flocculant, when combined with the inorganic flocculant polyaluminium chloride, has good flocculation and sterilization effects.

Using corn starch and 2,3-epoxypropyl trimethyl ammonium chloride as raw materials, and a water–ethanol mixed solvent as a dispersant, a highly efficient natural modified cationic flocculant can be prepared using a dry synthesis process. Experiments show that this cationic flocculant has a good flocculation effect when used for dyeing wastewater treatment, with a decolorization rate of 96.4%.

Using corn starch as a skeleton, an intermediate product is obtained by graft copolymerization with acrylonitrile, which can be used as a superabsorbent resin. Further saponification hydrolysis of its graft copolymer yields a weak anionic polymer, which can be further hydroxy methylated and sulfonated to obtain a strong anionic natural polymer flocculant. The synthesis process is as follows: the starch is first gelatinized, then graft copolymerized with acrylonitrile using Ce^{4+} as the initiator, hydrolyzed with NaOH to obtain a weak anionic polymer, and then further hydroxy methylated and sulfonated to obtain a strong anionic flocculant.

Using starch as a raw material, an amphiphilic polymer flocculant containing hydroxyl, cyano, amide, and quaternary ammonium salt groups was developed through chemical modification. Then, by adding sodium aluminate and sodium silicate in a certain ratio to make a compound, a composite polymer flocculant can be produced. Research shows that this composite flocculant has a strong flocculation and decolorization effect on dyeing, papermaking, leather, pharmaceutical and other industrial wastewater, and its flocculation performance is significantly better than cationic polyacrylamide, polyaluminium chloride, aluminum sulfate and other flocculants.

Starch can react with phosphates to produce phosphate ester starch, which can significantly alter the properties of the original starch even at very low degrees of substitution. Potato starch, which naturally contains phosphorus in its granules, is superior to other natural starches in industrial applications. Phosphoric acid can react with the three hydroxyl groups in starch molecules to produce mono-, di-, and tri-esters. Starch monophosphate, also known as monophosphate ester, is the most widely used phosphate ester starch in industry. Starch monophosphate is an anionic starch derivative, which, compared to the original starch, significantly improves the viscosity, transparency, and stability of the paste. Due to the ionizing properties of starch phosphates, they can serve as excellent emulsifiers, thickeners, and stabilizers in the food industry. Starch phosphates are also one of the most important starch derivatives used in papermaking. In water treatment applications, starch phosphates are good flocculants, suitable for wastewater treatment, flotation mining, and fine coal powder recovery from coal washing water. Adding starch phosphate to wastewater from meat, fish, fruits, vegetables, and brewing processing plants, followed by the addition of calcium salts, can facilitate recovery through precipitation. Starch phosphates can also be used in the flotation of bauxite ores to recover aluminum; they can be used in iron ore for the precipitation of iron oxide, and adding 4 ~ 10 mg/L to coal washing water can cause coal powder to flocculate and settle. Precipitating each ton of coal powder only requires 0.2 ~ 0.3 kg, and the settling effect of fluoride phosphate is even better. In addition, mixing starch phosphate with polyacrylamide in a 10:1 ratio can also improve flocculation. Since most fine particles and colloids in wastewater treatment have a negative charge, cationic modification of starch is an important research direction.

The coal washing and mineral processing industries need flocculants with good flocculation effect and cheap price. This is because the mineral processing industry discharges a large amount of mineral washing wastewater, which contains a lot of fine coal tailings. If wastewater containing fine coal tailings is directly discharged, it can cause serious environmental pollution and waste of water resources. The role of flocculants is to cause the fine powder to form flocs, settle quickly, and facilitate the recovery of fine coal tailings in the mineral washing wastewater. The water from which the fine coal tailings have been removed can be reused.

Polyacrylic acid and starch are currently two very widely used flocculants in the mineral processing industry, both of which can be used as selective flocculants for iron ore clay. Due to the special affinity of iron ore for starch, since 1975, starch has been successfully used as a selective flocculant for hematite in Tilden iron ore mining. S. A. Ravishankar et al. compared the performance of these two flocculants in selectively separating iron oxides from clay, and found that the natural polymer starch has better selectivity than polyacrylic acid.

The hydroxyl groups on starch molecular chains can be chemically modified to produce various ethers, which are important starch derivatives. Carboxymethyl starch is the largest produced starch ether in industry, and it belongs to anionic polymer electrolytes due to its anionic groups. Carboxymethyl starch is prepared by etherification of starch under alkaline conditions with monochloroacetic acid or its sodium salt, and industrial production is mainly for low substitution products. The gel of carboxymethyl starch is transparent, delicate, high in viscosity, strong in adhesion, good in fluidity and solubility, and has good emulsification, stability and permeability. It is not prone to spoilage and mold, and therefore has a wide range of uses in many industries, making it an important type of starch derivative. Carboxymethyl starch is an anionic polymer electrolyte and can be used as a flocculant, chelating agent, and adhesive.

Soluble starch is used to produce cross-linked starch with epichlorohydrin, which then reacts with chloroacetic acid to produce a product containing carboxymethyl groups, which has a good flocculation effect on direct dye wastewater.

Using corn starch, chloroacetic acid, and sodium hydroxide as raw materials, carboxymethyl starch sodium is synthesized by the solvent method. This anionic polymer flocculant has a good treatment effect on domestic sewage and oil refining wastewater.

Starch can undergo graft copolymerization with monomers such as acrylonitrile, acrylamide, acrylic acid, vinyl acetate, methyl methacrylate, and styrene to form graft copolymerized starch. The grafting methods include chemical grafting and radiation grafting. There are many water-soluble graft copolymers, the most common of which are starch grafted with acrylamide, acrylic acid, and several amino-substituted cationic monomers. The resulting copolymers have thermal water dispersibility and can be used as thickeners, flocculants, and absorbents. These copolymers can be used for flotation of ores or treatment of industrial wastewater, such as a settling agent for silica in the flotation of phosphate ores, and a flocculant in water suspensions of soil and coal dust.

Cai et al. used potassium permanganate as an initiator to graft copolymerize starch or microcrystalline cellulose with acrylamide, preparing cationic polyacrylamide and amphoteric polyacrylamide. The operation is as follows: the graft copolymer is hydrolyzed with sodium hydroxide, converting part of the amine in acrylamide into a negatively charged carboxyl group or sodium acrylate, thus forming a graft polyacrylamide containing acrylamide and sodium acrylate units; the aforementioned product reacts with diethylene glycol amine to further introduce cationic groups into the hydrolyzed polyacrylamide, resulting in a cationic/amphoteric graft polyacrylamide; at least one additive is added, which contains water-soluble inorganic salts and non-ionic surfactants, to improve the viscosity of the product. This method is easy to obtain raw materials, low production cost, simple process, mild reaction conditions, the product has good flocculation performance and wide adaptability, and it is an ideal sludge dewatering agent and wastewater treatment agent.

Using Co_{γ} pre-irradiation method to prepare starch-acrylamide graft copolymer, used as a flocculant to treat domestic sewage, the treatment effect is better than domestic polyacrylamide.

Researchers have synthesized a starch-acrylamide graft copolymer that has a significant flocculation purification effect on neutralized phosphorus wastewater. The reason is that the graft copolymer has a larger specific surface area in its spatial structure than the homopolymer, which is beneficial for bridging.

Using potato starch as a raw material, after harsh treatment, it is grafted with acrylamide to introduce a tertiary amine group to prepare a modified cationic flocculant. It has a good flocculation effect when used to treat kaolin suspension, sugar making, and tanning wastewater.

Using the product of granular starch and acrylamide graft copolymerization as a flocculant, the deinking wastewater from waste paper can be decolorized with good results. The flocculant made by graft copolymerization of taro starch (rich in branched starch) and acrylamide under the initiation of cerium salt has a high COD and suspended solids removal rate when used for the treatment of alcohol tank liquid.

Under the initiation system of $Ce(IV)/HNO_3$, the flocculant obtained from acrylamide-starch graft copolymerization has a good treatment effect on industrial wastewater such as oily wastewater, milk wastewater, printing and dyeing wastewater, papermaking wastewater, and electrophoretic dyeing wastewater. The test results show that the COD removal rate and sludge settling speed are not inferior to the polyacrylamide 3 million product.

Using Fe^{2+}-H_2O_2 as an initiator, soluble starch, acrylamide (AM), and dimethyl diallyl ammonium chloride as raw materials, a series of ternary graft copolymers synthesized have a flocculation effect. Studies have found that the flocculation effect is closely related to the chemical composition structure and dosage of the graft copolymer, and as the grafting rate increases, the flocculation performance of the starch graft copolymer generally improves. The starch/DMDAAC-AM graft copolymer has a good flocculation effect on artificially prepared high-turbidity kaolin water. Especially when the component ratio is properly matched, the flocculation effect is better than polyacrylamide and polydimethyldiallylammonium chloride. The reason why this graft copolymer has a good flocculation effect is mainly due to the grafting of polyacrylamide and polydimethyldiallylammonium chloride branches with flocculation functions on the starch. The cationic polydimethyldiallylammonium chloride branch is a quaternary ammonium salt, which is almost completely ionized in aqueous solution, so it has a high positive charge density, and bridging is not necessary under high turbidity, so it can capture negatively charged suspended particles; while the anionic polyacrylamide branch can promote the sedimentation of inorganic suspended matter, playing the role of a flocculation aid, therefore, the two fully exert their different group adsorption flocculation properties, achieving an ideal flocculation effect.

Starch xanthate is a starch ester compound that was developed in the 1970s. It was made shortly after the invention of the cellulose xanthation reaction. Starch xanthate is a reaction product of carbon disulfide and starch at room temperature. The water-soluble xanthate monoester is unstable and needs to be stored at low temperature (5 °C) to prevent decomposition. The non-water-soluble crosslinked starch xanthate diester can be used for the removal of heavy metal ions in industrial wastewater such as electroplating, mining, and brass smelting, with good results, such as the removal of cadmium, chromium, nickel, copper, zinc, mercury and other heavy metals, which not only recovers heavy metals but also reduces environmental pollution.

Research from the United States Department of Agriculture's Peoria III indicates that starch xanthate can be used to precipitate heavy metals such as Cd, Cr, Cu, Fe, Pb, Mn, Hg, Ni, Ag, Zn. When used in conjunction with polyamines, most of these heavy metals can be easily removed, with better results than using lime or alkali alone. Compared to the precipitation method, the volume of sludge is smaller and the gel-like state is less noticeable. The effective pH range is between 3 and 11. U.S. Patent 3,979,286 describes the use of insoluble starch xanthate esters prepared from crosslinked starch for heavy metal ions in industrial wastewater. U.S. Patents 4,051,316 and 4,083,783 describe the use of alkali metal-magnesium starch xanthates for removing most heavy metal ions from industrial wastewater.

Using crosslinked starch xanthate esters to treat nickel-plating wastewater can achieve a removal rate of over 95%, with a residual nickel concentration of less than 0.2 mg/L. Therefore, crosslinked starch xanthate esters are low-cost, non-secondary polluting, simple equipment, easy-to-operate, and easy-to-promote nickel ion remover. When treating chromium-containing electroplating wastewater with crosslinked starch xanthate esters, tests show that they have a high removal effect on various valence states of chromium ions, including cations and chromate anions, with a removal rate of over 99% and a residual concentration of less than 0.1 mg/L. Moreover, the residue is stable and will not cause secondary pollution.

Yong et al. have prepared a variety of modified starch flocculants. They found that hydrolyzed cassava or taro starch is a good flocculant for sludge suspension when insoluble metal salts are present, and it is very effective in treating asphalt sand tailings and phosphate slime. Comparative tests found that its treatment effect is far superior to hydrolyzed wheat starch.

Currently, starch is also widely used in the pottery industry. By mixing potato branch starch into 40% slurry, adding sodium sulfate, and adjusting the pH to 11.5 with soda solution. Adding a cationic reagent to cationize it, reacting at 34 °C for 18 h, terminating the reaction, washing the cationic starch with water, and drying it, it is necessary to ensure that the cationic starch contains very little alkali. The prepared product can be used as a flocculant and binder in the pottery-making process.

Furthermore, starch can also be used for settling red mud. The main source of aluminum is bauxite, which can be used to produce alumina through the Baeyer process (high temperature, strong alkali). The alumina is then mixed with cryolite and electrolyzed to produce aluminum cathodes. During the production of alumina, solid–liquid separation is required to remove the filtrate containing sodium aluminate, caustic soda, and silicates from the residual solid particles of ferric oxide, certain dolomites, and various siliceous rocks and silicates. Because ferric oxide is reddish, the residue is called red mud. The main polymer for separating red mud is starch. By changing the preparation time, temperature, and alkalinity, almost all types of starch can be used, including corn, potato, sorghum, and cassava starch. Due to price and applicability, corn starch is most commonly used in the United States. In iron ore processing, lime can cause the colloidal ferric oxide and silicates to coagulate, while starch makes them grow into flocs. Various synthetic substances compete in terms of performance and ease of operation (less manual labor required in preparation). Starch requires alkali treatment and sometimes needs to be heated (up to 82 °C) during preparation. Generally, starch is prepared into a 1 ~ 5% solution and diluted with process water before use. The role of starch includes rapid precipitation of solids and purification of aluminate solution, which helps to recover alkali through 6 ~ 9 stages of backwashing, thus minimizing the harm of discharged alkali to the environment. When processing Suriname bauxite, the required amount of starch is 1 ~ 2 pounds/ton (red mud), and for Jamaican ore, the required amount of starch is 40 pounds/ton (red mud). Although the amount of starch used is large, it is the only acceptable flocculant. Recently, synthetic substances (such as sodium polyacrylate) have begun to challenge the status of starch with their low usage and ability to handle higher concentrations of red mud. However, due to its poor water clarity and high price ($1.50/pound), starch is still in use. This starch is generally produced on reclaimed land after the removal of aluminate. Energy is needed to heat the starch. The concentration is generally about 1%, and the time to gel and fully mature is 1 ~ 2 h. The reserve solution should be diluted before being added to the sedimentation tank and concentration tank. Barham et al. have invented a method to improve the purification effect of red mud, which uses dextrorotatory glycoside starch and a water-soluble synthetic flocculant (such as polyacrylic acid, acrylamide and acrylic acid copolymers, etc.) to separate the red mud in the Bayer process. After this composite flocculant is mixed with the solution of the Bayer process, the red mud contained in the solution can be removed by sedimentation, centrifugation, or filtration. In conclusion, the rise of green chemistry provides a good opportunity for the application research of starch products in the field of water treatment. With the deepening of theoretical research on starch modification, the technology of modified starch products is gradually maturing, and it has become possible for modified starch products to partially replace environmentally harmful water treatment agents.

6.2 Galactomannan

6.2.1 Overview

Galactomannan is a storage polysaccharide in plants, mainly found in the seeds of leguminous plants with endosperm. In addition, it is also found in *Annona muricata* of the Annonaceae family, *Convolvulus tricolor*, *Ipomoea muricata*, and Leucacna. The seeds of *Leucaena leucocephala* also contain galactomannan. Galactomannan is also found in the stem and leaf tissues of red clover (*Trifolium pratense*). Immature seeds of monocotyledonous palms also contain galactomannan.

The study and utilization of galactomannan has a history of over seventy years abroad, originally derived from the seeds of the carob tree (*Ceratonia siliqua*), known as carob gum. Carob gum is an evergreen tree growing in southern Europe and has now been introduced worldwide. The use of carob gum gradually expanded from the 1920s and was widely used in many fields by the 1940s. According to 1975 statistics, the world sales of galactomannan gum: carob gum was about 12,500 tons, and guar gum was about 37,000 tons. The approximate industrial usage is as follows: the food industry accounts for about 40%, the paper industry 30%, the textile industry 20%, and petroleum, coatings, explosives, and adhesives account for about 10%. China started to study and use galactomannan gum relatively late. With the needs of China's industrial development, a certain amount of guar gum has been imported since the 1970s, which has led to domestic research and application of galactomannan gum and its resources.

6.2.2 Structure and Properties of Galactomannan

Galactomannan is a natural high molecular weight polysaccharide, and its molecular weight varies depending on the source. Its structure is formed by D-mannose connected by β-1,4 glycosidic bonds to form the main chain, and D-galactose forms side chains on some mannose through α-1,6 glycosidic bonds to form a multi-branched polysaccharide.

In the endosperm of leguminous plants, galactomannan can serve as a reserve food. Galactomannan is deposited on the outside of the protoplasmic membrane, and its potential functions are not only to nourish the protoplasts but also to provide protection. Galactomannan has strong water absorption and retention capabilities, acting as a "reservoir" during seed germination, maintaining a suitable microenvironment for seed germination, and resisting drought conditions. At the same time, when galactomannan is completely degraded into monosaccharides, it provides nutrients and energy for the early germination process.

Different leguminous plants have different degrees of galactose substitution in their galactomannan, ranging from 25% (low substitution) to about 95% (high substitution). The degree of substitution reflects the classification of leguminous plants. In the galactomannan stored in the seeds of leguminous plants, the ratio of galactose to mannose can also show the classification basis of this family. The basic structure of galactomannan in various plant seeds is not much different. This basic linear and branched structure determines its characteristics are significantly different from those of unbranched, fibrous, water-insoluble mannans and glucomannans. Galactomannan gum has good water solubility and cross-linking, and can form a high viscosity stable water solution at low concentrations.

The ratio of galactose to mannose varies with different species, for example, in guar gum the ratio of galactose to mannose is 1/2, in acacia gum it is 1/4, and in tara gum it is 1/3.

The most important natural galactomannan used in water and wastewater treatment is guar gum. Figure 6.2 shows the structure of guar gum. Guar gum is formed by connecting β-1,4-D-mannopyranose units as the backbone to form the main chain, with α-1,6-linked galactopyranose units as side chains for every two mannopyranose units, with a relative molecular mass of about 220,000. Derivatives of guar gum include hydroxypropyl, hydroxyethyl, carboxymethyl, carboxymethyl hydroxypropyl, and cationic guar gum.

Guar gum is a plant gum extracted from the legume plant—guar beans, grown on the Indian subcontinent. Its unique molecular structure and physicochemical properties make it a promising natural polymer. Since guar beans grow in high-temperature, drought, and water-deficient environments, they are not water-resistant, which is a significant advantage compared to starch that needs to be gelatinized before use.

The initial appearance of guar gum was as a substitute for locust bean gum. Before this, locust bean gum was widely used in industrial production and caused a demand crunch. Later research showed that although both guar gum and locust bean gum are galactomannans, they have significant differences in chemical composition and behavior. Locust bean gum needs to be boiled in hot water to reach maximum viscosity, while guar gum can be hydrated in cold water. In terms of chemical composition, locust bean gum has only 1.5 lactose branches for every 4 mannose units. Therefore, the number of branching units in guar gum is twice that of locust bean gum. This is considered the main reason why guar gum is easier to hydrate and has

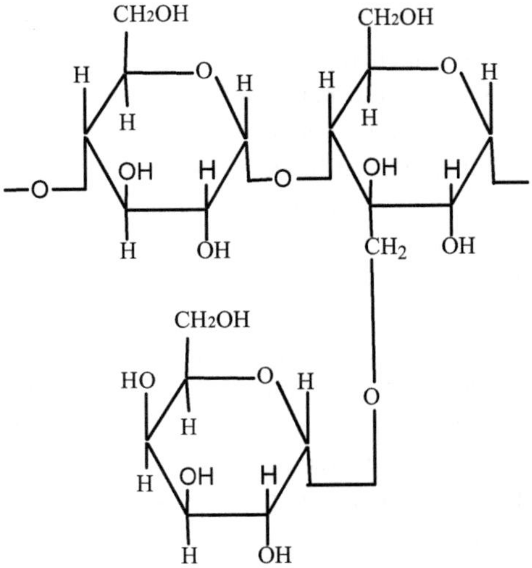

Fig. 6.2 Structure of guar gum

greater hydrogen bonding activity than locust bean gum. In addition, the cost of guar gum is only half that of locust bean gum.

There are no non-polar groups on the straight chain of guar gum, most primary and secondary hydroxyl groups are on the outside, and the galactose branches do not cover the active alcohol hydroxyl groups. Therefore, guar gum has the maximum hydrogen bonding area, and when it binds with fibers, the hydrogen bonding distance is short, and the binding force is strong.

Guar gum has good water solubility and thickening properties and is widely used as a thickener, emulsifier, etc. However, the original powder gum dissolves slowly, the content of insoluble matter in water is high, and the viscosity is not easy to control. People often use chemical means to change its physical and chemical properties to meet the actual needs of industrial production. There are four main directions for the modification of guar gum: (1) Functional group derivation. This method is based on the fact that there are three hydroxyl groups on the sugar unit of guar gum, which can undergo etherification, esterification, or oxidation reactions under certain conditions to produce ethers, esters, and other derivatives. (2) Graft polymerization. Under certain conditions, some initiators can cause guar gum or vinyl monomers to generate free radicals, thereby undergoing polymerization reactions, such as: acrylic acid, acrylamide, methyl acrylamide, acrylonitrile, etc. grafting. (3) Enzymatic method. This method uses enzymes to degrade and change the properties of guar gum. (4) Metal cross-linking method. Mainly use the cross-linking property of guar gum. The adjacent cis hydroxyl groups on the main chain of guar gum can form a gel with boron and some transition metal ions, such as titanium, zirconium, etc. In addition, there are modifications using a combination of multiple methods, such as hydroxypropyl guar gum and acrylamide grafting.

In addition to guar gum, our country also has strong advantages in the production and research of fenugreek gum, the structure of which is shown in Fig. 6.3.

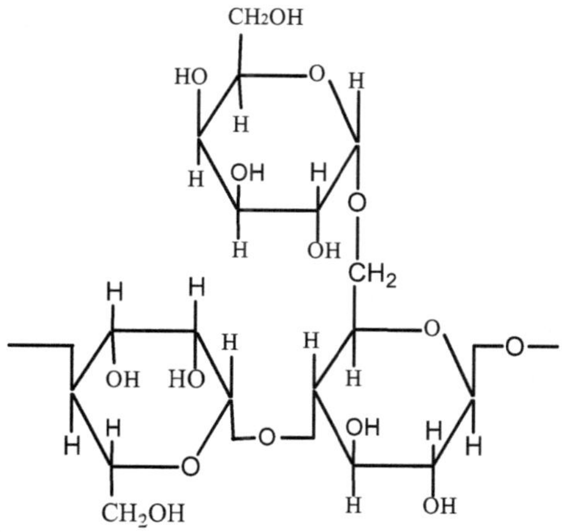

Fig. 6.3 Structure of carrageenan

6.2.3 Applications

Galactomannan, due to its good water solubility and cross-linking properties, can be used as a thickener, stabilizer, and adhesive, and is widely used in industries such as petroleum drilling and extraction, food and medicine, textile dyeing, mining, military explosives, daily-use ceramics, architectural coatings, wood processing, papermaking, and pesticides. Foreign research on galactomannan gum has a history of nearly a hundred years, and China began to study and use it in the 1970s. In recent years, due to the improvement of plant gum extraction technology and the continuous expansion of application technology and fields, this research field has attracted much attention.

Most coagulants are mainly used to treat suspended solids and colloidal impurities in water, but it is difficult to achieve a high removal rate for dissolved solids in water. In industrial water treatment to remove metal ions, ion exchange technology is an economical and effective method compared to chemical precipitation and solvent extraction technologies. Chelating resins have excellent ion exchange capabilities and are widely used. However, due to the rising prices of products such as petroleum, some researchers have begun to use galactomannan as a raw material, introducing sulfonic acid functional groups to make cheap and readily available cation exchange resins, which are also hydrophilic flocculants. The cationic polysaccharide obtained by introducing quaternary ammonium groups into galactomannan can be used as a flocculant for mineral processing and water treatment.

Galactomannan gum can be used to remove the sludge from potash mines. This process removes sludge in saturated brine, and it takes about 0.005 ~ 0.5 pounds of galactomannan gum to treat each ton of ore, and about 0.01 ~ 1.0 pounds of polyamine is needed as a collector to collect diatomaceous earth. The treated ore can be removed from the floating diatomaceous earth by froth flotation.

Rutenberg invented a method for preparing guar gum. Guar splits (containing 78 ~ 82% galactomannan, a small amount of protein, inorganic salts, water-soluble gums, cell membranes, and residual seed coats and embryos) are hydrated with sufficient water at 5 ~ 95 °C for at least 15 min, so that its water content reaches 25 ~ 80%. The hydrated guar splits are extruded and ground into a dispersed state in a hammer mill, and then dried in an air dryer at 100 ~ 125 °C for 10 ~ 15 min, so that the water content is less than 20%, to prepare fine-grained, high-viscosity guar gum. The resulting product can be used as a thickener and stabilizer in the textile, food, cosmetics, pharmaceutical and other industries, a pulping additive in the paper industry, and a fluid loss control agent and flocculant in drilling.

Guar gum is widely used in the paper industry. Because guar gum has a structure similar to cellulose, it can quickly adsorb on the surface of fibers and provide a large number of hydroxyl groups for the formation of hydrogen bonds between fibers, thereby effectively promoting the binding between fibers. In addition, guar gum can also improve the retention of fillers and fine fibers, while avoiding poor flocculation between fibers and improving the uniformity of the paper sheet. In the paper industry, the use of gelatinous borosilicate in combination with high molecular weight synthetic polymers and cationic starch or guar gum can improve paper performance. Guar gum used in cigarette paper has excellent retention, filtration, and enhancement effects, and the paper made does not have an odd smell when burned, meeting the needs of cigarette paper. At present, in major domestic cigarette paper factories in China, guar gum has basically replaced starch as a strengthener, retention aid, filter aid, and paper surface performance improver. The etherification reaction of guar gum with tertiary amines or quaternary ammonium salts can give guar gum a certain charge density, making it easier to interact with negatively charged fibers and fillers. Mixing cationic guar gum and cationic starch in a certain proportion in advance, and using them after pasting, the enhancement of cationic starch and the retention and filtration effects of cationic guar gum can be effectively exerted, and it is more important that it eliminates the harm of "dust" (guar gum powder) to the human body and the disadvantage of large water consumption when using cationic guar gum alone. Mixing cationic or amphoteric guar gum with silicic acid colloid in an appropriate proportion becomes a good paper strengthener, while also significantly improving the retention and filtration effects, reducing the pollution problem of papermaking white water.

The United States has a long history of using guar gum to treat wastewater from paper mills. In the process of producing products such as tissues, toilet paper, cigarette paper, and facial tissues, the products generally contain fibers from wood pulp, do not contain clay fillers, and may contain calcium carbonate only when producing cigarette paper. In the papermaking process, white water circulates in the device, and a considerable part of the fiber material is recovered through the saving device. The excess white water is factory wastewater. In order to protect water resources and reduce the high SS and BOD in these wastewaters, these factories must treat the wastewater before discharging it. Treatment methods include dissolved air flotation, and sometimes the Sven-Pedersen method is used to remove fine fibers from wastewater. Cationic guar gum has a good effect on the removal of this type of wastewater.

Cationic guar gum can cause fiber flocculation through hydrogen health and electrostatic attraction adsorption bridging. The anionic and hydration characteristics of the fiber also promote this. The resulting hydrated flocs are subjected to the pressure of saturated air water, and bubbles are released under normal pressure and float up. After this step of treatment, the suspended solids in the wastewater are generally less than 20 mg/L, and the BOD is also within an acceptable range. The cationic guar gum used is in powder form. For example, there are two commercial products, Jaguar C-13 (produced by Celanese Plastics and Specialties. Company) and Gendriv(produced by Henkel Company), comply with the regulations of the U.S. Food and Drug Administration (FDA). Depending on the volume of wastewater to be treated, the dosage of these products is generally 1 ~ 5 mg/L. The powder is usually prepared into a 0.25 ~ 0.50% concentration using softened water or factory supplemental water. It should be hydrated for at least 1 h before use, and then diluted to 0.05% or lower concentration with circulating water or process water before treating wastewater. When purifying wastewater, natural and synthetic polymers and other products are not as effective as cationic guar gum.

Most iron ore processing involves the use of large amounts of water. The processing techniques include gravity separation, magnetic separation, and froth flotation. The resulting tailings must be separated from the water, allowing 85 ~ 95% of the water to be recycled, while minimizing pollution caused by surface runoff. In the United States and Canada, most processes add polymers to the tailings thickener to recycle water and discharge fine solid particles and coarse tailings in the tailing's disposal area. Around 1977, a large iron ore plant in Canada began using a cationic guar gum with low charge density and very low molecular weight (below 200,000 Da). This product was replaced by a high-charge density polyamine, used at about 0.5 ~ 1.0 mg/L. Its use is not only for solid–liquid separation of mucus but also effective enough for water purification, allowing the iron ore concentrate to be well filtered without debris on the filter membrane. Therefore, liquid polyamine replaced it due to its easy operation and low cost. Most existing processes in the United States and Canada use synthetic substances such as polyamines and polyacrylamides. Some factories in the Mesabi Range combine lime and alkalinized starch. The dosage of starch is 0.2 ~ 0.5 pounds/ton (solid).

In the wastewater treatment process of natural soda ash production, guar gum is also used. Natural polyelectrolytes such as non-ionic guar gum can be used, and sometimes cationic guar gum is used to flocculate clay impurities, purifying the saturated carbonate wash solution before filtration and recrystallization. After roasting and digestion, heat is added, and flocculant is added to the feed of the purifier, where the clay flocculates and settles out. The unwashed residue is concentrated and transported to the disposal area as waste. The digested liquid is recrystallized to form soda ash. The key point is that the flocculant used for purification does not interfere with filtration or recrystallization. This is an advantage of using guar gum, while many synthetic substances can have adverse effects on certain minerals. But a major drawback of these natural polymers is that they cannot be prepared in saturated brine. They are generally prepared in 0.1 ~ 0.5% softened water or partially saturated brine and then used at a dosage of 4 ~ 10 mg/L based on the digestate. This results in a lower concentration of soda ash solution, so further evaporation is needed in the recrystallization process, consuming more energy.

In the past few decades, the dispersed air flotation method has been increasingly valued for the production of smaller coal particles. Correspondingly, water recycling is increasingly carried out in coal washing plants to minimize the amount of water discharged and meet increasingly stringent standards for pH value, BOD, iron ions, ferrous ions, and turbidity. In addition, tailings need to be dewatered to reduce their volume and the amount of leachate. Since the early 1960s, the application of polymers has become increasingly widespread, especially as flocculants in the refinement of tailings, filtration, and centrifugation processes. In the early 1960s, in West Virginia and Pennsylvania, flocculation mainly used coarse corn and potato starch, which needed to be heated to 82 °C and prepared into a concentration of 2.0 ~ 5.0%. The dosage is 0.25 ~ 1.0 pounds/ton (dry solids). Its performance still has many areas for improvement. In the early and mid-1960s, coarse starch was replaced by pregelatinized potato starch, which was later mixed with polyacrylamide. Later, most coal washing plants used anionic and cationic polyacrylamide. Due to the use of higher concentration low-flow concentrators, synthetic substances only need 1/5 or less of the original amount, and the settling speed is faster, so the use of natural organic matter gradually decreases. In the western United States, due to the increase in open-pit mines containing a large amount of clay, reports on the use of non-ionic guar gum and anionic polyacrylamide have also increased, claiming to reduce costs and have better treatment effects on fine clay. These systems require preparation and dispersion of two sets of devices.

In the United States, uranium ore comes from sandstone and shale, such as in Colorado, Wyoming, New Mexico, Texas, and Utah. In South Africa, uranium can be obtained from gold mines, and in Canada, uranium can be obtained from titanium uranium mines. After being washed with acid or carbonate, solid–liquid separation is carried out, such as countercurrent decantation (CCD) in a thickening pool or filter pool. Solid residues can be separated from the uranium-containing mother liquor for further treatment to recover uranium. Solid residues are disposed of at tailings points, and radioactive waste may be lost due to seepage. The use of guar gum and its derivatives alone or in combination with polyacrylamide can achieve solid–liquid separation. Other natural polymers are not as effective as polyacrylamide, and only Canada, the United States, and South Africa use gelatin, which has good clarification effects. Since gelatin can promote fungal growth, care should be taken when using gelatin. Guar gum is very effective as a filter aid because the flocs it forms are very regular in shape. Synthetic substances produce a large amount of rapidly settling flocs that do not leak water, so they can be used as good filter materials for uranium mines. The dosage used in concentration is between 0.1 and 1.0 pounds/ton, and the dosage in filtration is 0.1 ~ 1.5 pounds/ton (dry solids). These substances are generally prepared in softened water, with a concentration of 0.25 ~ 0.50%, hydrated for at least 1 h before use, and diluted to 0.1% for filtration, or diluted to 0.05% or lower concentration before use for concentration. Process water is generally used for dilution. In many factories, guar gum and polyacrylamide are mixed to improve flocculation effects or reduce costs. Some factories add guar gum and polyacrylamide in sequence. This is a very effective method, but it requires two systems to prepare the flocculant.

In the textile industry, dyeing and printing concentrates help suspend dyes, producing clear imprints and hues. These concentrates are dispersed and enter factory wastewater along with other chemicals. The wastewater contains a large number of suspended solids and a high concentration of organic matter, which must be treated. In a factory in the United States, alum and anionic polyacrylamide are used together before filtering in a mixed media gravity filter pool, causing the activated sludge to flocculate for clarification. The BOD and suspended solids indicators of the effluent achieved ideal results. It has been found that replacing aluminum salt/anionic polyacrylamide with a cationic guar gum has superior performance and overcomes the problem of aluminum salt damaging anaerobic microorganisms.

Seawater is an indirect natural raw material for producing magnesium metal in refractory materials. Various methods can be used to precipitate magnesium hydroxide at high pH values. If it is precipitated only in the form of magnesium hydroxide, the settling rate is too slow, and it is difficult to produce the density required for the preparation of ideal refractory materials through sintering. Therefore, natural polymers such as Jaguar MD7A (a non-ionic guar gum) can be used, with a dosage of 0.1 ~ 0.3 pounds/ton (solid). The overflow liquid in the concentration pool needs to be returned to the sea, so it should be clarified. In recent years, high molecular weight anionic polyacrylamide has replaced guar gum. In the production process of magnesium metal, the concentrated sludge of $Mg(OH)_2$ is redissolved and electrolyzed to produce magnesium cathode. In this case, seawater can be considered as waste, and the recovery of magnesium from seawater is considered as waste treatment.

In the process of treating copper oxide or roasted copper ore with leaching, purification, and electrolytic metallurgy, a waste product accumulates in the tailing's disposal area. This waste comes from the acid leaching waste liquid of copper oxide, and solid–liquid separation is carried out in the concentration pool and filter pool after leaching. The use of natural polymers such as guar gum alone or in combination with polyacrylamide can achieve these purposes. Using guar gum in the concentration pool can produce clear overflow liquid. This overflow liquid can be used for electrolytic copper after iron removal by lime precipitation (more modern methods use solvent extraction to purify the solution). The use of guar gum and polyacrylamide to filter high-density low-flow liquid can achieve good results, and the residue is deposited in the tailings area. One advantage of using guar gum for filtration is that it can completely wash the filter cake, thereby reducing the metal content in the tailing's leachate. Many processes use iron filings to recover copper. The dosage of guar gum used in the concentration pool is 0.05 ~ 0.3 pounds/ton (dry solid), and the dosage in the filter pool is 0.1 ~ 0.2 pounds/ton. The concentration of the reserve solution prepared on the recycled wasteland is 0.50%. It needs to be diluted before use, diluted to 0.05% in the concentration pool, and diluted to 0.1% in the filter pool.

In the process of alkali metal ore treatment, guar gum is often used in the removal of talc and carbonaceous shale. When concentrating alkali metal ores containing copper, lead, zinc, nickel, and iron (pyrite), the presence of talc and carbonaceous shale vein ore makes it difficult to achieve the desired metal content through current treatment methods (for example, using thiol-containing substances for froth flotation). The use of natural polymers and their derivatives such as guar gum, locust bean gum, carboxymethyl cellulose, and starch, or their combined use, is effective in reducing the talc and carbonaceous shale components in the ore in most cases. In some cases, they are not completely equivalent, but the use of one or a combination of these substances can work in all situations. By reducing the vein content and adding it to the waste, they can be placed in the tailings area. There is currently no method to predict which product works best in each unit consumption. It can generally be estimated through laboratory flotation tests, and then attempted to be applied to the factory. In most cases, mechanical stirring in water should be performed to prepare the product. Starch, guar gum, and carboxymethyl cellulose are generally prepared into 0.5 ~ 1.0% solutions at room temperature, allowing hydrolysis for several hours before use. Add 0.1 ~ 1.5 pounds/ton (dry material) to the coarse hopper, and 0.1 ~ 0.5 pounds/ton to the fine hopper. There is usually no need to dilute the reserve solution. The feeding point is generally set before the collector. The adjustment time is 0 ~ 5 min. The effect of synthetic materials is not as good as natural polymer products.

There is a special problem in the process of transferring radioactive waste to the landfill for disposal. Because these wastes are usually in the form of acidic liquids, there is a potential danger of leakage on the way. These liquids need to be partially solidified with stable liquid gels. Since natural polymers such as guar gum can form stable colloids, they can be considered for use. These gels may not remain stable, but they are easy to transport when necessary.

In drilling engineering, there are many plant gums used at home and abroad to improve the viscosity of non-clay washing fluids, such as: guar gum, hydroxypropyl guar gum, konjac gum, tianjing, drilling powder, etc. At the same solution concentration, the thickening effect of hydroxypropyl guar gum is the best, and its funnel viscosity, apparent viscosity, and plastic viscosity greatly exceed those of other plant gum solutions. There are already some domestic companies in China that grow and produce hydroxypropyl guar gum, with abundant raw material sources and stable and reliable performance.

The seeds of the *Strychnos patatorum* plant can be used as flocculants for water treatment in some third world countries. Researchers have extracted two polysaccharides from the plant that can produce flocculation effects, namely a mixture of galactomannan and galactose 1:1.7, and used them for the study of flocculation performance of artificial kaolin blended water. However, it is not yet conclusive which of these two polysaccharides is the active ingredient.

China has also made significant progress in the production and research of tianjing gum. Huang Qihua et al. first studied the chemical modification of tianjing gum and prepared carboxymethyl type, hydroxyethyl type, and carboxymethyl hydroxyethyl tianjing gum. Later, a series of modified tianjing gums were produced, which were used in petroleum fracturing fluid, papermaking, wastewater treatment, ceramics, printing and dyeing industries etc.

6.3 Cellulose

6.3.1 Overview

Cellulose is one of the earliest studied and most abundant natural polymers closely related to humans. Cellulose mainly comes from plants and is the primary structural material in higher plants. This polymer accounts for over a third of the total composition of higher plants. Through photosynthesis, plants can produce billions of tons of cellulose each year. Besides the plant kingdom, cellulose is also found in the animal kingdom, for example, some marine organisms have animal cellulose in their outer membranes. There are also reports of cellulose existing in the outer space. China was the first to recognize the important functions of cellulose and invented papermaking, which greatly promoted human culture.

6.3.2 Structure and Properties of Cellulose

Cellulose is a carbohydrate composed of glucose units with the empirical formula $C_6H_{10}O_5$. It is the primary structural unit of cell walls or wood or other higher plants.

Cellulose is a highly dispersed, high molecular weight polymer, connected by long chains of D-glucose units using β1-4 glycosidic bonds. Figure 6.4 shows the structure of cellulose. Each glucose unit contains three polar hydroxyl groups—one primary group and two secondary groups—and has a multi-ring structure, making the molecular chain rigid. It has a high degree of regularity in its structure and forms hydrogen bonds between molecules, filling the gaps. The large molecules are amorphous in equilibrium, but after orientation, they can have a considerable degree of regular crystalline structure. The polymer has a high density of convergence, so it is insoluble in organic solvents and can only be dissolved in special solutions like cuprammonium. Although it can also be dissolved in concentrated solutions of sulfuric acid or zinc antimony chloride, hydrolysis reactions occur simultaneously.

Cellulose derivatives include nitrocellulose, cellulose acetate, methyl cellulose, carboxymethyl cellulose, hydroxyethyl cellulose, and hydroxypropyl cellulose. The most commonly used in water treatment is carboxymethyl cellulose.

6.3.3 Applications

Cellulose can be esterified or etherified to produce many widely used cellulose chemicals. For example, nitrocellulose can be used to manufacture photographic film, spray paint, and explosives; cellulose acetate is used to make photographic film, magnetic

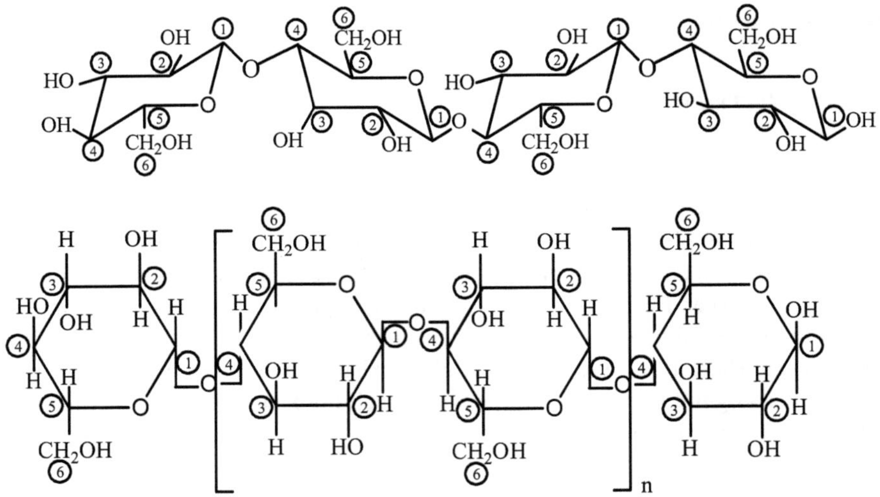

Fig. 6.4 Structure of cellulose

tape, and cigarette filter materials; cellulose xanthate is used to make so-called "cellophane," which is widely used as a packaging material in the food and light industries. It is also the raw material for the so-called "artificial silk" or "artificial cotton" (i.e., viscose fiber); cellulose ethers, commonly used are methyl or ethyl cellulose, carboxymethyl cellulose, and hydroxyethyl cellulose. Especially the latter two cellulose derivatives, due to their unique viscosity, emulsification, and surface activity, have been widely used in many sectors such as detergents, oil drilling, papermaking, textiles, food, paint, cosmetics, environmental protection, and more.

The most commonly used in water treatment is carboxymethyl cellulose, a very important water-soluble cellulose derivative. Because its acid solubility is poor, the commonly sold product is sodium carboxymethyl cellulose, also known as CMC, which was initially used in the food industry and is a water-soluble cellulose anionic ether.

Carboxymethyl cellulose was first produced by Jansen of Germany in 1918 and patented in 1921. During this period, its application research was quite active, and several quite enlightening patents were published. During World War II, Germany used carboxymethyl cellulose to synthesize detergents. The Hercules Company produced carboxymethyl cellulose for the first time in the United States in 1943 and produced refined products in 1946, which were recognized as safe food additives. This led to companies like Dupont, Buekye, and Wyandott joining the product line. Japan began producing carboxymethyl cellulose in 1944, and China started production in Shanghai in 1958. Carboxymethyl cellulose is a product with large production, multiple specifications, and rapid development among China's cellulose ether products. Currently, it is widely used in industries such as petroleum, textiles, papermaking, food, medicine, and daily chemicals. Carboxymethyl cellulose is obtained by the reaction of chloro (substituted) acetic acid or chloro (substituted) acetic acid sodium salt with alkali cellulose. There are currently three main production methods: water medium, solvent, and solution methods.

Researchers have made flocculants by compounding polyferric sulfate with carboxymethyl cellulose to treat hospital wastewater, which has better effects in flocculation, adsorption, and deodorization than using alone. Under the same experimental conditions, it has earlier flocculation time, larger flocs, faster settling speed, and higher removal rates of suspended solids and COD than other compounded products. The compounding of PFS and CMC, due to the linear structure of PFS, can adsorb colloidal particles and other impurities in the wastewater, and CMC can form a transparent gel-like substance in water, which can wrap the polymer formed by PFS, forming larger alum flowers, which helps coagulation.

The wastewater produced in the CMC production process contains a certain amount of CMC. If it is recycled, it can not only artificially reduce the COD and BOD of the wastewater but also improve economic benefits. Some researchers chose trivalent aluminum ions to separate CMC from wastewater, and then applied it to industrial wastewater treatment. They found that when it was compounded with a certain cationic organic high polymer flocculant at a certain ratio, it showed a certain synergistic effect and showed good application prospects.

Cationic carboxymethyl cellulose is a cationic quaternary ammonium salt polymer generated by esterifying carboxymethyl cellulose as a raw material under acid catalysis, and the esterification product reacts with iodoethane. The use of cationic carboxymethyl cellulose products and a small amount of aluminum sulfate and polyacrylamide can effectively remove COD, color, and suspended solids in drilling wastewater. The treatment of mining drilling wastewater with this synthetic substance using an intermittent coagulation device has a significant effect and greatly reduces costs.

Modified cellulose can be produced by reacting etherified sesame stalk cellulose with a quaternary salt, as shown in Eq. 6.8. The modified cellulose has adsorption properties for aniline in wastewater, and the removal rate of aniline can reach 93.2%.

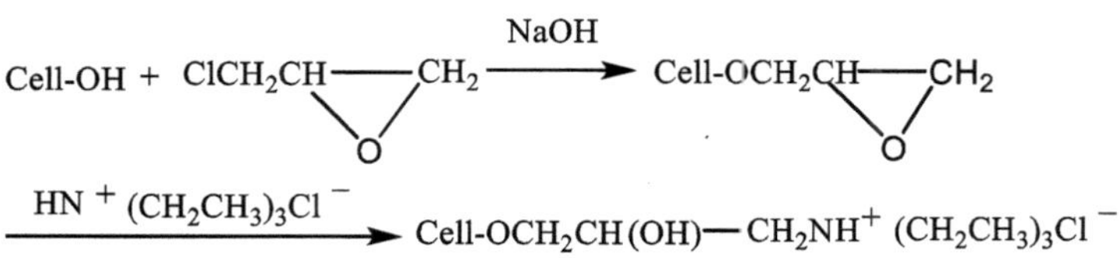

(6.8)

Hydroxypropyl methyl cellulose can be used as a coating film material for flocculants such as polyaluminum chloride. The coated flocculant has good moisture resistance and rapid solubility in water and still maintains a good flocculation effect on organic pollutants in wastewater.

Cai et al. used potassium permanganate as an initiator to graft copolymerize starch or microcrystalline cellulose with acrylamide, preparing cationic polyacrylamide and amphoteric polyacrylamide. The operation is as follows: the graft copolymer is hydrolyzed with sodium hydroxide, converting part of the amino groups in acrylamide into negatively charged carboxyl groups or sodium acrylate, thus forming a graft polyacrylamide containing acrylamide and sodium acrylate units; the aforementioned product is reacted with ethylene diamine alcohol, further introducing cationic groups into the hydrolyzed polyacrylamide, thus obtaining cationic/amphoteric graft polyacrylamide; at least one additive is added, which contains water-soluble inorganic salts and non-ionic surfactants, to improve the viscosity of the product. This method is easy to obtain raw materials, low production cost, simple process, mild reaction conditions, the product has good flocculation performance and wide adaptability and is an ideal sludge dewatering agent and wastewater treatment agent.

For a long time, the removal of heavy metal ions in wastewater has been a hot topic of research. In recent years, researchers have gradually begun to focus on the use of naturally modified high polymer compounds to remove heavy metals in wastewater, which is low cost and does not cause secondary pollution, and some progress has been made. The use of insoluble starch xanthate and cellulose xanthate to remove heavy metal ions in water has a good effect. The insoluble starch xanthate synthesized with highly cross-linked starch as the skeleton has good performance, a simple preparation process, large chelating capacity, and can treat a variety of metals, but its properties are unstable, difficult to regenerate, and the cost of using highly cross-linked starch is still high, so it is limited in application.

The basic principle of preparing cellulose xanthate is the reaction of alkali cellulose with carbon disulfide. Due to the existence of various parallel reactions, the reaction mechanism of xanthate is very complex. Firstly, carbon disulfide with low reactivity reacts with sodium hydroxide to produce highly reactive ionized water-soluble substances—disulfide carbonate, which spontaneously reacts with cellulose to produce cellulose xanthate.

During the process of treating heavy metals with cellulose xanthate, the amount of cellulose xanthate, pH value, and reaction time all have a significant impact on the treatment effect, and these factors should be controlled during use. The cellulose xanthate residue after wastewater treatment can be landfilled, or it can be soaked with sulfuric acid or nitric acid for the recovery of heavy metals, achieving the purpose of comprehensive utilization of resources. The cellulose xanthate synthesized from soybean dregs as the main raw material has a high removal rate for heavy metal ions such as Cu^{2+}, Pb^{2+}, Ni^{2+}, Fe^{2+}, Hg^{2+}, Cr^{6+} under appropriate pH values, and the residual concentration of each ion in the wastewater after treatment can meet the national industrial wastewater discharge standards.

Cellulose was prepared using straw as the skeleton material, modified, grafted with ethylenediamine, and a nitrogen-containing cellulose ion exchanger was synthesized. This modified cellulose has a good removal effect on copper ions, and overcomes the difficulties of regeneration and instability of insoluble starch xanthate, and the raw material source of modified cellulose is very wide, various crop stalks, bamboo, sawdust, waste paper scraps can be used, the preparation process is simple, the investment is small, and it is easy to promote.

Using a 5% sulfuric acid solution as a coagulant, a mixture of cellulose and chitosan is coagulated in a 6% sodium hydroxide and 5% thiourea aqueous solution, preparing a biodegradable cellulose/chitosan bead. The adsorption performance of these beads for heavy metal ions in aqueous solutions was studied. Experimental results show that these cellulose/chitosan beads can effectively adsorb Pb^{2+}, Cd^{2+}, and Cu^{2+}. At low concentrations, the adsorption selectivity order of these heavy metal ions is $Pb^{2+} > Cd^{2+} > Cu^{2+}$, the adsorption equilibrium time is about 4 ~ 5 h, and the desorption time is 5 ~ 15 min. Moreover, after treatment with hydrochloric acid solution, the regeneration rate of these beads can reach 98%. The adsorption mechanism is mainly based on the coordination adsorption model and the affinity of the hydroxyl groups on the beads for metal ions. The production process is simple, and the product is an environmentally friendly water treatment product for removing and recovering heavy metal ions.

Using epichlorohydrin as a crosslinking agent, β-cyclodextrin can be loaded onto viscose fibers in an alkaline medium to synthesize functional cellulose fibers loaded with β-cyclodextrin, which can enrich inorganic heavy metal ions, aniline, phenol, and hydroquinone in simulated water samples. The cellulose fibers loaded with β-cyclodextrin have good enrichment effects on inorganic heavy metal ions (Cu^{2+}, Cd^{2+}, Pb^{2+}), aniline, phenol, and hydroquinone, with enrichment capacities of 0.2428, 0.2954, 0.3438 mmol/g for Cu^{2+}, Cd^{2+}, Pb^{2+} respectively, and 1.154 mmol/g, 1.117 mmol/g, 0.9576 mmol/g for aniline, phenol, and hydroquinone respectively.

Humic acid substances extracted from weathered coal, containing active groups such as hydroxyl, carboxyl, and quinone, can be widely used in various fields such as industry, agriculture, medicine, and environmental protection. However, the effect of using humic acid substances alone as a treatment agent for heavy metal ions in wastewater is not ideal. Firstly, the sedimentation rate is slow, and secondly, the formed sediment has a high-water content and is not easy to separate from the water, affecting the treatment effect of heavy metal ions in wastewater. Some researchers have reacted sodium humate with cellulose to form a composite, and studied the complexation performance of the composite for harmful heavy metal ions in wastewater. It was found that the complexation effect of the composite of humic acid and cellulose is better than that of humic acid alone. This composite has a fast sedimentation rate and good complexation performance. Further research is expected to become a valuable heavy metal wastewater treatment agent.

6.4 Chitosan

6.4.1 Overview

Chitin is a type of organic matter that is abundant in nature, mainly derived from animals. Chitin is a biopolymer similar in structure to cellulose and is an important component of the exoskeletons of many lower animals, especially shrimp, crabs, and insects. It also exists in lower plants, such as fungi and algae, and is widely distributed in their cell walls. According to research data, the amount of chitin synthesized by organisms in nature each year can reach 10 billion tons. Chitin is the largest nitrogen-containing natural organic compound on Earth, second only to cellulose in terms of storage in natural polymers, and therefore holds a very important position. Chitin can also be referred to as chitinous, chitin, chitosan, crab chitin, and chitin protein. The utilization and research of chitin began nearly a hundred years ago. Since the 1970s, research, production, and application of chitin and its derivatives have become very active. From 1977 to 1988, four international academic conferences on chitin and chitosan were held consecutively in Japan, the United States, and Norway. These conferences discussed the current status of chitin production, the prospects and existing problems of products and derivatives made from it, thus confirming the wide application of chitin and consolidating its highly regarded position in the world market.

Chitin is an amino polysaccharide synthesized by organisms in nature. It is abundant in storage and can be biodegraded because it is a natural polymer synthesized by organisms. It is safe, non-toxic, has good biocompatibility, and has stable chemical properties. It has many unique advantages and is highly valued for its applications in medicine and functional foods. It is known as the "sixth life element" of the human body.

6.4.2 *Structure and Properties of Chitin*

Chitin is a polysaccharide formed by the connection of N-acetyl-2-amino-2-deoxy-D-glucose in the form of a β-1-4 glycosidic bond. Its structure is shown in Fig. 6.5. It is a polysaccharide of N-acetyl-D-glucosamine. From the structural formula, it can be seen that chitin is very similar to cellulose. If the acetylamine (CH_3CONH–) on the sugar base is replaced with a hydroxyl group (HO–), it becomes cellulose. Due to the action of O…H–O– type and O…H–N– type hydrogen bonds, there is an orderly structure between the large molecules of chitin. Due to different crystalline structures, chitin exists in α, β, and γ polymorphs. α-chitin is the most abundant and stable, but it transforms into other forms during the processing and separation process. Due to the strong hydrogen bond interaction between large molecules, chitin begins to decompose only when heated above 200 °C. If the acetyl group in the chitin molecule is removed, it becomes chitosan (or chitooligosaccharide), also called deacetylated chitin, as shown in Fig. 6.6. Generally speaking, if more than 55% of the N-acetyl group is removed, it can be called chitosan.

Chitin and chitosan can be considered as products where the –OH hydroxyl group at the C_2 position of cellulose is replaced by –CH_3CONH acetylamine (chitin) or –NH_2 amine (chitosan). The reason why chitin and chitosan have important theoretical research significance and commercial value lies in their molecular structure and composition characteristics. It is the only alkaline polysaccharide in nature and is also the most abundant nitrogen-containing organic matter besides proteins. Its nitrogen content (6.89%) is about 5 times higher than that of artificially synthesized nitrogen-containing cellulose derivatives (1.25%).

Chitosan is a white or grayish-white, semi-transparent flake-like solid. Due to the very stable crystal structure of the large chitosan molecule, the rigidity of its

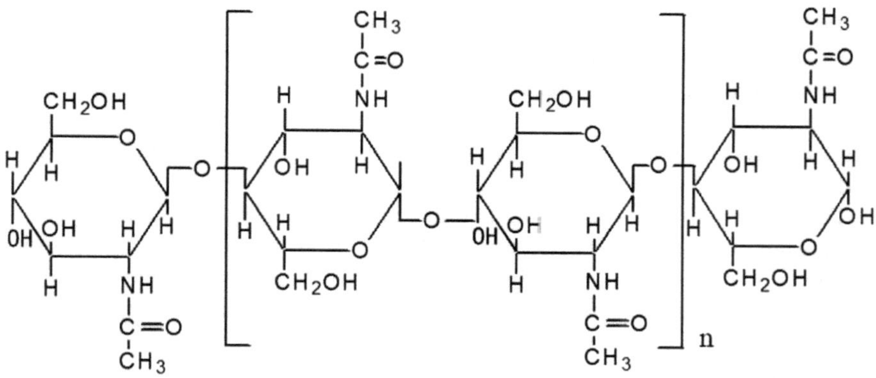

Fig. 6.5 Structure of chitin

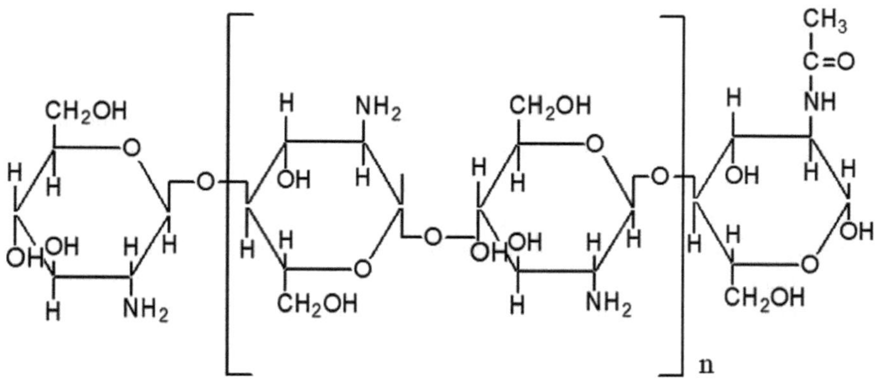

Fig. 6.6 Structure of chitosan

molecular chain and its strong hydrogen bonds gives it stable physical and chemical properties. Chitosan is insoluble in water, dilute acids, and general organic solvents, but can dissolve in certain coordination compound solvents. It does not swell in alkali, swells violently in acid, and degrades when dissolved in concentrated inorganic acid. Chitin is a white or grayish-white, slightly pearlescent, semi-transparent flake-like solid. It is insoluble in water and alkali solutions. Chitin can dissolve in low-acidity aqueous solutions, such as hydrochloric acid, acetic acid, benzoic acid, cycloalkanoic acid, etc., so it is also called soluble chitosan, while chitosan does not have this solubility and is also called insoluble chitosan. However, even in dilute acid, chitin will slowly hydrolyze, the viscosity of the solution gradually decreases, and finally completely hydrolyzes into glucosamine, so the chitin solution is best used as soon as it is prepared and should not be stored for a long time. Moreover, once chitin encounters an alkaline solvent, it will immediately form a gel, and chitin cannot dissolve in organic solvents, which brings some difficulties to research and application. Whether it is chitosan or chitin, it completely hydrolyzes into glucosamine in hydrochloric acid at 100 °C, and hydrolyzes into glucosamine, chitobiose, chitotriose… and other low molecular weight polysaccharides under milder conditions. Because chitin contains free amino groups, it can bind acid molecules. It is the only alkaline polysaccharide in natural polysaccharides, so it has many special physical and chemical properties and physiological functions. As an industrially valuable product, chitin, the N-deacetylation degree must be above 70%.

The essence of chitin dissolving in dilute acid is that the chitin molecular chain has free amino groups. There is a pair of unbound electrons on the free amino group. This amino group is weakly alkaline in aqueous solution and can bind a hydrogen proton from the solution, thereby making chitin a positively charged polyelectrolyte, or a high molecular cation, which is also a high molecular salt. These cations destroy the hydrogen bonds between chitin molecules and within molecules, making them soluble in water. Therefore, it is not actually chitin that dissolves in dilute acid, but chitin salt (polyelectrolyte) that dissolves in water.

The derivative reactions of chitin and chitosan molecules mainly occur on these functional groups, due to the presence of functional groups such as –OH hydroxyl, $-NH_2$ amino, pyran ring, oxygen bridge and other functional groups. Because chitosan molecules have hydroxyl, amino, imino and other functional groups, they are easy to undergo chemical reactions, which provides convenience for the modification of chitosan and chitin. Sulfation, grafting, cross-linking, hydroxyethylation, carboxymethylation, cyanoethylation, alkylation, nitration, phosphatization, deamination, etc. are performed on chitosan and chitin, giving them different special properties, thereby expanding the application field of chitosan and chitin.

The main quality indicator of chitin products is viscosity, and products with different viscosities have different uses. At present, domestically and internationally, products are divided into three categories based on viscosity: 1% chitin dissolved in 1% acetic acid aqueous solution, high-viscosity chitin has a viscosity greater than 100 cPa s; medium-viscosity chitin has a viscosity between 10 and 20 cPa s; low-viscosity chitin has a viscosity between 2.5 and 5 cPa s.

Chitosan and its derivatives have extremely strong hygroscopicity. The hygroscopicity of chitosan can reach 400 ~ 500%, which is more than twice that of cellulose. The hygroscopicity of chitosan is stronger than that of chitosan, second only to glycerin. During the hygroscopic process, the amino groups, hydroxyl groups, and other polar groups in the molecules interact with water molecules to hydrate, and the molecular chains gradually swell. With different pH values, the molecular chains change from spherical micelles to linear shapes. Chitosan and its derivatives have excellent oxygen permeability. They will degrade under various conditions and can be degraded to form amino glucose in concentrated inorganic acids. They can undergo biodegradation under the action of lysozyme, and hydrolysis will occur if left in water for a long time, some groups will fall off, and the glucose ring will open. The neighboring position of the free amino group of chitosan is a hydroxyl group, so it has the function of chelating divalent metal ions and shows various colors. It is a high-performance heavy metal trapping agent. The free amino group of chitosan has a high affinity for various proteins and can be used as a carrier for the immobilization of enzymes, antigens, antibodies, and other physiologically active substances.

Shrimp, crab shell washing → decalcification and inorganic salt removal → protein and fat removal → alkali removal, rinsing → washing, drying → Chitosan product → concentrated alkali treatment → washing, drying → crude chitosan product → purification → chitosan product

Fig. 6.7 Preparation process flow of chitosan and chitosan

The traditional preparation methods of chitosan and chitosan are shown in Fig. 6.7. Due to different raw materials and preparation methods, the relative molecular weight of chitosan and chitosan varies from hundreds of thousands to millions.

The production of chitosan generally adopts the traditional basic process of decalcification, deproteinization, and deacetylation using shrimp and crab as raw materials, which consumes a large amount of alkali and causes serious pollution. With the deepening of research, in order to solve a series of problems such as single raw materials and serious pollution, microbial methods, microwave methods, and biological preparation methods have been developed to produce chitosan and chitosan and their derivatives. Chitosan is the main component of the cell walls of most fungi and the main component of the extension part of the fungal hyphae. The formation of chitosan is closely related to the growth of fungal hyphae. Existing research mainly extracts from the waste mycelium of citric acid fermentation and can also be extracted from the mycelium filtered out of the waste liquid of penicillin and streptomycin fermentation. In recent years, domestic research and development have directly cultured fungi to produce chitosan or directly extracted chitosan technology.

Microwave radiation to promote organic synthesis is a new organic synthesis technology that emerged in the late 1980s. In the process of chitosan deacetylation to produce chitosan, microwave heating can be used instead of electric heating or steam heating to produce chitosan. In addition, enzymatic hydrolysis can be used, that is, specific or non-specific enzymes are used to degrade chitosan to produce chitosan oligosaccharides.

6.4.3 Applications

Chitosan and its derivatives, due to their excellent biocompatibility, biodegradability, and bioactivity, as well as their unique physical and chemical properties, have a wide range of applications in chemical engineering, pharmaceuticals, cosmetics, papermaking, printing and dyeing, agriculture, environmental protection, and as carriers for enzyme immobilization. The research and development of chitosan has become a globally recognized high-tech field and a promising emerging industry.

In 1977, Japan first used chitosan as a flocculant to treat wastewater. In the same year, the first international conference on chitosan was held in Boston, USA.

Since then, the application of chitosan in various fields such as food, environmental protection, and medicine has seen greater development.

In recent years, significant progress has been made in the application of chitosan as a flocculant and adsorbent in water treatment. Conventional organic or inorganic flocculants often have problems such as large usage, complicated operation, large amount of sludge production, high treatment cost, and the possibility of toxic residues after treatment, which can easily cause secondary pollution. Using chitosan to treat wastewater is safe, non-toxic, biodegradable, without secondary pollution, and effective, making it an ideal water treatment agent. Chitosan is a good complexing agent for many metal ions, phenolic compounds, natural and synthetic polyanions, and also has appropriate antibacterial effects. According to reports, Japan uses about 500 tons of chitosan for water treatment each year, mainly for water and sludge treatment, while the United States mainly uses chitosan for water supply and drinking water purification.

In the treatment of water and wastewater, due to the large number of free hydroxyl and amino groups distributed on the chitosan molecular chain, the amino groups are easily protonated in some dilute acid solutions, making the chitosan molecular chain carry a large amount of positive charge, becoming a soluble polyelectrolyte, acting as a cationic flocculant. The flocculation mechanism of chitosan mainly includes: (1) Bridging action: the flocculant molecule binds multiple particle molecules at the same time through ionic bonds and hydrogen bonds, thus playing the role of an "intermediate bridge", linking these particles together to form a network structure and precipitate. (2) Charge neutralization: the colloidal particles in water carry a negative charge. When the chain-like biomacromolecule flocculant or other hydrolysis products with a positive charge approach these colloidal particles, they neutralize some of the charges on their surface, causing the colloidal particles to destabilize and collide with each other and precipitate. (3) Group reaction: some active groups in the flocculant macromolecule react chemically with the corresponding groups of the flocculated substance, aggregating into large molecules and precipitating. There are many factors affecting the flocculation effect of chitosan, such as its properties (degree of deacetylation), the properties of the treated wastewater, including pH value, content of soluble and insoluble substances, ion strength, etc.; as well as the amount of chitosan used in wastewater treatment, the contact state between molecules, and the spatial resistance to flocculation.

Jiang Ting et al. used suspensions (including kaolin, montmorillonite, and natural clay) as research subjects, comparing chitosan with strong cationic polymer flocculants (C-109P polyacrylamide, Sanyo Chemical Industries, Japan), weak cationic polymer flocculants (C-809P polyacrylamide, Sanyo Chemical Industries, Japan), and anionic polymer flocculants (polyacrylamide, Lanzhou) under high, medium, and low turbidity. The experimental results found that the turbidity removal effect of chitosan is similar to that of strong cationic polymer flocculants. From the perspective of polyelectrolytes, chitosan is a weak cationic type, but it has the coagulation effect of strong cationic coagulants, which may be the result of its unique structure.

Chitosan has a strong coagulation effect on suspended substances. Compared with traditional alum and polyacrylamide as flocculants, chitosan has a better clarification effect. Zhou Yaping and others studied the flocculation behavior of three chitosans with roughly the same molar mass but different degrees of deacetylation (93, 78, 62%) on soapstone suspensions in the pH range of 3 ~ 6, and compared it with the non-ionic flocculant polyacrylamide, proving that chitosan with free amino groups has a good flocculation ability for soapstone suspensions, and the bridging mechanism may play a leading role in this flocculation process.

Research by Ravi Divakaran et al. shows that it is relatively difficult to flocculate kaolin suspensions with chitosan alone, but if the suspension contains a small amount of soluble humic substances, chitosan can easily flocculate and settle kaolin. This flocculation reaction is very sensitive to pH changes. In addition, they also reported for the first time the removal effect of chitosan on suspended titanium dioxide particles when the water contains humic acid.

Murcott et al. used chitosan or cationic starch as natural cationic coagulants for drinking water treatment, and added bentonite to the water to enhance conventional coagulation. This method has a good removal effect on particles and color in the water, and it was also determined that the optimal ratio of natural polymers to clay is between 1:5 and 1:20.

In drinking water treatment, chitosan has a good purification effect. When chitosan is used in combination with soapstone, a smaller amount can precipitate particles and harmful substances in drinking water, quickly clarifying the water quality.

Since humic substances in water can have a negative impact on water quality, it is crucial to remove soluble humic acids and their complexes with metal ions and organic pollutants by flocculation in drinking water treatment. Some researchers have used natural cationic polymers to neutralize the negative charge of humic colloids and form polyelectrolyte complexes. The interaction of humic acid with two soluble chitosan salts (chloride and glutamate) was studied through the charge of the formed complex and the effect of coagulation color removal under different pH values and polymer ratios. Within a wide concentration range, when the pH is close to the isoelectric point of the polyelectrolyte complex, both chitosan salts will flocculate, but if chitosan glutamate is used, less dosage is required.

Kind Code et al. have invented a method of treating drinking water by using water-soluble multivalent inorganic salts as coagulants, anionic or non-ionic polymers with a molecular weight higher than 2 million and good water solubility and dispersibility for bridging flocculation, and chitosan as a coagulant aid. During the treatment, microbial disinfectants, water-soluble alkalis, silicates extracted from clay or zeolites, food additives or nutrients are added, thereby quickly and efficiently purifying the water quality.

H. Ganjidoust et al. compared the removal effects of several synthetic polymers (non-ionic polyacrylamide, polyethyleneimine, hexamethylene bischlorohydrin polycondensate), natural polymers (chitosan), and inorganic coagulants (alum) on industrial papermaking wastewater, mainly using the color and TOC of papermaking black liquor as measurement indicators. The test results showed that the coagulation effect of non-ionic polymer PAM is not as good as that of cationic polymers. At a pH value of about 6, polyethyleneimine and hexamethylene bischlorohydrin polycondensate can produce good brown precipitates. After gravity settling for

30 min, their color removal rate reached 80%, while the TOC removal rate reached 30%. Alum has a better precipitation effect, its color removal rate is the same as that of polyethyleneimine and hexamethylene bischlorohydrin polycondensate, but it can remove 40% of TOC. Compared with synthetic polymers and alum, natural polymer chitosan has the highest color and TOC removal rates, reaching 90% and 70% respectively.

Jill Ruhsing Pan used powdered chitin separated from crab shells, deacetylated with 45% sodium hydroxide in a 100 °C water bath for 60 min, and the reaction was stopped in an ice bath. Subsequently, it was washed several times with deionized water until the pH of the suspension reached 7. The suspended particles were collected by filtration with a filter membrane and dried at 80 °C for 48 h. This method of preparing chitosan powder is different from the past, so it has unique properties. The modified chitosan was added to a 1% hydrochloric acid solution and stirred at 100r/min for 60 min until dissolved to make a 1% reserve solution. Before each experiment, the reserve solution was added to deionized water and gently stirred with a magnetic stirrer to obtain a coagulant solution of the required concentration. In addition, a coagulant mixture was prepared, that is, chitosan and polyaluminum chloride (PAC) were mixed in different mass ratios. The coagulation test compared the coagulation effects of modified chitosan, PAC, and a mixture of chitosan and PAC on turbid water. The results show that the modified chitosan requires the smallest dosage, and the flocs produced are large and settle quickly. Lowering the pH will reduce the optimal dosage, but the flocs produced will also become smaller and the settling speed will slow down. When the dosage is lower than the optimal dosage, increasing the intensity of rapid stirring will improve the coagulation effect.

Using chitin as a raw material, under alkaline conditions, carboxymethyl is introduced by reacting with monochloroacetic acid, and deacetylation is carried out simultaneously by hydrolysis to produce amphoteric chitosan that can be dissolved in dilute acid, dilute alkali, and water. Or, taking advantage of the property of the amine group in chitosan reacting with the aldehyde group to form Schiff base, an aldehyde containing a carboxyl group in the molecular structure is selected to produce amphoteric chitosan. These two types of amphoteric chitosans have many similar properties, and N-[(3′-hydroxy-2′,3′-dicarboxy)-ethyl] chitosan amine can also be produced by reacting the lactate of deacetylated chitosan under alkaline conditions with epoxyethane dicarboxylic acid.

Zeng Defang et al. have used discarded shrimp (crab) shells as the main raw material to prepare chitosan composite flocculants, and applied them to the primary enhanced treatment of several typical industrial wastewaters such as electroplating, tanning, printing and dyeing, and dye wastewater, with significant effects. This flocculant has a good flocculation effect when the pH value is 6 ~ 7. Compared with traditional flocculants, it can increase the removal rate of COD, SS and heavy metal ions by 10 ~ 20%, and reduce the cost by 40 ~ 60%. Its main economic and technical indicators are significantly better than traditional flocculants, and it has a good application prospect in industrial wastewater treatment. The researchers analyzed its flocculation mechanism. The linear molecular chain of chitosan contains multiple hydroxyl and amino groups, which can provide electrons to metal ions M^{n+} with empty d orbitals to chelate into stable internal salts, thereby removing harmful metal ions such as Al^{3+}, Zn^{2+}, Cr^{6+}, Hg^{2+}, Pb^{2+}, Cu^{2+} from the water. Moreover, the amino group can protonate with H^+ in the water to form a cationic polyelectrolyte, which can aggregate coarse and fine ions in the water into large flocs and settle down through electrostatic attraction and adsorption, thereby achieving the purpose of removing COD and SS in the water. In addition, adding an appropriate amount of sulfate to the chitosan composite flocculant can help the colloid destabilize and accelerate the sedimentation speed. Because the cationic chitosan can form a chemical bridge with the sulfate group, it neutralizes and reduces the surface charge of the colloidal particles, compresses the double electric layer of the colloidal particles to destabilize the colloidal particles, and is more prone to flocculation and sedimentation.

Li Yuling et al. used $CaCl_2$, $MgCl_2$, $Al_2(SO_4)_3$, polyvinyl alcohol, polyacrylamide and homemade acetic acid chitosan flocculant to flocculate and purify bentonite slurry. The results show that the homemade acetic acid chitosan solution has the best purification effect on bentonite. The whiteness of the purified bentonite reaches 78.6%, and the montmorillonite content reaches 90.1%.

Chitosan is obtained by chemically treating crude chitin, and it is further processed into a protein flocculant. This protein flocculant is used to treat protein impurities in starch saccharification solution, and the effect is good. Compared with the traditional neutralization technology, the removal rate of protein increases from 29.8 to 79.3%, and no impurities such as chlorides are introduced during the treatment process, which improves the purity of the purified sugar solution.

In addition, Japanese researchers have found that using chitosan as a flocculant can effectively remove carcinogenic and mutagenic substances such as halogenated hydrocarbons in the purification of tap water, and its removal effect is much higher than that of activated carbon and rayon. Using chitosan as a clarifying aid for hard water treatment, its performance is better than alum and polyacrylamide.

Chitosan and its derivatives can adsorb many metal ions and non-metal substances such as dyes, phenols, halogens, proteins, nucleic acids, amino acids, etc., through coordination and ion exchange. Studies have shown that the hydroxyl, amino, and other active groups in chitosan molecules, when the pH value is above weakly acidic, the adsorption and chelation of heavy metal ions increase with the increase of pH value. At a pH of 9, the adsorption rate of chitosan to Pb^{2+} can reach 100%. Therefore, it can be used for the enrichment, recovery, separation, and treatment of metals. The application of chitosan to recover copper from industrial wastewater has been industrialized. Proteins can be recovered from the wastewater of poultry, protein processing plants, starch factories, and MSG factories as feed for livestock and poultry. The suspended solids in the wastewater of the fishmeal industry are mainly composed of proteins, which can be recovered to optimize the production process. Using chitosan to coagulate fishmeal wastewater, followed by centrifugal separation, at a pH of 7.8, the TSS removal rate can reach 75%. Chitosan is also an effective flocculant for activated sludge and proteins. It is worth noting that the coagulation effect of chitosan is related to both the molecular weight and the object of coagulation. When treating activated sludge, high-viscosity chitosan has a good coagulation effect, and when treating suspended substances in cheese whey, low-viscosity chitosan has a good coagulation effect. Chitosan is used in sludge dewatering, and the effect is better when used with other coagulants.

Laurent et al. used chitosan solution in combination with another anionic natural polymer (such as xanthan gum, guar gum, alginate, etc.) to treat the processing wastewater of livestock, poultry, and fish, with significant effects. This type of wastewater contains a large number of fatty acids, proteins, fats, etc., which can be removed with natural flocculants, and the resulting sludge can be used as a safe animal feed, fertilizer, and fuel source, or it can be landfilled.

The application of chitosan as a flocculant in the food industry has its unique advantages. Most of the cationic flocculants currently used in industry are synthetic polymers with high toxicity and are generally not suitable for use in the food industry. Chitosan is non-toxic and tasteless, biodegradable, and will not cause secondary pollution, making it very suitable for use in the food industry. Turbid fruit juice mainly contains pectin, tiny solid particles, etc. Pectin is a polysaccharide aldehyde acid, which can produce electrostatic attraction with positively charged macromolecules and precipitate (the pH value can be adjusted with hydrochloric acid to facilitate pectin precipitation). The high-viscosity chitosan is a natural cationic flocculant, which has a strong coagulation ability for pectin and also has a strong adsorption effect on pigments, thereby clarifying the fruit juice and making the color lighter, which is very beneficial for improving the light transmittance of fruit juice. Currently, enzymatic methods or the addition of coagulant aids are also used to treat fruit juice, but this method not only has high production costs and long production cycles, but also produces secondary precipitation if the treatment of pectin is incomplete, making the actual operation process quite complicated. Using chitosan as a flocculant not only reduces production costs and shortens the production cycle, but also increases output and reduces unnecessary losses. At the same time, using chitosan as a processing aid in clarifying fruit juice can promote solid–liquid separation, remove acid and suspend solid particles from turbid fruit juice, enhance transparency, and improve the quality of fruit juice products. In addition, chitosan can also be used for fruit juice deacidification and removal of sediment in vinegar.

The adsorption capacity of chitosan is several times that of granular activated carbon, and the raw material is non-toxic, there is no secondary pollution, and the adsorption cost is low. Its phosphate ester derivative, phosphoric acid chitosan, can recover uranium from seawater, with an adsorption capacity of up to 2.6 mg/L. After adsorption, uranium can be easily recovered by dissolution with dilute sodium phosphate solution. Chitosan can also effectively adsorb radioactive nuclides in nuclear industry wastewater. Chitosan can also effectively remove CH_3HgCl, CH_3HgOAC, etc. from some industrial wastewater.

Using a 5% sulfuric acid solution as a coagulant, the mixture of cellulose and chitosan can coagulate in a 6% sodium hydroxide and 5% thiourea aqueous solution, and biodegradable cellulose/chitosan particles can be prepared. These cellulose/chitosan particles can effectively adsorb Pb^{2+}, Cd^{2+}, and Cu^{2+}. At low concentrations, the adsorption selectivity order of these heavy metal ions is $Pb^{2+} > Cd^{2+} > Cu^{2+}$, the adsorption equilibrium time is about 4 ~ 5 h, and the desorption time is 5 ~ 15 min. Moreover, after treatment with hydrochloric acid solution, the regeneration rate of these particles can reach 98%. Its adsorption principle conforms to the coordination adsorption model and the affinity of the hydroxyl group on the particles for metal ions. The production process of this method is simple, and its product is an environmentally friendly water treatment product for removing and recovering heavy metal ions.

Qu Rongjun et al. used diethylene glycol bisglycidyl ether (DGBE) as a cross-linking agent to synthesize a gel-type chelating resin with chitosan as the matrix, and studied its adsorption performance for metal ions such as copper, nickel, and cobalt. The results showed that this resin can selectively adsorb Cu^{2+} ions when the three ions coexist, and its selectivity coefficients are $K_{Cu(II)/Ni(II)} = 6.28$ and $K_{Cu(II)/Co(II)} = 4.23$ respectively.

Chitosan and chitosan come from nature, are safe, non-toxic, tasteless, biodegradable, will not cause secondary pollution when used for wastewater treatment, and have many unique excellent properties. The application research of them as flocculants and adsorbents in wastewater treatment has made great progress. China has a long coastline and abundant shrimp and crab resources. The raw materials for producing chitosan are sufficient. It can be expected that the development and utilization of this renewable resource will bring huge benefits to China's economy and society. Chitosan and chitosan have a broad application prospect in water treatment.

6.5 Lignin

6.5.1 Introduction

In 1938, French agronomist P. Payen isolated cellulose from wood, and also found a compound with a higher carbon content. Later, F. Schulze isolated this compound and named it lignin.

Lignin is one of the basic chemical components of coniferous trees, broad-leaved trees, and herbaceous plants, and is also present in all vascular plants (except tree ferns). Lignin is a complex aromatic polymer in plant cells, whose main function in the plant body is to lignify polysaccharides such as cellulose, thereby increasing the mechanical strength of the plant body. Lignin, along with cellulose and hemicellulose, are the main components of the plant skeleton. In terms of total amount, it is estimated that 1.5 trillion tons of lignin can be produced worldwide each year from plant growth. China's forest resources are not very rich, but there are 500–600 million tons of crop straw each year. Lignin is abundant in nature and is always associated with cellulose. Humans have been using cellulose for thousands of years, but the real study of lignin began after 1930.

6.5.2 Structure and Properties of Lignin

The structural unit of lignin is propylbenzene, and a methoxy group is present on the benzene ring. As the main structure of lignin, it is currently believed that there are three basic structures (non-condensed structures) with propylbenzene as the main structure, namely guaiacyl structure, syringyl structure, and p-hydroxyphenyl structure, showed in Figs. 6.8, 6.9 and 6.10. From the study of the biosynthesis process, it is known that these three basic structural units are all formed by the aromatization reaction of glucose to form coumaric acid, and then coumaric acid synthesizes these three basic structures of lignin.

There are many hydroxyl and carbonyl groups in the lignin structure, and the hydroxyl groups exist in the form of alcohol hydroxyl and phenolic hydroxyl. The phenolic hydroxyl in the lignin structure is a very important structural parameter. The amount of phenolic hydroxyl will directly affect the physical and chemical properties of lignin, such as reflecting the etherification and condensation degree of lignin and can also measure the solubility and reactivity of lignin.

The physical properties of lignin are related to the type, structure, and part of the plant, and also to the separation and extraction method. Originally, lignin was a white or nearly colorless substance with a relative density of about 1.35 ~ 1.50. There is no asymmetric carbon in its structure, so there is no optical activity. Lignin is a polymer with many polar groups in the structure, especially many hydroxyl groups, resulting in strong intra- and intermolecular hydrogen bonds, so raw lignin is insoluble in any

Fig. 6.8 Guaiacyl structure

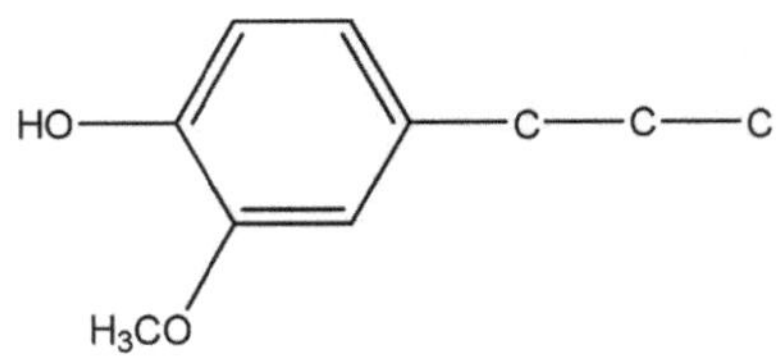

Fig. 6.9 Syringyl structure

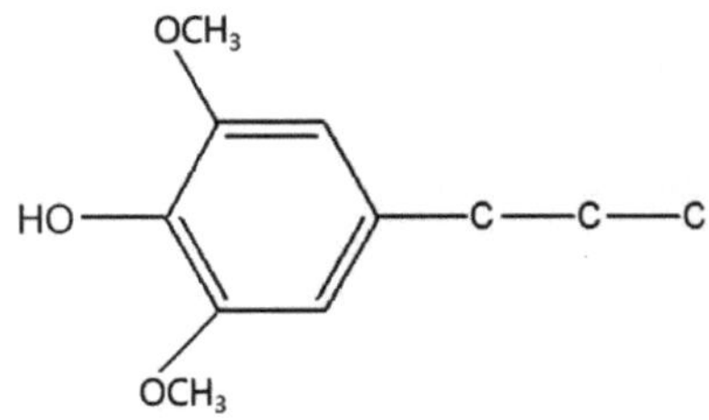

Fig. 6.10 p-Hydroxyphenyl structure

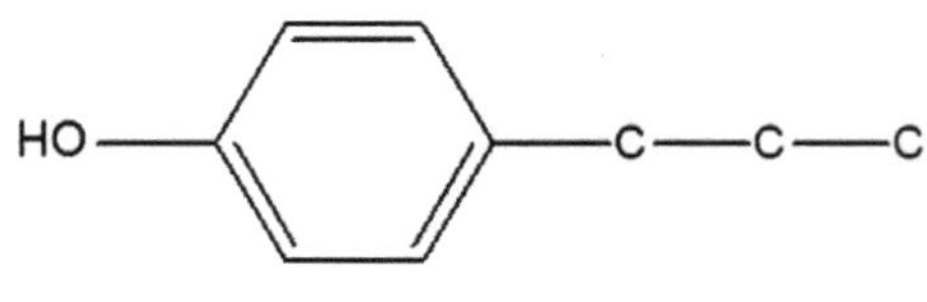

solvent. After separation, the lignin undergoes condensation or degradation, many physical properties change, and the solubility also changes, so that there are soluble lignin and insoluble lignin. The former is an amorphous structure, and the latter is the morphological structure of the raw fiber. The presence of phenolic hydroxyl and carboxyl groups makes lignin soluble in concentrated strong alkali solution. The relative molecular weight of raw lignin can reach hundreds of thousands, but the relative molecular weight of separated lignin is much lower, generally a few thousand to tens of thousands. The high and low relative molecular weight is related to the separation method.

The molecular structure of lignin contains active groups such as aromatic base, phenolic hydroxyl, alcohol hydroxyl, carbonyl, methoxy, carboxyl, conjugated double bond, etc., which can undergo many chemical reactions such as oxidation, reduction, hydrolysis, alcoholysis, acidolysis, photolysis, acylation, sulfonation, alkylation, halogenation, nitration, condensation or graft copolymerization.

6.5.3 Applications

Currently, people have begun to gradually recognize the important uses of lignin, such as for the production of synthetic resins and adhesives, rubber reinforcing agents, oilfield chemicals, building material additives, etc. It is also used in light industry and other industries and can be used in agriculture as a fertilizer, pesticide slow-release agent, plant growth regulator, soil amendment, liquid film, etc.

Industrial lignin mainly comes from the cooking wastewater of the paper pulping industry. According to the cooking method, industrial lignin can be mainly divided into lignosulfonates and alkali lignin, which can be used industrially as concrete water reducers, dispersants, slurry treatment agents, soil stabilizers, surfactants, water treatment agents, adhesives, etc. However, its performance is not ideal, and there are obvious disadvantages in performance compared to synthetic compounds. Therefore, the recovery of industrial lignin and further modification to improve its application performance can effectively solve the problem of paper wastewater polluting the environment and bring significant economic benefits. The application of lignin in water treatment is also increasingly attracting attention, and lignin is now used as a flocculant, corrosion inhibitor, rust inhibitor, scale inhibitor, dispersant, ion exchange resin, etc.

Lignin is an amorphous polymer with a large networked spatial structure, containing various polar and non-polar functional groups, the most abundant of which are phenolic hydroxyl and free hydroxyl groups. These functional groups carry a negative charge, so lignin is an anionic polymer flocculant. The reactive functional groups present on the lignin macromolecule make lignin easily form chemical bonds during the flocculation process, which plays an important role in promoting the adsorption of soluble organic matter and the net capture of colloids and suspended particles.

Jantzen and Vincent F. Felicetta et al. found that low molecular weight lignosulfonates have the effect of precipitating proteins after multiple separations and purifications. Later, Striker Joerg, Arnold Manfred, Jost Renate et al. found that moderately sulfonated high molecular weight lignosulfonates can be used as flocculants for protein wastewater. Low molecular weight lignosulfonates react with proteins to form insoluble complexes in acidic solutions; while the ones with high molecular weight bridge proteins to form flocs. The spatial configuration of lignin molecules is an intermediate form between spheres and densely curled bodies, with a lower average molecular weight and fewer active adsorption points, which directly affect the flocculation performance. People improve the flocculation performance of lignin by changing the spatial configuration of lignin, increasing the molecular weight, and introducing functional groups with flocculation performance through cross-linking reactions and condensation reactions. The cross-linking reaction links lignin molecules together at a certain distance to form a large molecule with a soft bond. When this product is dissolved in a water medium, it can form a loose and soft molecule, thereby increasing the area of lignin molecules to adsorb and capture small solid particles or colloid particles in the suspension, improving the flocculation effect. The condensation reaction can attach specific functional groups to the lignin molecule, such as $-CH_2N(CH_3)_2$, $-CONH_2$, $-C(=NH)NH-$, $-OSO_3CH_3$, $-N(CH_3)_2$, $-NH-R$, etc., thereby increasing the active adsorption points on the lignin molecule and improving the coagulation performance.

Metal ions also have a certain impact on the flocculation performance of lignosulfonates. Research shows that when the molecular weight is very high, the flocculation action of lignosulfonates containing chromium is stronger than that of lignosulfonates containing other metal ions.

In order to obtain better performing flocculants, there are many types of chemicals that react with lignin. In 1988, Erkki Pulkkinen et al. from Finland proposed that alkali lignin cationic flocculants could be produced by reacting alkali lignin with quaternary amine reagents such as trimethylamine chloride and glycerol trimethylamine chloride. These flocculants have good flocculation performance.

Miller et al. used lignin as a flocculant, in combination with natural polymers or synthetic polymers such as polysaccharides, to treat food processing wastewater and recover proteins, fats, and oils from it. This method has good flocculation effects at low pH values. Hoftiezer et al. heated alkali lignin with formaldehyde and polyamine in an aqueous solution to prepare a cationic lignin flocculant through the Mannich reaction. By adjusting the ratio of lignin to polyamine, a product with optimal flocculation performance can be prepared. This new type of modified cationic product can dissolve in both acidic and alkaline solutions and has a good effect on suspended particles in water when used as a flocculant to treat kaolin suspensions. Schilling et al. reacted lignin, aldehydes, and polyamines to synthesize lignin amines with a high nitrogen content that can dissolve in both acidic and alkaline solutions. This product can be used as a flocculant, filter aid, precipitant, corrosion inhibitor, asphalt emulsifier, oil well additive, cationic dispersant, etc. When used as a flocculant, it can form a better flocculation state. The synthesis reaction can be carried out in an organic solvent, and the molecular weight of the lignin amine can be controlled by adding a certain amount of crosslinking agent, so that the modified lignin amine flocculant exhibits amphoteric characteristics and can dissolve in a wider pH range.

Lei Zhongfang et al. conducted early research on lignin flocculants in China. They used lignin extracted from black liquor before and after anaerobic treatment as a flocculant to treat montmorillonite suspensions, monosodium glutamate wastewater, and printing and dyeing wastewater, achieving good results. Experiments found that lignin flocculants can effectively reduce the turbidity of water at low pH values. When the lignin dosage is 25 mg/L, the turbidity removal rate for montmorillonite suspension is 83.8%. This flocculant has a good effect on the removal of turbidity and COD from monosodium glutamate wastewater, and has a good decolorization effect on active dye wastewater, with a decolorization rate of up to 87%.

The lignosulfonate produced from the waste liquid of wheat straw pulping by the sulfite-formaldehyde-anthraquinone method has the characteristics of high sulfonation degree, low molecular weight, good water solubility, and good surface activity, and has a certain dispersibility. It can be used as a diluent, foaming agent, and flocculant for treating waste drilling fluid for drilling fluids. Adding a certain amount of this lignosulfonate to the drilling fluid has a significant flocculation effect on the solid particles in the slurry.

Liu Qianjun et al. used calcium lignosulfonate as a raw material to prepare lignin-based flocculant decolorizers through grafting and aminomethylation, and combined them with aluminum sulfate to examine their decolorization treatment effects on reactive orange K-G and reactive yellow X-R. The experiment proved that the effect of using lignin-based flocculant alone is not good, but the decolorization performance is significantly enhanced after being combined with aluminum sulfate, the dosage is reduced, and under appropriate conditions, this flocculant has a good decolorization effect. In addition, calcium lignosulfonate can also be used as a filter aid for magnetite slurry due to its flocculation action and surface activity.

Using lignin as a raw material, lignin and formaldehyde are reacted in a cross-linking reaction to obtain a lignin solution with a water-soluble polymer skeleton. Then, sodium bisulfite, ferric chloride, and sodium persulfate are added to carry out a sulfonation reaction, preparing an anionic lignin polymer flocculant. This flocculant is used to treat the wastewater from a silkworm boiling workshop, recovering the protein in it. It is determined that the optimal dosage of the flocculant for treating this wastewater at pH = 3.0 is 30 mg/L. After stirring at a speed of 24 r/min for 2 min, the COD_{Cr} removal rate of the wastewater can reach 62.3%. After drying the flocculated sediment, the protein content can reach over 60%, which can be used as animal feed. This indicates that this product has practical significance for recovering protein from silkworm cocoon wastewater.

Wu Bingyan a et al. used trimethylamine, epichlorohydrin, sodium hydroxide, lignin extracted from black liquor from papermaking, and self-made quaternary ammonium salt monomer $[(CH_3)_3(CH_2)_2CHON]^+Cl^-$ as raw materials to synthesize lignin quaternary ammonium salt flocculant. Through orthogonal experiments, the effects of reaction material ratio, activation time, catalyst concentration, reaction time, and other factors on the synthesis were studied, and the optimal conditions for the synthesis of the flocculant were optimized. The flocculation decolorization performance of lignin quaternary ammonium salt was studied with high concentration, high color (3000 ~ 5000 times) dye intermediate wastewater. The optimal dosage of lignin quaternary ammonium salt flocculant is 20 mg/L, and the color removal rate of wastewater is 85%. And through infrared spectroscopy analysis, it is speculated that the flocculation of lignin quaternary ammonium salt reduces the molecules with sulfonic acid groups in the wastewater, thereby achieving color removal.

Zhang Zhilan et al. used lignin as a coagulant to treat monosodium glutamate liquid and dye synthesis wastewater, and compared it with polyaluminum chloride and polyacrylamide. The results prove that lignin coagulant has superior coagulation performance. The main mechanism of lignin treating monosodium glutamate wastewater is electrostatic attraction and electropositive neutralization, hydrophobic sweeping and net trapping sedimentation, while the mechanism for dye synthesis wastewater is hydrogen bond attraction, sweeping and net trapping sedimentation.

Fang Guizhen et al. synthesized a cationic lignin quaternary ammonium salt flocculant used lignin in black liquor from sulfate pulping as raw material. The factors affecting the synthesis such as reaction temperature, stirring speed, feeding method, etc. were quantitatively analyzed, and the effects of reaction material ratio, activation time, reaction time, and water addition on the synthesis of flocculant lignin quaternary ammonium salt were studied by orthogonal experiments. The results show that the suitable conditions for the synthesis of flocculant lignin quaternary ammonium salt are: ammonium persulfate catalysis, reaction temperature 70 °C, stirring speed 500 r/min, one-time feeding, activation time 3 min, lignin to monomer mass ratio of 1:2.5, reaction time 4 h, lignin to water mass ratio of 5:1. The lignin quaternary ammonium salt prepared by this method is a cationic flocculant, which is more suitable for use under weakly acidic conditions. This is because under acidic conditions, the molecular conformation of the cationic flocculant tends to stretch, fully exerting the bridging effect of large molecules. When the synthesized lignin quaternary ammonium salt is used for flocculation decolorization of acidic black ATT, the optimal pH value of the solution is between 2 and 3, and the decolorization rate can reach 94.02%.

Huang Minsheng et al. used sodium polyacrylate as the main flocculant, lignin as the coagulant, and natural zeolite as the adsorbent to pretreat monosodium glutamate concentrated wastewater, achieving very good results. The removal rates of COD, SS, and SO_4^{2-} in the pretreatment process reached 69%, 91%, and 43% respectively, and the cost of pretreatment reagents was about 6.24 yuan/ton of wastewater, and the economic benefit of the separated SS protein was about 27 yuan/ton of wastewater.

Lignin is a precious resource given to mankind by nature, but it has not been widely used so far, which often result in not only a waste of resources but also environmental pollution. China has abundant lignin resources, but research on it is far from meeting practical needs. We should make full use of this natural high polymer resource and develop new and efficient lignin coagulants.

6.6 Tannin

Tannin is a general term for plant extracts that have tanning, astringent, and water-soluble properties. This component is widely found in the bark and seeds of plants and serves as a defense against external enemies for plants. Tannin is a type of polyphenol, with a molecular weight generally limited between 400 and 3000. Tannin is soluble in water, alcohol, and ketones, but insoluble in ether, chloroform, and petroleum ether. Tannin has a strong interaction with proteins and can form precipitates with various alkaloids and heavy metal ions. Tannin turns black-green when reacting with ferrous salts, and most tannins have phenolic hydroxyl groups and are weakly acidic. These characteristics of tannin have been widely used in industry.

Tannin is so closely related to people's lives that has been studied since ancient times. However, because its separation and refinement are relatively difficult, it has become one of the plant components whose chemical structure is not yet clear. Since the 1980s, structural chemistry has made rapid progress, and many chemists have conducted systematic research on tannin compounds. So far, about 500 types of tannin structures have been identified. Tannins can be classified into condensed tannins, hydrolysable tannins, and new tannins from a chemical structure perspective. In addition to the flavan-3-ol structure (showed in Fig. 6.11), condensed tannins also have the structure unit of catechol-β-phenol. Condensed tannins can be divided into dimers, trimers, etc. based on the number of structural units. Treating pure condensed tannins with acid can produce anthocyanins, known as pre-anthocyanins. Hydrolysable tannins, hydrolyzed to acid, include gallotannins that only produce gallic acid, and ellagitannins that produce ellagic acid. New tannins are new types of tannins with c-glucosyl flavones as the parent nucleus. Figure 6.12 showed the commercially available tannin m-digallic acid used for immobilization experiments. In this way, tannin is a group of compounds with diverse chemical structures, but the common point in known types of tannins is the presence of multiple phenolic -OH hydroxyl groups in the molecule.

Fig. 6.11 Flavan-3-ol

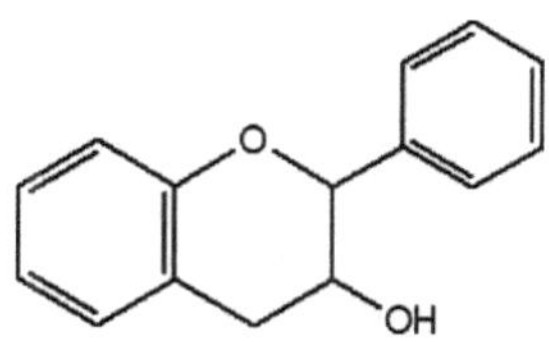

Fig. 6.12 Digallic acid

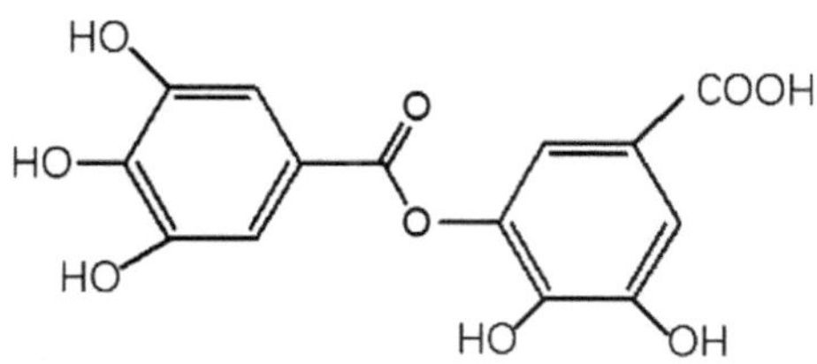

Since the 1980s, people have begun to propose the use of cationic products of tannin compounds as flocculants for water treatment. Kelly et al. reacted condensed tannins with aminated compounds (such as aminoethanol, methylamine, ammonium chloride, etc.) and formaldehyde to produce a tannin flocculant. which was used in conjunction with inorganic iron or aluminum salts to flocculate suspended solids in water. This tannin flocculant has a pH of less than 3.0, and its tannin raw materials are derived from the bark of white quebracho or acacia trees. The preparation process of the tannin flocculant is as follows: under weakly acidic conditions with a pH slightly below 7.0, tannin, aminated compounds, and formaldehyde react in the liquid phase, with a molar ratio of NH_2– on the aminated compound to the tannin structural unit of 1.5 ~ 3.0; the reaction mixture is heated until the temperature rises to 65 ~ 93 °C, until the reaction product is formed. The viscosity of the intermediate should be properly controlled during the reaction process, so that the generated product has a longer shelf life, for example, when the temperature is 82 °C, the viscosity of the key intermediate in the system measured by the Brookfield viscometer should be between 2 and 100 cps; when the viscosity of the intermediate product reaches this range, the reaction is terminated, the solid content of the solution is adjusted to 20 ~ 60%, and the pH value is adjusted to below 3.0 to obtain the modified cationic tannin flocculant, which contains about 40% of the modified cationic tannin component. When this tannin flocculant is mixed with inorganic flocculants, the mass percentage of the two is 9:1 ~ 2.33:1. The method described in U.S. Patent US4558080 is similar to this.

The hydroxyl group of tannin reacts with the group of an esterifying agent (such as acetic anhydride, dichloromethane, or N-(3-chloro-2-hydroxypropyl) trimethyl ammonium chloride, etc.) to form an ester, and a modified tannin compound can be obtained. Tannin can also react with formaldehyde or acetaldehyde and cyclohexylamine and other nitrogen-containing substances, and the resulting tannin derivatives are water-soluble or soluble at a pH value below 7 and insoluble in water at a pH value above 7. These two modified tannin compounds have a coagulating effect on particulate matter in water, can reduce the adhesion of particles, and can also have a demulsifying effect on oil/water emulsions.

Reed et al. have invented a method for preparing alkylated tannin-Mannich compounds. This method involves reacting tannin-Mannich condensates with an alkylating agent at pH of 5 ~ 14. The tannin-Mannich condensates are prepared by condensing tannin, amine, and formaldehyde. The alkylated tannin-Mannich polymer can further react with formaldehyde to increase its viscosity. The specific method is as follows: add tannin to softened water to form a uniform mixture, stir continuously until it dissolves, then add ethanolamine until the temperature reaches 350 ~ 400 °C; add formaldehyde to the mixture, then add hydrochloric acid to balance its pH value; add softened water and sodium bisulfite to increase its viscosity; cool the mixture, dilute; react the mixture with methyl chloride to increase its viscosity, forming a polymer that can be used for wastewater decolorization. Tannin can react with formaldehyde and amine to undergo aminomethylation, the resulting tannin-Mannich polymer has a high molecular weight due to the cross-linking of formaldehyde and Mannich base, and because both cationic amines and anionic phenols exist on the polymer chain, it has amphoteric characteristics. Due to this amphoteric characteristic, the tannin-Mannich polymer has high solubility at extreme pH values, but it is often insoluble when the pH value is close to 7, because at this time it is close to its isoelectric point (at the isoelectric point, the net charge on the polymer is zero). During the storage period of the tannin-Mannich polymer, it will continue to cross-link until it forms an insoluble gel, with a storage period of about 3 months. The equation for the tannin aminomethylation reaction is as follows:

$$CH_2{=}O + H^+ \longrightarrow CH_2{=}\overset{+}{O}H \tag{6.9}$$

$$CH_2{=}\overset{+}{O}H + (CH_3)_2NH \longrightarrow N(CH_3)_2{-}\overset{+}{C}H_2 + H_2O \tag{6.10}$$

$$N(CH_3)_2{-}\overset{+}{C}H_2 \longrightarrow \overset{+}{N}(CH_3)_2{=}CH_2 \tag{6.11}$$

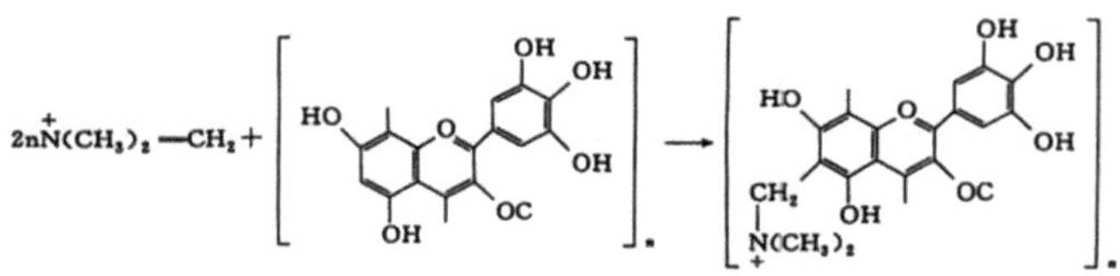

(6.12)

Using the cationic ortho-dihydroxyphenyl tannin in the aqueous extract of *Acacia mearnsii* bark as a raw material, a plant-based flocculant can be prepared, which can be used to treat drinking water, industrial water, boiler, and cooling tower water, as well as general domestic and industrial wastewater. Comparing this flocculant with other inorganic or organic flocculants, it was found that this flocculant has some special properties and can achieve better treatment effects. For example, the alkalinity of the water does not decrease during application, and there is no need to adjust for the activation of the flocculant. If a metal salt is used as a flocculant, the conductivity of the water often increases, but the modified tannin is a plant-derived organic polymer and does not increase the conductivity. In addition, this flocculant is active over a wide range of pH values and can promote the chelating ability with metals in water, causing the metal ions to react with the flocculant and settle.

Tannin is a biodegradable anionic polymer. Using tannin as a coagulant can make it a substitute for synthetic anionic polyelectrolytes. Tannin can avoid the health effects of residual aluminum from inorganic coagulants and synthetic polymers, and the resulting sludge can be biodegraded. Mahmut Özacar et al. have studied the coagulation properties of tannin in carob shells, determining its potential as a coagulant. They also studied its coagulation performance through beaker coagulation tests. This tannin comes from an oak tree that grows in Asia Minor, and the tannin content in the carob shell is about 35%. The hydrolyzed tannin content in the carob shell extracted in this study was 53.50%. Since tannin is an anionic polymer, a synthetic anionic polyelectrolyte AN913 (SNF) was selected for the experiment. Polysaccharides is a copolymer of acrylamide and acrylic acid, with a molecular weight between 1×10^6 and 2×10^6, and its free acrylamide content is about 0.1%.

It is one of the best anionic coagulant aids. Comparative coagulation experiments were conducted on artificial water with different pH values and turbidity. The results showed that the removal rate of turbidity was greatly improved when aluminum sulfate was used in conjunction with tannin or AN913, compared to using aluminum sulfate alone. However, tannin has better coagulation performance than AN913, forming larger flocs with faster settling speed, and can significantly reduce the amount of aluminum sulfate used. In addition, the study found that after using tannin as a coagulant, the resulting sludge is easier to filter, making the filtration process simpler and more economical, with reduced residual aluminum content in the treated water and no detectable residual tannin substances. The researchers also studied the effects of tannin, AN913, and clay as coagulant aids on the removal of phosphate from water, comparing six agents: alum, alum + clay, alum + tannin, alum + AN913, alum + tannin + clay, and alum + AN913 + clay. The results showed that for raw water with an initial phosphate concentration of 5 ~ 15 mg/L, the phosphorus removal effect was significantly improved after adding tannin or AN913 to the alum, compared to using alum alone. Using polymers or clay can improve the efficiency of phosphorus removal and reduce the amount of alum added. FTIR spectroscopy was used to detect the aluminum species formed during the phosphorus removal process, and it was found that hydroxy aluminum phosphate, hydroxy tannic acid aluminum, and aluminum complexes containing phosphorus, tannin, and AN913 were formed.

Researchers in China have also prepared cationic products of tannin and used them as flocculants to treat drilling wastewater, and evaluated their performance. Cationic tannin flocculants can also be prepared using larch rosin with a tannin content of 45% as raw material. Formaldehyde and dimethylamine react with tannin under weakly acidic conditions in a Mannich reaction, causing tannin to undergo amination methylation. At the same time, side reactions occur, with formaldehyde causing tannin molecules to form tannin gel through methylene linkage, and tannin molecules may also undergo irregular condensation to form a red powder that is insoluble in water. The study determined the optimal reaction conditions for the cationization of tannin, with the molar ratio of tannin structural units: dimethylamine: formaldehyde = 1:4:5, pH = 6.5, reaction temperature of 85 °C, and reaction time of 1.5 h. Beaker coagulation test results showed that by adding cationic tannin flocculant, the treatment of high COD_{Cr} and high color industrial wastewater can greatly reduce the amount of inorganic coagulant used, improve treatment effect, and reduce treatment costs.

The two important sources of tannin resources are gallnut and Chinese gall, and China is one of the main producers of gallnut, possessing rich tannin resources. The research and development of such resources have a very broad prospect in the water treatment industry.

6.7 Microbial Polysaccharides

Microbial polysaccharides are a new type of fermentation product developed in the 1960s and 1970s. They are high molecular weight polymers synthesized from monosaccharides (for example, glucose is fermented in a solution containing non-pathogenic bacteria, enzymes, nutrients, oxygen, and trace metal catalysts). Figure 6.13 shows a commonly used production and recovery process for microbial polysaccharides. The composition of microbial polysaccharides is related to the microbial strain. Table 6.4 lists anionic and extracellular microbial polysaccharides. Many microbial polysaccharides, compared to plant polysaccharides, have a porous structure and a larger specific surface area. For example, microbial cellulose has a net-like structure of single fibers and can be used as an adsorbent. Microbial polysaccharides can also be used as cleaning and descaling agents, and have applications in the textile, cosmetics, food, and pharmaceutical industries.

Xanthan gum is one of the most distinctive types of microbial polysaccharides currently being developed both domestically and abroad, and it is also the largest-scale and most widely used microbial polysaccharide in the world. The United States once evaluated nine types of microbial polysaccharides, and xanthan gum ranked first due to its comprehensive functions and wide range of uses. Xanthan gum is used in

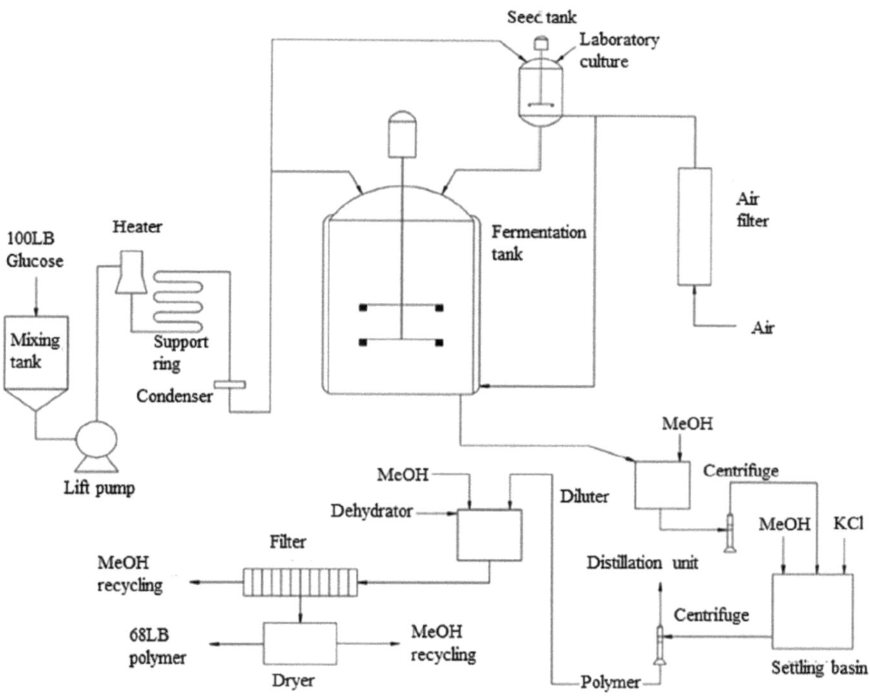

Fig. 6.13 Production and recovery process of microbial polysaccharides in a pilot plant

Table 6.4 Composition of anionic and extracellular microbial polysaccharides

Microorganism	Component	Molar ratio	Bond
Xanthomonas	D-Mannose D-Glucose	3.0	GA(β,1 → 2)M
	D-Glucuronic acid	3.0	M(1 → 4)G
	(Potassium salt)	2.0	G(1 → 4)GA
	O-acetyl	1.7	G(β,1 → 4)G
	Pyruvic acid	0.63	
Klebsiella	D-Glucose	1	MA(β,1 → 4)G
	D-Galactose	1	G(β,1 → 4)Gal
	D-(Potassium salt)	1	Gal(β,1 → 4)MA
	O-acetyl	4	
Cryptococcus	D-Mannose	4	M(1 → 3)M
	D-Xylose	1	X(1 → ^{c})M
	D-glucuronic acid (Potassium salt)	1	GA(β,1 → 2)M
	O-acetyl	1.7	
Hansenula	D-Mannose	5	M(α,1 →)POK(→)M
	Potassium	1	M(1 → 3)M
	Orthophosphate	1	M(1 → 2)M

the petroleum industry as the best mud treatment agent. The United States, the United Kingdom, France, Italy, and other countries are the main producers of xanthan gum, with an annual total production of more than 30,000 tons, and domestic consumption of more than 1000 tons per year.

Xanthan gum is a water-soluble gum, also known as Xanthomonas polysaccharide, Hansen gum, and Shanton gum, etc. It is an extracellular polysaccharide produced by Xanthomonas using starch or sucrose as the main raw material through modern biotechnology. Studies have shown that it is composed of sugar residues such as D-glucose, D-mannose, and D-glucuronic acid. The main chain is composed of D-glucose molecules connected by β-1,4 glycosidic bonds, similar to cellulose. Every other glucose monomer on the main chain is connected to a trisaccharide side chain. The trisaccharide side chain is composed of two mannoses and one glucuronic acid, each carrying an acetyl group and a pyruvate group. Xanthan gum forms a double helix structure in aqueous solution and forms a three-dimensional network with weak non-covalent bonds. The chemical structure and molecular conformation of xanthan gum determine its unique properties. The unique properties of xanthan gum are mainly reflected in its rheological properties, such as strong viscosity enhancement, good pseudoplasticity and stability, acid and alkali resistance, heat resistance, biodegradation resistance, good tolerance to high salt environments, excellent emulsion stability and particle suspension ability, and can be used in construction, coatings, papermaking, textiles, ceramics, petroleum, and water treatment industries. Xanthan gum mixed with other non-gelling polysaccharides easily forms a gel.

6.8 Animal Glue and Gelatin

The glue used for water and wastewater treatment comes from collagen—the main component of animal skin, tendons, and bone connective tissue. Animal glue is closely related to gelatin (glue gelatin), which is a hydrolysis product of collagen, containing many structural units of varying sizes, from simple peptides to large colloidal molecules, with molecular weights ranging from 3000 to 80,000. The functional groups that cause flocculation are amino and carboxyl groups. Animal glue is insoluble in cold water, but when the temperature is below its freezing point, it can absorb 6–8 times its weight in water. At around 60 °C, the swollen particles can dissolve in the solution. Heavy metal ions can react with animal glue to form irreversible gels. Glue is usually characterized by gel strength (measured in g in the Bloom test) and viscosity. Gelatin is a mixture of water-soluble high molecular weight proteins, derived from the hydrolysis of collagen.

Gelatin molecules are large and complex proteins, with molecular weights ranging from 15,000 to 250,000. They may be formed by 18 different amino radicals connected in a certain rule-based manner. Table 6.5 shows the amino acids obtained after complete hydrolysis of gelatin. The chain belongs to the polypeptide chain. Since the early development of the photographic industry, gelatin has been used as a colloidal protective agent. U.S. Patent US5908921 describes the preparation of gelatin, especially the method for producing high-purity gelatin for the photographic industry. Due to the gelation, bonding, and concentration characteristics of gelatin, it is still widely used in the food and pharmaceutical industries. Recently, FabroGen Inc. in the United States applied for a patent, providing animal collagen and gelatin and their combinations, as well as their production methods.

Gelatin can be halogenated to form a flocculant, with the molecular formula $[C_6H_{2n+1}O_nR_m]_x$, where R is a halogen, $n = 1 \sim 5$, $m = 1 \sim 5$, $x = 400 \sim 2000$. Using

Table 6.5 Amino acids produced by complete hydrolysis of gelatin

Amino acid	Mass percentage %	Amino acid	Mass percentage %
Alanine	11.0	Histidine	0.9
Arginine	8.8	Phenylalanine	2.2
Aspartic acid	6.7	Proline	16.4
Glutamic acid	11.4	Serine	4.2
Glycine	27.5	Threonine	2.2
Histidine	0.78	Tyrosine	0.3
Hydroxyproline	14.1	Valine	2.6
Leucine and isoleucine	5.1	Cysteine	Trace
Lysine	4.5		

rigid or semi-rigid gelatin and a 36 ~ 38% (weight) HR solution as raw materials, the ratio of the two feed amounts is 1:0.17 ~ 2. The gelatin is stirred in warm water at 25 ~ 55 °C to swell into a colloid, the colloid is heated to 40 ~ 100 °C and the HR solution is added dropwise, controlled to be completed within 30 ~ 60 min, kept warm for 2 ~ 4 h, and cooled to room temperature to discharge. This product is a highly efficient, non-toxic flocculant for water treatment, which can be used alone or as a coagulant for inorganic polymers, adding 2 ~ 6% can improve the flocculant effect by 3 ~ 4 times.

Some researchers have used flocculants for the clarification of traditional Chinese medicine water extracts, based on the co-polymerization and precipitation of the flocculant with the coarser particles in the traditional Chinese medicine extract. Adding chitosan to the extraction of traditional Chinese medicine can cause the colloidal protein in the solution and negatively charged suspended particles to aggregate into a co-polymer precipitate that is insoluble in water. The traditional Chinese medicine extract itself contains a small amount of tannin, which can react with gelatin to form a gelatin-tannin complex, which can co-precipitate with the suspended particles in the extract. Both chitosan and gelatin can form precipitates with tannins, proteins, waxes, etc. in the extract. If the two are used together, the small flocculent precipitates can quickly cross-link to form large flocs, accelerating precipitation, and the effect is good in the clarification process of traditional Chinese medicine water extract. Roberts et al. have invented a method to remove toxic fatty acids from pulp and paper wastewater, which is to add anionic gelatin (or albumin, alginic acid, alginate, etc.) to the wastewater before the traditional purification process, which can significantly improve the removal rate of fatty acids.

Compared with artificially synthesized organic flocculants, natural (modified) polymer flocculants have the advantages of low cost, wide source, low toxicity, and easy biodegradability, and attracting are increasingly attention. The research and application in water and wastewater treatment are also becoming more in-depth and extensive. The development of new and efficient natural polymer flocculants, as a substitute for synthetic organic polymer flocculants, is of great significance and has broad application prospects from the perspective of sustainable environmental development.

6.9 Moringa Seed Extracts

Moringa oleifera has been proven to be an effective, safe, and inexpensive traditional coagulant substitute, widely used in drinking water treatment. In some developing countries, such as Sudan, rural residents are accustomed to using Moringa seed powder to remove turbidity, which usually exists in the form of suspended organic and inorganic particles, such as clay, microbes, and other decomposed plant residues.

Table 6.6 Water sample characteristics

Parameter	Value
Turbidity	61 NTU
pH	7.4
Conductivity	529 μS/cm
TOC	4.143 mg/L
Temperature	28.5 °C

Some research results show that Moringa does not release toxic substances when treating water, has antibacterial effects against Salmonella typhi, Vibrio cholerae, and Escherichia coli, and is a promising natural antibacterial agent with potential applications in controlling waterborne diseases. Compared with chemical coagulants, its greatest advantage lies in the stability of pH during the coagulation and flocculation process. However, other studies point out that Moringa has a negative impact on the concentration of TOC in water, which may be one of the reasons hindering its large-scale use.

During water treatment, the formation of flocs and the reduction of turbidity are enough to indicate the coagulation activity of Moringa seed extract. Some researchers used a flocculator to compare the coagulation activity of Moringa, polyaluminum chloride (PAC), and their combination. The initial turbidity of the corresponding water sample was 61 NTU (Table 6.6 is showed for water sample characteristics). The results showed that the coagulation activity of Moringa seed extract was lower than that of PAC ($P \leq 0.01$). However, the combination of half doses had the same effect as PAC, with no significant difference ($P \leq 0.05$).

The efficiency of Moringa increases with the increase of water turbidity, and its performance is comparable to the inorganic coagulant PAC. Its performance is shown in Fig. 6.14. The turbidity removal rate is 68.6%, and the final turbidity of the water body is 19.2 NTU, which does not meet the standards set by the World Health Organization (WHO). The coagulation effects of the chemical coagulant PAC and the combination of Moringa and PAC are better, with turbidity removal rates of 96.1% and 95.7%, and final turbidities of 2.4 and 2.7 NTU, respectively.

In this study, TOC analysis found that the increase in Moringa dosage also increases the concentration of TOC in the water body (Fig. 6.15 is showed for the effect of different doses of Moringa extract on TOC), while the combination of Moringa and PAC significantly reduced the TOC content in the water sample (Fig. 6.16 is showed for the effect of Moringa and PAC on TOC).

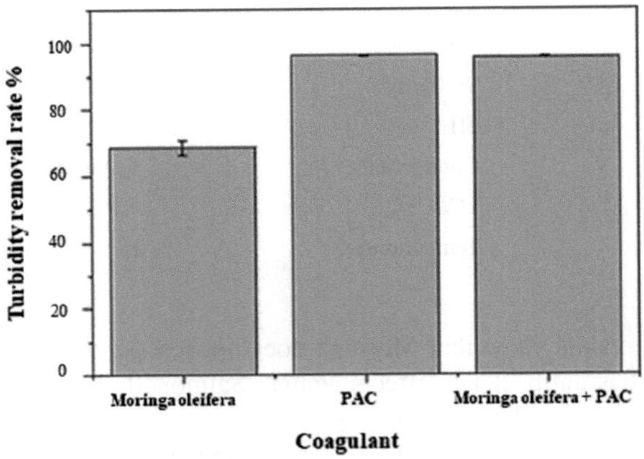

Fig. 6.14 The effect of Moringa, PAC, and their combination on turbidity removal

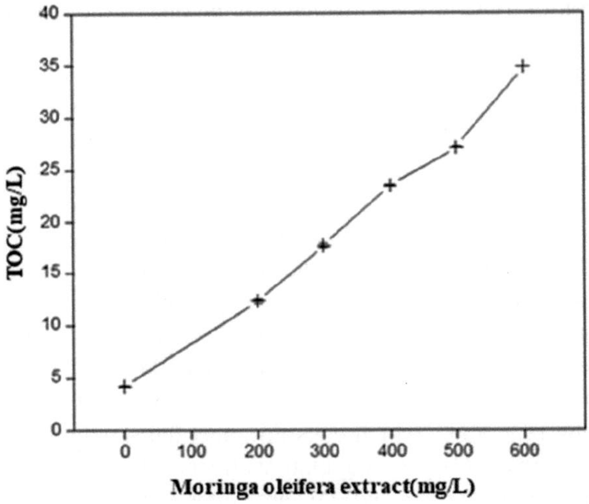

Fig. 6.15 The effect of different doses of Moringa extract on TOC

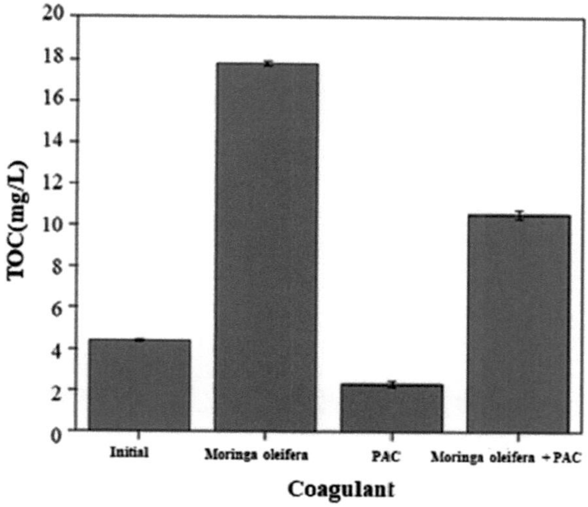

Fig. 6.16 The effect of Moringa and PAC on TOC

Chapter 7
Decolorizing Agent

Printing and dyeing wastewater refers to the wastewater discharged from the pretreatment, dyeing, printing, and finishing processes of textile products such as cotton, wool, and synthetic fibers. The wastewater contains dyes, pulp, auxiliaries, acids, alkalis, fiber impurities, inorganic salts, etc. The composition of dyes is very complex, containing many difficult-to-degrade components, and the color, ammonia nitrogen content, organic matter content, etc. are very high. The nitro and amino compounds in the dye structure and heavy metal elements such as copper, chromium, zinc, and arsenic have significant biological toxicity. The dyeing process uses water as a medium, requires multiple washings, uses a large amount of water, and 10 ~ 20% of dyes in the processing process are discharged into wastewater, seriously polluting the environment. With the development of the dye industry and the advancement of dyeing and printing processing technology, the types and functions of dyes are increasing, and the stability of dye structures has greatly improved. Therefore, printing and dyeing wastewater has characteristics of high concentration, high color, complex composition, large changes, and high treatment difficulty. The decolorization problem of printing and dyeing wastewater has become a major problem that needs to be solved urgently in wastewater treatment at home and abroad.

At present, most printing and dyeing factories adopt a combination of chemical treatment and biochemical treatment. Biochemical treatment uses microbial methods to degrade dye molecules and organic matter, but the degradation rate of harmful molecules in the biochemical treatment process is low, the equipment investment is large, and the operation cost is high. Therefore, choosing a simple, economical and effective treatment method has become the focus of research on the decolorization of printing and dyeing wastewater. In addition to the biochemical method, there are many studies and applications reported on other physical and chemical or chemical decolorization methods such as adsorption, oxidation–reduction, ion exchange, membrane, coagulation, etc. Among them, coagulation is widely used in the treatment of printing and dyeing wastewater because of its low cost, large treatment capacity, convenient operation and management, and high decolorization rate for

L. Fengting, *Water Treatment Chemicals: Coagulants and Flocculants*,
https://doi.org/10.1007/978-981-96-8221-8_7

hydrophobic dyes. It can effectively reduce the color of wastewater and remove various high-molecular substances, colloids, heavy metals, and soluble inorganic compounds that cause eutrophication of water bodies, such as phosphorus.

7.1 Coagulation Decolorization Treatment Technology

There are many types of dyes, which are divided into two types according to whether they are soluble in water: one is water-soluble dyes, which mainly include direct dyes, acid dyes, reactive dyes, and cationic dyes; the other is non-water-soluble dyes, which mainly include dispersed dyes, reduction dyes, sulfur dyes, insoluble azo dyes, etc. Water-soluble dye molecules contain more hydrophilic groups that can form hydrogen bonds with water molecules, such as $–SO_3H$, –OH, –COOH etc. Non-water-soluble dyes do not contain hydrophilic groups or contain a small number of hydrophilic groups but have a large molecular weight, and usually exist in wastewater in the form of hydrophobic suspended particles or colloids.

Coagulation decolorization refers to the aggregation and sedimentation of colloidal particles and tiny suspended matter in water under the action of coagulant hydrolysis and hydrolysis products, or by destroying carboxyl, sulfonic acid and other hydrophilic groups, making the dye molecules lose their hydrophilicity. Under the action of the coagulant, they coagulate to form flocs, thereby effectively reducing the color of the wastewater. The mechanism of coagulation decolorization mainly includes the following categories:

(1) Adsorption and Neutralization

The colloidal particles in the dye wastewater usually carry a negative charge and are always suspended in water in a state of mutual repulsion. When a coagulant with an opposite charge to the colloidal particles is added, the colloidal surface has a strong adsorption effect on the opposite ions, opposite colloidal particles, or chain ions with opposite charges, thereby neutralizing some of the charges, reducing the repulsion between the colloids, and the original stable state is destroyed, and the colloidal particles aggregate and settle.

(2) Compression of Double Electric Layer

When a coagulant is added to the wastewater, ions with opposite charges to the colloidal particles (counter ions) are formed in the solution. As the concentration of counter ions increases, the charge of the effective component in it neutralizes with some counter ions on the surface of the colloidal particles, exchanges occur, the double electric layer is compressed, the thickness of the diffusion layer becomes thinner, the number of charges on the colloidal particles decreases, the ζ potential decreases, and when the ζ potential reaches the critical potential, the colloid becomes unstable, forming gradually sinking colloidal particles.

(3) Adsorption Bridging

When the coagulant added to the wastewater is a high polymer or aluminum-iron compound, it can generate a linear structure of high polymer through hydrolysis and condensation reactions. Even if the distance between the gel particles is far, the high polymer can adsorb colloidal particles through electrostatic attraction, van der Waals forces, hydrogen bonds, coordination bonds, etc., to perform adsorption bridging. Due to the bridging effect, the gel particles are combined into a floc, gradually achieving sedimentation.

(4) Precipitation Net Capture

In the wastewater treatment process, coagulants with metal salts will generate precipitates due to hydrolysis, such as metal hydroxides, metal carbonates, etc. These precipitates have a large surface area. When stirring continues, the colloidal particles and suspended particles in the water will be swept by the hydrolyzed precipitates and sediment.

Due to the variety of dyes, it is not possible for a single coagulant to have a good decolorization effect on all dyes. It is necessary to choose the appropriate coagulant based on the characteristics of the dye.

7.2 Types of Decolorizing Coagulants

Coagulants used for dyeing wastewater treatment can be divided into two major categories: inorganic and organic. Inorganic coagulants include metal salts and inorganic high polymer electrolytes, compounds of iron salts, magnesium salts, aluminum salts, and silicon and calcium elements. Organic flocculants include surfactants, natural polymers and their modified flocculants, synthetic organic polymer flocculants, etc.

7.2.1 Inorganic Coagulants

(1) Aluminum Salt

Commonly used aluminum salts include aluminum sulfate and polyaluminum chloride. The hydroxyl compound with a high positive charge formed by the hydrolysis of aluminum salt has a good coagulation decolorization effect on large molecular weight hydrophobic dyes, such as insoluble dyes, direct dyes, basic dyes, reducing dyes, sulfur dyes, dispersed dyes, etc. However, it does not have a high decolorization rate for active dyes, acidic dyes, especially small molecular weight, monoazo bonds, and water-soluble dye wastewater containing several sulfonic acid groups.

(2) Iron Salt

Commonly used iron salts mainly include iron sulfate, ferrous sulfate, ferric chloride, polyferric sulfate, etc. The hydrolysis process and coagulation mechanism of iron salt are similar to those of aluminum salt, but iron has a stronger affinity for OH- than aluminum, and it hydrolyzes faster in solution, has strong coagulation ability, good decolorization performance, is safe and non-toxic, can avoid secondary pollution, and is especially superior to aluminum-based coagulants in decolorizing sulfur and dispersed dyes, so it is widely used.

Under slightly acidic conditions, trivalent iron salts undergo hydrolysis reactions, and the resulting iron hydroxide has a selective adsorption effect. It has a good adsorption effect on dye molecules with acidic groups such as sulfonic acid groups and carboxylic acid groups, thereby achieving a decolorizing effect.

Unlike the decolorization mechanism of trivalent iron salts, Fe^{2+} ions have reductive properties and have a strong selective reduction effect on oxidative nitrogen-containing groups such as nitro and azo groups. For dye molecules with $-SO_3^-$, $-OH^-$, $-NH_2$ and other groups, Fe^{2+} ions can react with the unshared electron pairs of these groups to form large molecular chelates, enhancing the hydrophobicity of the dye molecules. This makes the colloid particles easily adsorbed on the hydrolysis products produced by Fe^{2+} ions, achieving the effect of color removal. However, because the mechanism of ferrous sulfate decolorization is to reduce the chromophoric group, the reduction product is a small organic molecule that cannot be effectively removed by coagulation, so the COD removal rate is not high, and the consumption of alkalinity in the solution is large, and the amount of coagulant is also large.

(3) Magnesium Salts

Commonly used magnesium salts mainly include magnesium oxide, magnesium sulfate, etc. In China, magnesium salts and other magnesium compounds are not widely used alone in wastewater treatment, but a considerable amount of research in dyeing and printing wastewater shows that magnesium salts and other magnesium compounds have a large specific surface area, strong decolorization ability, and good effluent turbidity. They are excellent coagulating decolorizing agents, especially for water-soluble dyes containing acidic groups (carboxyl, sulfonic acid). At a pH of 11 ~ 12, magnesium salts are converted into magnesium hydroxide precipitates, and the newly formed precipitates have strong adsorption, which can adsorb dye molecules in the solution, thereby playing a decolorizing role. Therefore, while decolorizing, its COD removal rate is also high, which is suitable for some wastewaters with a high original pH. The treatment of reactive dye and disperse dye wastewater with $MgCl_2$ and $Ca(OH)_2$ is better than $Al_2(SO_4)_3$, PAC, $FeSO_4/Ca(OH)_2$. The mechanism is that Mg^{2+} reacts with hydroxyl, carboxyl or sulfate ions to form stable chelates, which can be removed from wastewater by coagulation. However, magnesium salts also have the disadvantage of a narrow pH range.

(4) Other Inorganic Coagulants

Many regions have acetylene production companies. The waste residue formed after the hydrolysis of calcium carbide contains a large amount of lime, ferrous and silicon, which is a very good decolorizing agent for dyeing and printing wastewater.

Activated silicic acid is made from water glass as a raw material and treated with various activators (generally sulfuric acid). It can be regarded as an anionic inorganic high polymer electrolyte. It is usually used together with conventional coagulants such as aluminum salts and iron salts to promote the coagulation decolorization process.

Inorganic composite coagulants are formed by compounding various composite high polymer coagulants such as inorganic iron salts, silicates, and inorganic aluminum salts. Such as polyaluminum ferric sulfate, polyaluminum ferric chloride, which can overcome the disadvantages of slow alum flower formation, light alum flower, slow settling in aluminum salt treatment, and can overcome the disadvantages of unclear effluent and high color of iron salt, the decolorization effect is significantly better than traditional inorganic coagulants.

7.2.2 *Organic Flocculants*

(1) Surfactants

Due to the polar groups of long-chain cationic surfactants carrying positive charges, they can neutralize the negative charges of dye molecules. At the same time, the core atoms with lone electron pairs can complex with the polar groups of dye molecules. Their non-polar ends are hydrophobic and can be adsorbed on the hydrophobic groups of the flocculent, thereby removing the dye in the form of flocculent. Therefore, the larger the molecular weight and the longer the carbon chain, the higher the surface free energy, and the more likely it is to undergo complexation reactions.

There are many reports on the use of surfactants for dyeing wastewater treatment. Octadecyl amine acetate can be used to treat insoluble dyes, such as treating cotton fabric wastewater containing sulfur black B dye, with a dye removal rate of up to 99.2%. Stoical uses octadecyl trimethyl ammonium chloride and hexadecyl pyridinium bromide in combination with $Al_2(SO_4)_3$ to treat silk dyeing wastewater containing acid and direct dyes at a pH of 4 ~ 11 through coagulation and air flotation, with a decolorization rate of 90 ~ 100%. However, the complexation of cationic surfactants with dye molecules has strong selectivity, so it is often difficult to achieve good results when used alone, and it needs to be used in combination with aluminum salts.

(2) Natural polymers and their modified coagulants

Natural organic polymer flocculants show good application prospects due to their wide source of raw materials, low cost, non-toxicity, and easy biodegradability. Natural polymer flocculants used for dyeing wastewater treatment mainly include chitosan derivatives, natural starch and its derivatives, and lignin derivatives.

Chitosan is the second largest natural polymer after cellulose, widely found in the exoskeletons of crustaceans such as shrimp and crab, and in the cell walls of some fungi. Chitosan is a deacetylated product of chitin, with many hydroxyl and amino groups distributed on its molecular chain, as well as some residual N-acetyl groups. Their presence allows chitosan to adsorb dyes in the solution through electrostatic attraction, hydrogen bonding, and van der Waals forces. At the same time, the molecular structure of chitosan contains $-NH_2$ and –OH groups, both of which have strong reactivity and can undergo various chemical modifications under appropriate conditions. For example, the water-soluble carboxymethyl chitosan obtained by carboxylation modification of chitosan has excellent decolorization effects on water-soluble dye wastewater, especially for water-soluble anionic dyes.

Starch is a type of sugar substance present in the bodies of natural plants, mainly deposited in the seeds, tubers or roots, fruits, and leaves of plants, and is spread throughout the plant world. Starch is a polysaccharide high polymer compound composed of glucose, and each unit of its large molecule contains three free hydroxyl groups, which can undergo various chemical modifications. Natural lignin is a type of amorphous polymer with a large network space structure and is also the main component of the cooking waste liquid in the pulp and paper industry. Lignin molecules contain phenolic hydroxyl, alcoholic hydroxyl, methoxy, aldehyde, ketone, carboxyl, and other active groups, which can undergo various chemical modifications. Most of the fine particles and colloids in wastewater treatment carry negative charges. To enhance the action of starch and lignin molecules on these small molecules, cationic modification is an important research direction. Cationic ionized starch and lignin can be used to treat cationic dye, direct dye, and acidic dye wastewater, with a decolorization rate of over 90%.

(3) Synthetic organic polymer flocculants

Synthetic organic high molecular weight flocculants have high molecular weights and carry many active functional groups in their molecular chains. Therefore, they have a large extension in water, good flocculation performance, require less usage, have a wide pH range, and simultaneously have superior performance in filtration, dewatering, and other solid–liquid separation operations. Depending on the charge they carry, synthetic organic high molecular weight flocculants can be divided into cationic, anionic, amphoteric, and non-ionic flocculants. Since colloids and suspended substances often carry negative charges, cationic flocculants are often used in the treatment of printing and dyeing wastewater.

Polyacrylamide (PAM) is the most widely used synthetic organic high molecular weight flocculant, accounting for about 80% of the total production of synthetic organic high molecular weight flocculants. The main types used in the treatment of printing and dyeing wastewater are cationic PAM and amphoteric PAM.

Polydimethyldiallylammonium chloride is a water-soluble cationic polymer. The homopolymers and copolymers of dimethyldiallylammonium chloride have the advantages of high positive charge density, good water solubility, easy control of relative molecular mass, high efficiency, and non-toxicity. They can be used in oil extraction, papermaking, textile printing and dyeing, daily chemical industry, and water treatment fields.

Currently, the most effective high molecular weight flocculant is PAN-DCD, which is a water-soluble, amphoteric polyelectrolyte with various active groups, modified from polyacrylonitrile as the main chain using dicyandiamide under alkaline conditions. PAN-DCD has good decolorization effects on neutral dyes, reactive dyes, and acid dyes, with a decolorization rate of over 90%. It has a dual effect on printing and dyeing wastewater, decolorizing and removing COD. If used in combination with polyaluminum, the removal effect is even better, with the highest COD removal rate being 63%. Another type of decolorizer worth noting is the dicyandiamide formaldehyde condensate that has appeared in recent years. It has excellent decolorizing effects on printing and dyeing wastewater, but a large dosage will increase the treatment cost.

7.3 Decolorizing Effect and Mechanism of Dicyandiamide Formaldehyde Condensate

Dicyandiamide formaldehyde condensate is a viscous transparent gel-like liquid formed by the condensation of dicyandiamide and formaldehyde in the presence of strong acid or salt. Its viscosity depends on the degree of polymerization of dicyandiamide and formaldehyde. The dicyandiamide formaldehyde condensate has been reported since 1891 by E. Benberge and has a history of over 100 years. It can be used as a fusion agent, paper and glass fiber lubricant, light leather agent, color fixative, electroplating additive, etc. In the early 1950s, it was widely used as a printing and dyeing fixative.

In the development and application of flocculants, it was found that dicyandiamide formaldehyde condensate has good flocculation ability under certain conditions. In addition to having a treatment effect on dyed wastewater, it also has a good treatment effect on oily wastewater, papermaking wastewater, and slaughterhouse wastewater, which has attracted people's attention. In the 1960s, there were many patents that used dicyandiamide formaldehyde condensate for the treatment of papermaking wastewater and dyeing and finishing wastewater, such as Japanese Patent Showa 35-6652, which reported the use of dicyandiamide formaldehyde condensate for the treatment of pulp wastewater. Japanese Patent Showa 36-23231 elaborated on the use of dicyandiamide formaldehyde condensate for the treatment of wastewater from the paper industry, meat processing plants, and oil factories.

The application of melamine formaldehyde resin as a decolorizing agent began in the early 1990s, and it is now widely used as a decolorizing agent and paper sizing agent both domestically and abroad. Melamine formaldehyde resin provides a large number of cations, causing the negative charges carried by dye molecules to be neutralized and destabilized. At the same time, the added melamine formaldehyde resin hydrolyzes to produce a large amount of flocculent material, which can adsorb and trap the destabilized dye molecules, separating them from the water and achieving the purpose of decolorization.

Japanese Patent Sho 51-29116 reports the production of melamine formaldehyde resin using a "one-step condensation" process, in which the reactants are added all at once and the reaction is carried out quickly. However, this method has serious problems such as rapid temperature rise, difficult control of the reaction, and easy boiling. In addition, the product obtained from the reaction of melamine and formaldehyde has a relatively low molecular weight, and the flocs formed with dye molecules are small and settle slowly. Therefore, in recent years, extensive research has been conducted on the synthesis, modification, and compounding of melamine formaldehyde with other coagulants.

7.3.1 Preparation Method

It is generally believed that when melamine reacts with formaldehyde, formaldehyde first reacts with amino ($-NH_2$) or imino ($=NH$) to form hydroxymethyl ($-CH_2OH$), then hydroxymethyl dehydrates with hydrogen on amino or imino to form ether bond (–C–O–C–) and form linear or branched high molecular weight polymers. This process can be represented by the following equations.

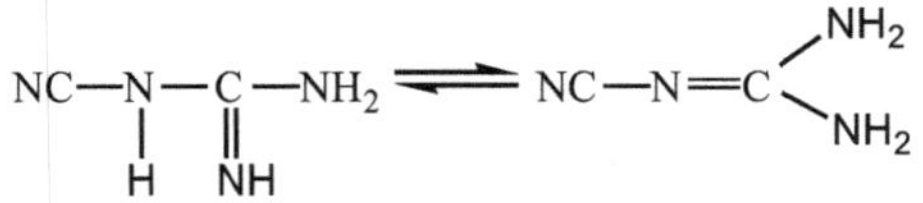

(7.1)

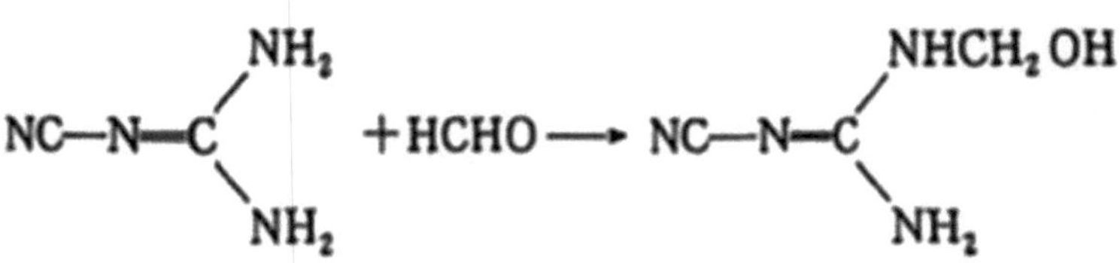

(7.2)

During the reaction, the addition of ammonium chloride reacts with hydroxymethyl melamine as follows:

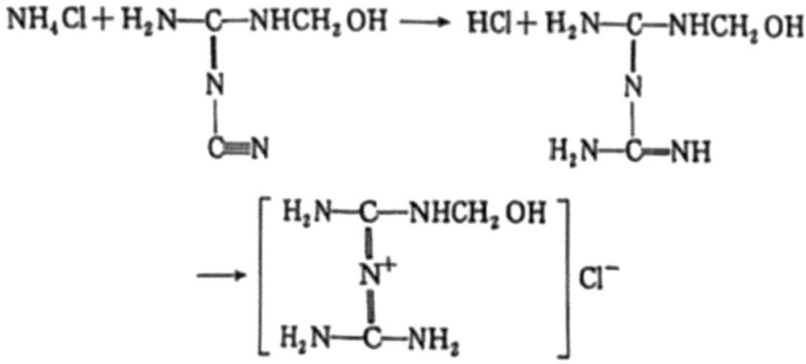

(7.3)

Further reaction with formaldehyde produces:

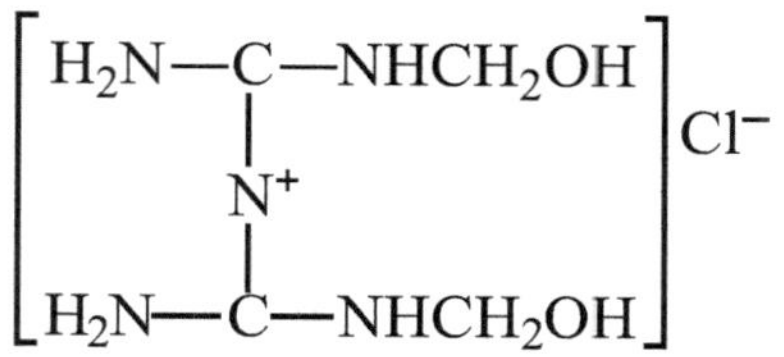

The hydroxymethyl on the above intermediate product reacts with the amine and amino on another molecule to form a secondary methyl bond, forming a polymer with the following structure:

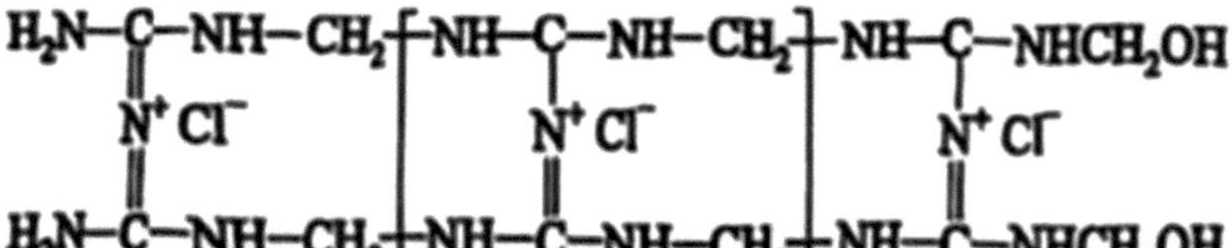

At present, the "two-step condensation" process is commonly used in China, the product quality is relatively stable, easy to control, and through the selection of raw materials and optimization of process conditions, the prepared melamine formaldehyde has a broad-spectrum and efficient application effect. The two-step method uses formaldehyde as the main raw material, Due to the potential harm of residual formaldehyde to aquatic organisms. Therefore, glyoxal, melamine, and ammonium chloride can be used as raw materials to produce polymers.

7.3.2 Production Process

In industrial production, the "one-step" reaction was once used, that is, all the raw materials were put into the reaction kettle at once. Since this reaction is a strong exothermic process, the temperature can rise by more than 10 °C per minute at the peak of heat release, and the temperature can rise to close to 100 °C, making operation difficult and prone to boiling and spraying accidents, making production extremely unsafe. The improved "two-step method" adds formaldehyde or ammonium chloride in two steps to alleviate the exothermic process. That is, more than half of the materials are added first, and after the reaction reaches a certain stage, the remaining part of the materials is slowly added. This is the production process widely used at present. Because the price of melamine is relatively high, some companies use urea instead of melamine to reduce costs. The product has poor low-temperature resistance and becomes turbid or even stratified when the winter temperature is low. Although it can be restored after heating and stirring, it does not affect the treatment effect, but it still brings inconvenience to the user. After improvement, solid polymers can be produced, which are not affected by temperature and storage time, and the solubility of solid polymers is very good, and they can all dissolve within 10 min.

Melamine formaldehyde resin is generally used under neutral or weakly alkaline conditions. Under certain conditions, the flocs produced during the decolorization process are small, affecting their subsequent separation. Nonionic or cationic polyacrylamide flocculants can be added to increase the particle size of the flocs. In addition, this polymer can also be added with polyacrylamide after dosing to increase the particle size and settling speed of the flocs. During use, it can also be treated with aluminum salt coagulant first, and then treated with polymer coagulation, which can greatly reduce the dosage of decolorizer.

7.3.3 Comparison with Other Coagulants

The difference between melamine formaldehyde resin and commonly used polyacrylamide (PAM) and inorganic coagulants can be roughly summarized as shown in Table 7.1. Users can choose according to the characteristics of the wastewater to be treated.

Common inorganic coagulants used in the treatment of printing and dyeing wastewater include iron salts, aluminum salts, and magnesium salts. In the following experiments, four widely used and representative inorganic coagulants were selected, namely aluminum sulfate, ferrous sulfate, polyaluminum chloride (PAC), and polyferric sulfate (PFS). Among them, $Al_2(SO_4)_3 \cdot 18H_2O$ and $FeSO_4 \cdot 7H_2O$ are analytically pure reagents. The main parameters of PAC and PFS are as follows: PAC is a liquid with an Al_2O_3 content of 18% and a basicity of 60%; PFS is a liquid with a Fe^{3+} content of 11%, a basicity of 12%, and a density of 1.47 g/cm^3. The above four coagulants were compared with the melamine formaldehyde decolorizing flocculant, and the optimal coagulation conditions and corresponding decolorization rates of the five coagulants are shown in Table 7.2.

As can be seen from Table 7.2: Comparing the decolorization rates of aluminum and iron coagulants, it is not difficult to find that the coagulation effect of $FeSO_4$ is better than that of PFS, aluminum sulfate, and PAC. When the pH value of the dye solution is 9 ~ 11, adding iron salt coagulant has a certain decolorization effect, and the effect of ferrous sulfate is better than that of polyferric sulfate. In the experiment,

Table 7.1 Comparison of the performance of melamine formaldehyde resin with PAM and inorganic flocculants

Item	Melamine formaldehyde resin	PAM series	Inorganic coagulant
Relative molecular mass	$10^3 \sim 10^4$	$10^6 \sim 10^7$	10^2
Solubility in water	Easily soluble	Difficult to dissolve	Easily soluble
Working solution concentration	Low	Low	High
Suitable pH range	Neutral to alkaline	Acidic to alkaline	Weakly acidic to weakly alkaline
Treatment effect on water-soluble dye wastewater	Good	Poor, can be used in combination with melamine formaldehyde resin or inorganic coagulant	Has some effect, worse than melamine formaldehyde resin, can be used in combination with it
Treatment effect on wastewater containing suspended, colloidal dyes	Good	Has some effect, can be used in combination with melamine formaldehyde resin or inorganic coagulant	Has some effect, worse than melamine formaldehyde resin, can be used in combination with it
Price	High	Highest	Low
Dosage	Medium amount	Trace amount	Large amount
Amount of slag	Small amount	Small amount	Large amount
Possibility of waste slag incineration	Yes	Yes	Not suitable, and will release acidic gas

Note The applicable dyes are reactive, acidic, direct, and dispersed dyes, excluding cationic dyes

Table 7.2 Comparison of optimal coagulation conditions and decolorization rates of five coagulants

Coagulants	$Al_2(SO_4)_3$	PAC	$FeSO_4$	PFS	Melamine formaldehyde decolorizing agent
pH Value	6.5 ~ 7	6.5 ~ 7	11	6.5 ~ 7	7 ~ 10
Dosage (ratio to dye concentration)	1:8	1:2	1:4	1:2	1:1
Decolorization rate (%)	33.1	66.8	90.5	45.4	99.9

it was also observed that the dye water sample after the addition of $FeSO_4$ changed color, the solution changed from red to light yellow, which may be due to the strong reductive ferrous hydroxide reducing the dye. However, the sample after using the decolorizing agent can achieve complete colorlessness, and at the same dosage, the decolorization effect increases with time. Therefore, in actual production and application, a composite method can be adopted to prepare a composite coagulant of polyaluminum chloride and ferrous ions to improve the decolorization effect.

7.4 Factors Affecting the Coagulation Decolorization Effect of Melamine Formaldehyde Resin

7.4.1 Influence of Dye Molecular Structure

The decolorization experiment of simulated wastewater of Acid Red 1, Reactive Black 5, Van Yellow RK and Disperse Blue 60 dyes using melamine–formaldehyde decolorizer was carried out, and the decolorization rate can reach more than 99%. However, due to the different structures of various dyes, the coagulation decolorization effect has certain differences. The structural formulas of the dyes used in the experiment are shown in Fig. 7.1.

Apart from reactive dyes, all acidic and disperse dyes have good water solubility due to the presence of sulfonic acid groups in their molecular structures. Because the backbone of the dye molecule is hydrophobic, without the presence of carboxyl or sulfonic acid groups, the solubility of the dye molecule is very poor. Weak acid dyes are generally monoazo or diazo types, with complex structures, containing $-SO_3H$, –OH, and other hydrophilic groups in the molecule, and exist in a state close to a colloid in aqueous solution at room temperature.

The basic skeleton of the reducing dye molecules is a large molecular weight polycyclic aromatic compound, containing multiple –C=O and –NH– groups. Due to the large number of hydrophobic aromatic rings and fewer hydrophilic groups, they mainly exist as hydrophobic suspended particles, with poor stability. When general coagulants, such as aluminum sulfate, are added, they easily agglomerate and are removed.

Reactive dyes, direct dyes, and disperse dyes generally belong to diazo or triazo structures, with a higher content of $-SO_3H$, –OH, and other hydrophilic groups in the molecule, good water solubility, high solubility, and often exist in a state close to a true solution. Even if a large amount of general coagulant is used, the decolorization rate is very low.

7.4.2 The Effect of the pH Value of the Dye Solution or Wastewater

The positive charge on the molecular chain of the melamine–formaldehyde decolorizer comes from the imino group. The OH– in the solution affects the probability of the decolorizer binding with the negative charge in the dye molecule. In addition, the optimal pH value is also related to the molecular weight of the decolorizer. Figure 7.2

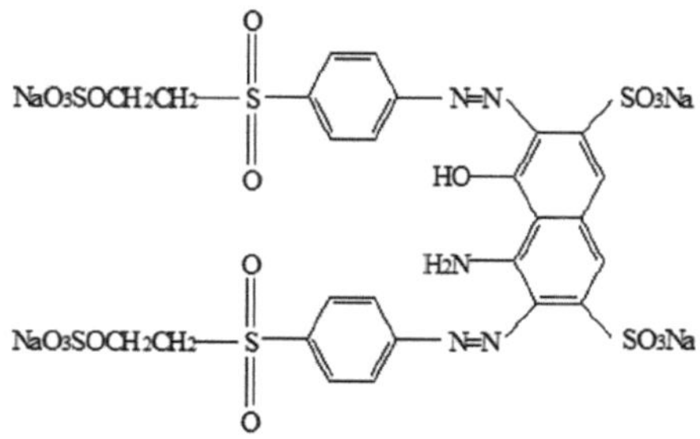

Reactive Black 5

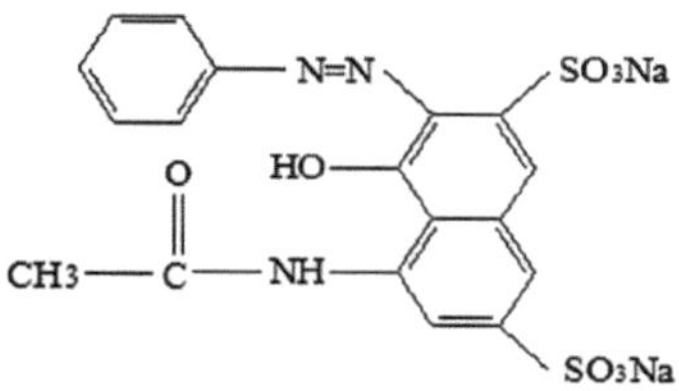

Acid Red 1

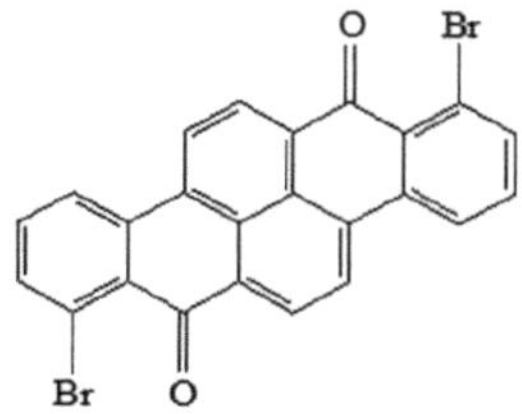

Van Yellow RK

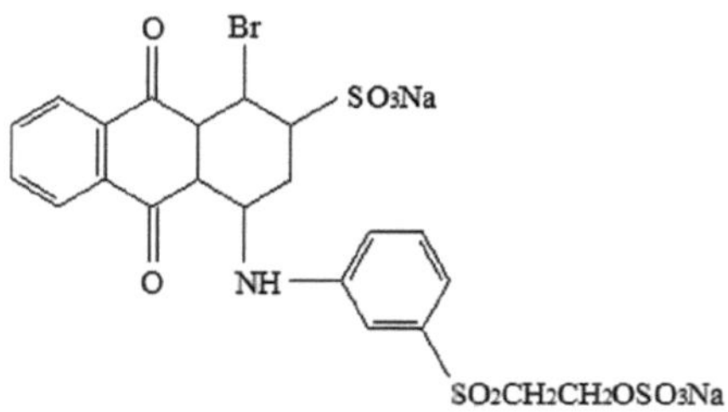

Disperse Blue 60

Fig. 7.1 Structural formulas of the dyes used in the experiment

is the UV–visible spectrum of the mixture of Reactive Red 2 solution and melamine–formaldehyde decolorizer at pH 6.0. Curve A is the dye solution, and curve B is the mixed solution after adding the decolorizer. It was found in the experiment that when the pH value is 6.0, the Reactive Red 2 dye solution remains a solution after mixing with the decolorizer, and no precipitation appears in the treated dye solution. However, during the mixing process (dye/decolorizer = 200 mg/mL/200 mg/L), the color of the mixture can be observed to gradually change from vermilion to pink. As

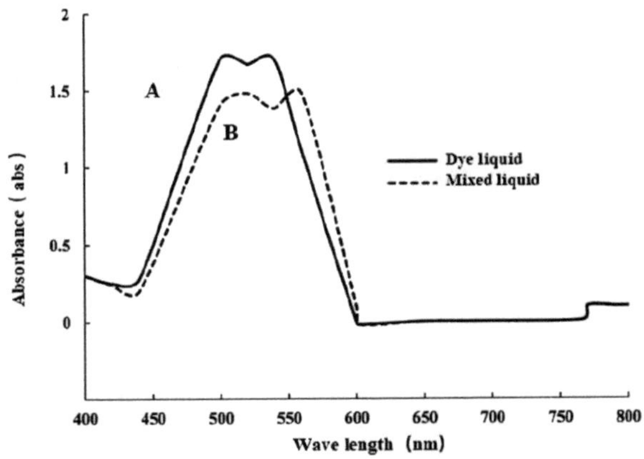

Fig. 7.2 Visible spectrum of the mixture of Reactive Red and decolorizer

can be seen from Fig. 7.2, the maximum absorption wavelength of the mixed solution after adding the decolorizer changes from 535 to 525 nm. It was also found in the experiment that under this pH condition, increasing the amount of decolorizer added does not affect the reaction between the dye solution and the decolorizer. Changes in the maximum absorption wavelength can also be observed after other dye solutions are mixed with the decolorizer. If the pH of the dye solution is adjusted to 8.0 before adding the decolorizer, small alum flowers will be produced after stirring with the decolorizer. This shows that the reaction of Reactive Red 2 with the decolorizer is related to the pH value. When treating other dye solutions containing sulfonic acid groups, such as Reactive Black 5 and Acid Red 1, with melamine–formaldehyde decolorizer, the same results can be obtained.

From the above experiments, it can be seen that the pH value has a significant impact on the decolorization effect of this cationic flocculant. This may be related to the molecular structure of the polymer. The polymer is relatively stable under weakly alkaline conditions, while under strongly acidic and strongly alkaline conditions, some groups in the polymer may dissociate and become unstable, resulting in a less-than-ideal removal rate of the dye solution's color. The pH value not only affects the cationic flocculant but also affects the hydrogen bonds between the dyes. The dyes used in the experiment, Reactive Red 2, Reactive Black 5, and Acid Red 1, all contain –OH or $-NH_2$ groups in their molecular structures, and there are also molecular hydrogen bonds when interacting with dye molecules. Due to the influence of pH value on the hydrogen bonds between molecules, the hydrogen bond interaction between molecules weakens in acidic media, making it difficult for the small molecules produced by the reaction to aggregate and settle, resulting in a lower decolorization rate of the dye solution than in neutral and slightly alkaline media. However, if the alkalinity is too high ($pH > 11$), the decolorization rate will decrease. The likely reason is that a high pH value reduces the $-NH_3^+$ in the flocculant, affecting its binding with the sulfonic acid group.

Research shows that under weakly alkaline conditions, the decolorizing agent molecules exist in a relatively stable cationic form, react with the carboxyl or sulfonic

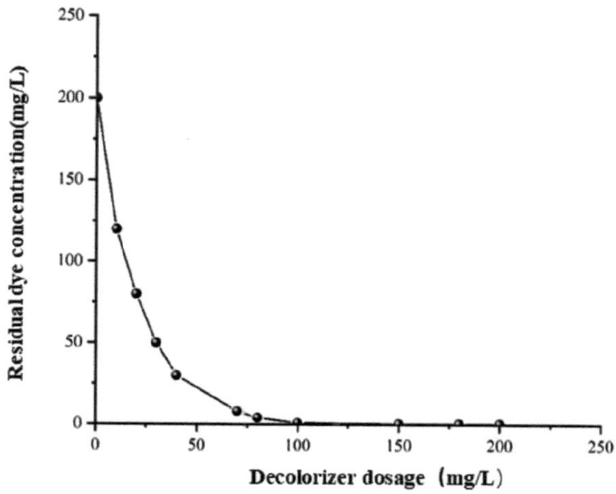

Fig. 7.3 Relationship between the residual concentration of Reactive Red 2 dye solution and the dosage of decolorizing flocculant at pH 8

acid groups contained in the dye molecules, destabilize the soluble dye molecules, and further flocculate and precipitate. Figure 7.3 shows the relationship between the residual concentration of Reactive Red 2 dye solution and the dosage of decolorizing flocculant when the pH value is 8. The curve in the figure shows that this decolorizing agent can effectively remove the color of the Reactive Red 2 solution, with a decolorization rate of over 99%.

In the experiment, it was found that the residual color of the Reactive Red 2 solution after treatment was difficult to measure with a spectrophotometer. In the experiment, it was easy to distinguish the solution of Reactive Red 2 with a concentration of 0.5 mg/L and deionized water with the naked eye and spectrophotometer, but it was difficult to distinguish the color difference between the dye solution after treatment with the decolorizing agent and deionized water. Figure 7.4 is the wavelength-absorption value relationship of the dye after treatment with the decolorizing agent and the standard solution. Curve A is the blank sample, curves B and C are solutions of Reactive Red 2 dye with concentrations of 0.5 mg/L and 1.0 mg/L, respectively, and curve D is the solution after treatment with the decolorizing agent. Although the peak bottom of the treated sample in the figure is higher than the standard 0.5 mg/L Reactive Red 2 solution, it can be observed in the experiment that the former is almost colorless, while the standard solution still has color. The high peak bottom of curve D may be caused by residual salts and sodium chloride. When the pH value is 8, the removal rate of the color of the Reactive Red 2 solution by the melamine formaldehyde decolorizing agent can reach more than 98%. The same effect can be achieved by treating other reactive and acidic dye solutions with this decolorizing agent.

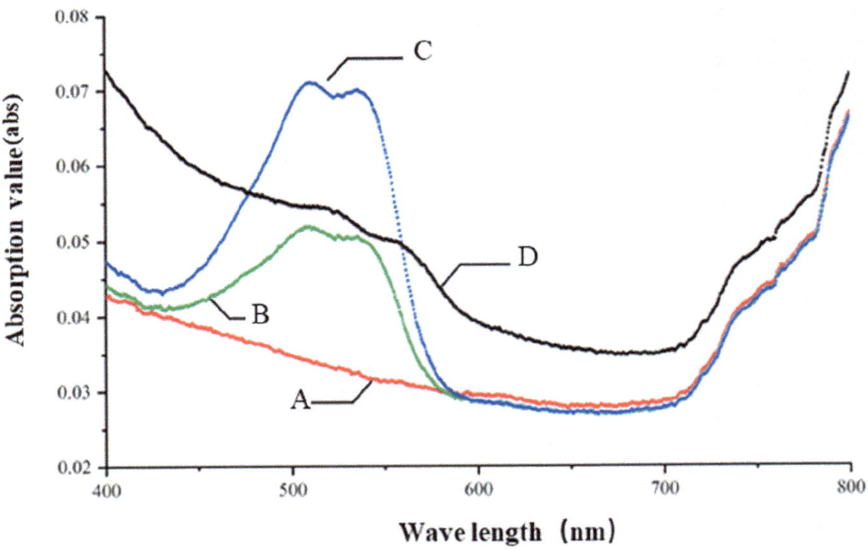

Fig. 7.4 Wavelength-absorption value relationship of dye solution after treatment with decolorizing agent and standard solution

7.4.3 The Impact of Stirring Speed

The coagulation process is divided into two stages: mixing and flocculation. The purpose of mixing is to rapidly and evenly disperse the reagent in the water, so as to react with the dye molecules and create good flocculation conditions. At the same time, the colloid begins to destabilize, and coagulation occurs through particle Brownian motion and turbulent action, but it does not require the formation of large flocs. The purpose of flocculation is to separate the coagulated particles from the water through inter-particle bridging, net trapping, etc., and form good flocs. The requirements for stirring are different for different purposes. The mixing stage requires intense and rapid action, with a short completion time, while flocculation requires a gradual decrease in stirring speed as the floc grows, and a longer reaction time. Because the melamine formaldehyde decolorizer has both coagulation and flocculation functions, its use method is the same as that of ordinary coagulants, that is, rapid stirring is required in the early stage of the reaction to destabilize the colloid; slow mixing is required in the later stage to allow the floc to grow and produce a good sedimentation effect. Generally, fast stirring (250 rpm/min) for 1 min and slow stirring (50 rpm/min) for 5 min are chosen.

7.4.4 *The Impact of Decolorizer Dosage*

According to the relationship between the dosage of melamine formaldehyde decolorizer and the removal rate of Reactive Red 2 dye wastewater mentioned earlier, when the dosage of decolorizer is 0.6 ~ 0.9 times the dye concentration, the decolorization rate can reach more than 95%, and the COD removal rate is greater than 70%. When the dosage of decolorizer is about 1 time the dye concentration, the decolorization rate can reach 99 ~ 100%. Therefore, it is speculated that the decolorizer reacts with the dye solution in an equivalent complexation reaction. However, because this decolorizer contains a large number of double bonds and amino groups, if the dosage exceeds the amount required for flocculating the dye, it often causes the COD of the solution to increase, so the dosage must be properly controlled during use.

7.4.5 *Preliminary Exploration of the Structure of the Decolorizer and the Flocculation Mechanism*

By measuring the Raman spectra of the decolorizer, Reactive Red 2 dye, and the precipitate formed after their reaction, it was found that the spectral lines of the precipitate and the dye are basically similar, indicating that the structure of the dye molecules did not change before and after decolorization, and there are no dye molecules in the solution after complete removal, and no chemically degraded substances were produced during the reaction. This shows that the dye is removed by flocculation to form insoluble macromolecules, and no reduction reaction occurs. Because Reactive Red 2 and other dye molecules contain hydrophilic sulfonic acid groups, they have excellent water solubility, and polar groups will form a water film on the surface of the gel particles after adsorbing a large number of polar molecules, which is usually very stable. This type of dye can only be destabilized and removed by electrostatic neutralization and compression of the double electric layer, which can be disproved by the fact that high cationic degree polyacrylamide and diallyl dimethylammonium chloride copolymer have no decolorizing effect on active dyes. In addition, the precipitate formed during the coagulation process is insoluble in strong acid and strong alkali, which further proves that this coagulation is a chemical reaction. If the decolorization mechanism of the aforementioned Reactive Black 5, Acid Red 1, Disperse Blue 60, and active dyes is completely the same, but due to the difference in hydrophilicity or the number of hydrophilic groups of different dyes, the amount of coagulant needed for destabilization and decolorization still has a certain difference. For reductive dyes, such as Van Yellow RK, because of its poor hydrophilicity, general aluminum salt coagulants can achieve very good results. Although the use of decolorizers can achieve very good results, the treatment cost is higher.

During the synthesis process of the decolorizing agent, acrylamide is added to form a polyacrylamide molecular chain that connects cationic units, linking together smaller molecular weight cationic ammonium salt polymers, thereby giving the decolorizing agent good bridging properties. It is precisely because of this special structure that the decolorizing agent has good precipitation and flocculation effects. Therefore, it can be used alone or in combination with aluminum salt coagulants.

7.5 New Developments in Coagulant Decolorizing Agent Research

Given the characteristics of strong hydrophilic sulfonic and carboxylic groups in the molecular structure of reactive dyes and direct dyes, which carry negative charges when dissolved in water, the key is to destroy or seal the hydrophilic groups of the dyes, reduce their hydration, and then destabilize, coagulate, and flocculate under the action of flocculants to achieve the purpose of separation from the solution. At the same time, in response to problems such as the relatively small molecular weight of the product obtained by the reaction of melamine and formaldehyde, the small size of the flocs formed with dye molecules, slow sedimentation, and poor flocculation performance, scholars at home and abroad have conducted a series of modification studies. For example, by adding a certain amount of polyamine crosslinking agent, the relative molecular weight of the product and the flocculation effect can be improved. By adding aluminum chloride, aluminum sulfate, acrylamide, urea, and other catalysts and modifiers, as well as compounding with polyaluminum chloride, magnesium chloride, etc., the flocculation effect can be greatly improved. The State Key Laboratory of Pollution Control and Resource Reuse at Tongji University has successfully developed the TJ series of decolorizing agents, which have both flocculation and precipitation effects, and can effectively remove various active, acidic and other soluble dyes, with a decolorization rate of over 98%, providing a good approach for the treatment of printing and dyeing wastewater in China; this series of decolorizing agents use guanidine polymers to seal the hydrophilic groups of dyes and precipitate the dyes. This decolorizing agent can be used in biochemical treatment or pretreatment processes, and there are no specific requirements for water treatment facilities.

In addition to coagulation decolorization, adsorption and advanced oxidation technologies are also commonly used decolorization methods. Common adsorbents mainly include activated carbon, coke, silicon polymers, diatomaceous earth, kaolin, and industrial slag. The adsorption method has problems such as the adsorbent is easily saturated, the treatment effect decreases over time; the regeneration of the adsorbent is difficult, and the regeneration wastewater and saturated waste adsorbent can easily cause secondary pollution. Therefore, it is generally used for the treatment of low-concentration printing and dyeing wastewater or advanced treatment. Fenton oxidation technology based on hydroxyl radicals with high oxidation–reduction potential is usually used for efficient pretreatment or advanced treatment of printing and dyeing wastewater. The Fenton method has a large amount of iron sludge production, and the amount of acid and alkali used to adjust the pH value significantly increases the operating cost. In recent years, achieving heterogeneous Fenton oxidation under neutral or near-neutral pH conditions, reducing the production of iron sludge and the consumption of acid and alkali has received widespread attention. The method of combining ozone oxidation and bio-aerated filters (BAF) has been used for advanced treatment and reuse of printing and dyeing wastewater in China, with an organic matter removal rate of 30 ~ 40% for secondary biochemical effluent of printing and dyeing wastewater. The future development direction is to enhance ozone oxidation with catalysts. In addition, domestic and foreign research has been conducted on ion exchange decolorization, ultrafiltration membrane decolorization, and biological decolorization technologies. Among them, the treatment of water-soluble dyes that are difficult to decolorize by conventional methods using ion exchange methods has been studied and made some progress. The research focuses on the development of ion exchange resins and ion exchange decolorizing fibers.

For slightly soluble dyes and dye components with larger molecular weights, ultrafiltration or reverse osmosis technology can be used for decolorization treatment, but considering economic feasibility, ultrafiltration technology is currently mostly used for high-concentration dyes and dyeing wastewater treatment, especially for the recovery and reuse of insoluble dyes.

Due to the poor biodegradability of dye components in printing and dyeing wastewater, conventional biochemical methods have always been unsatisfactory in terms of decolorization. The current solutions, besides improving the biodegradability of wastewater through pretreatment, mainly involve the selection of excellent decolorizing bacteria and the enhancement of the biological treatment process. Enhanced biological processes, optimized biochemical processes, and other methods represented by anaerobic–aerobic systems, biochar methods, bio-iron methods, and forced aeration have improved their decolorization efficiency to a certain extent. In addition, every city currently produces a large amount of acetylene waste liquid. Enzyme-catalyzed methods can effectively decompose it, but the degradation speed is slow, and there is no foreseeable application prospect in the near future. Overall, there has been no breakthrough progress in biological decolorization, and it must still be used in conjunction with other treatment methods.

Chapter 8
Adsorbents

8.1 Principles of Adsorption

8.1.1 Phenomenon of Adsorption

Under normal circumstances, when any two phases come into contact, the forces at the interface are unbalanced, and interactions occur between the molecules, causing the molecules at the interface to lose balance and undergo adsorption. This phenomenon, where certain components in the liquid accumulate or concentrate automatically on the surface of the solid when the solid comes into contact with the liquid, is called adsorption. Substances with adsorption capabilities are called adsorbents, and the substances being adsorbed are called adsorbates. For example, in wastewater, activated carbon is used to adsorb suspended particles. The activated carbon is the adsorbent, and the dissolved substances or finer suspended particles that are adsorbed are called adsorbates.

The adsorption process can be divided into: ① The adsorbate first transfers from the solution to the outer surface of the adsorbent in the form of molecules or ions through convective diffusion, which is the external mass transfer process; ② Then the adsorbate transfers from the outer surface of the adsorbent to the inner surface of the microporous structure through pore diffusion, which is the internal mass transfer process; ③ The adsorbate diffuses along the surface of the pore; ④ Finally, the adsorbate is adsorbed on the surface of the pore.

Solid-phase adsorbents have a wide range of applications in production and scientific experiments. In wastewater treatment, the adsorption effect of the surface of solid substances on substances in wastewater is mainly used to remove trace pollutants in wastewater, achieve the purpose of purifying water quality, such as the removal of trace heavy metal ions in wastewater, the removal of trace harmful biodegradable organic substances, decolorization, deodorization, etc.

L. Fengting, *Water Treatment Chemicals: Coagulants and Flocculants*,
https://doi.org/10.1007/978-981-96-8221-8_8

8.1.2 *Physical Adsorption and Chemical*

Adsorption According to the different nature of the interaction between the adsorbent and the adsorbate, adsorption can be divided into physical adsorption and chemical adsorption. During physical adsorption, the adsorbent and the adsorbate molecules interact with each other through van der Waals forces; during chemical adsorption, the adsorbent and the adsorbate molecules undergo a chemical reaction and bind together through chemical bonds. Because physical adsorption and chemical adsorption have essential differences in intermolecular forces, they exhibit many different adsorption properties, as shown in Table 8.1.

Since physical adsorption is caused by intermolecular forces, the heat of adsorption is small, generally within 41.9 kJ/mol. Because intermolecular forces are ubiquitous and non-selective, one adsorbent can adsorb multiple adsorbates. However, due to the different polarities of the adsorbent and the adsorbate, the amount of adsorption of a certain adsorbent for various adsorbates is different. Molecules can separate from the adsorbent through thermal motion, so physical adsorption is reversible.

Chemical adsorption generally occurs at higher temperatures, with larger adsorption heat, equivalent to the heat of chemical reactions, generally 83.7 ~ 418.7 kJ/mol. One adsorbent can only chemically adsorb a certain or several adsorbates, so chemical adsorption is selective. Because chemical adsorption relies on the chemical bond force between the adsorbent and the adsorbate, adsorption can only form a single molecular adsorption layer. The bond energy of the chemical bond is large and generally cannot spontaneously break, so chemical adsorption is irreversible.

Physical adsorption and chemical adsorption are not isolated and often occur together. Most of the adsorption in water treatment is the result of the combined action of the two types of adsorptions. Due to the influence of adsorbates, adsorbents, and other factors, a certain type of adsorption may be dominant.

Table 8.1 Comparison of the properties of physical adsorption and chemical adsorption

Property	Physical adsorption	Chemical adsorption
Adsorption force	Van der Waals force	Chemical bond force
Adsorption layers	Single or multiple layers	Single layer
Adsorption heat	Small (close to liquefaction heat)	Large (close to reaction heat)
Selectivity	None or poor	Strong
Reversibility	Reversible	Irreversible
Adsorption equilibrium	Easy to reach	Not easy to reach

8.1.3 *Adsorption Isotherm and Adsorption Isotherm Equation*

8.1.3.1 Adsorption Isotherm

In physical adsorption, after the wastewater is fully in contact with the adsorbent, on the one hand, the adsorbate is adsorbed by the adsorbent, on the other hand, some of the adsorbed adsorbate can leave the surface of the adsorbent due to molecular thermal motion and return to the liquid phase. The former is called the adsorption process, and the latter is called the desorption process. When the adsorption rate and the desorption rate are equal, that is, the amount of adsorption per unit time is equal to the amount of desorption, the concentration of the adsorbate in the solution and the concentration on the surface of the adsorbent no longer change and reach equilibrium. At this time, the concentration of the adsorbate in the solution is called the equilibrium concentration, that is, the so-called adsorption equilibrium.

The adsorption capacity of the adsorbent is represented by the adsorption amount q. The so-called adsorption amount refers to the mass of the adsorbate adsorbed by the unit mass of the adsorbent (g). Take a certain volume V(L) of water sample with an adsorbate concentration of C_0(g/L), and add activated carbon with a mass of W(g). When the adsorption equilibrium is reached, the remaining adsorbate concentration in the wastewater is C(g/L), then the adsorption amount q can be calculated by the following formula:

$$q = \frac{V(C_0 - C)}{W} \tag{8.1}$$

where q is adsorption capacity, g/g;

V is wastewater volume, L;

W is amount of activated carbon added, g;

C_0 is original water adsorbate concentration, g/L;

C is remaining adsorbate concentration in water at adsorption equilibrium, g/L.

Under constant temperature conditions, the adsorption amount increases with the increase of the equilibrium concentration. The curve of the change of the adsorption amount with the equilibrium concentration is called the adsorption isotherm. There are two common types of adsorption isotherms, as shown in Fig. 8.1.

In order to better describe the adsorption isotherm and obtain the maximum adsorption amount and other information, various models are used to fit the adsorption isotherm data. Common adsorption isotherm models are shown in Table 8.2.

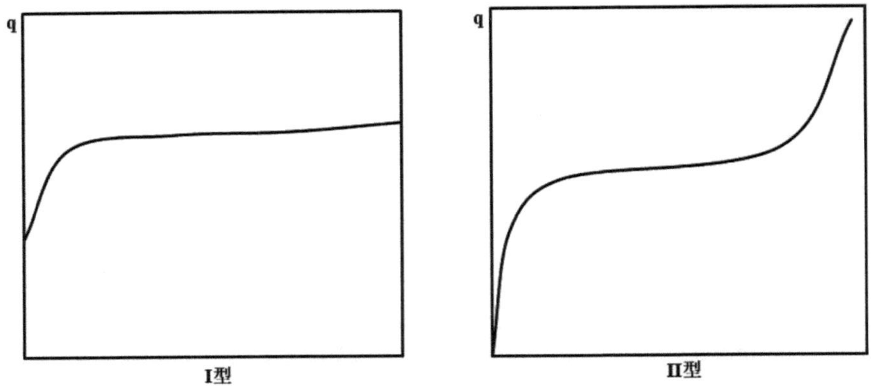

Fig. 8.1 Form of adsorption isotherm

Table 8.2 Common adsorption isotherm models

Model name	Formula	Parameter meaning
Linear model	$q_e = kC_e$	K is the adsorption affinity coefficient (L/g)
Langmuir model	$q_e = \frac{q_m C_e}{1/b+C_e}$	q_m is the maximum adsorption capacity (mg/g); b is the adsorption equilibrium constant (L/mg)
Freundich model	$q_e = K_f {C_e}^{1/n}$	K_f is the Freundich affinity coefficient ((mg/g)/(mg/L)$^{1/n}$); n is the Freundich constant
Langmuir-Freundich model	$q_e = \frac{q_m (bC_e)^n}{1+(bC_e)^n}$	q_m is the maximum adsorption capacity (mg/g); b is the adsorption equilibrium constant (L/mg); n is the Freundich constant
BET model (gas adsorption)	$\frac{V}{V_m} = \frac{cp}{(p_0-p)[1+(c-1)p/p_0]}$	V is the volume of the adsorbed gas; V_m is the adsorption amount when a single molecular layer is adsorbed; p_0 is the saturated vapor pressure of the adsorbate at the adsorption temperature is a constant related to the heat of vaporization of the adsorbate

8.1.3.2 Langmuir's Monolayer Adsorption Theory and Adsorption Isotherm

In 1916, Langmuir proposed a theory of solid–gas adsorption based on a large number of experimental facts and from a kinetic point of view, known as the monolayer adsorption theory. The Langmuir model is the most commonly used adsorption isotherm equation, with the following assumptions: ① Adsorbates can only adsorb in a monolayer on the solid surface; ② The adsorption action on the solid surface is uniform; ③ There is no interaction between the adsorbed molecules. ④ Adsorption equilibrium is dynamic equilibrium. That is to say, the Langmuir adsorption model can simulate experimental results well when the adsorption action on the solid surface is quite uniform and adsorption is limited to a monolayer. The assumptions of this model are not strictly valid, it is sensitive to changes in experimental conditions and has considerable limitations. In the actual adsorption process, especially in liquid phase adsorption, there are many influencing factors, and there is often a force between the adsorbed molecules, causing many adsorption processes to not be well fitted by this model. Therefore, this model can only explain the situation of monolayer adsorption. Nevertheless, the Langmuir model is still an important adsorption isotherm model, its derivation process considers the adsorption mechanism, laying the foundation for the establishment of other adsorption models.

Because liquid phase adsorption is very complex and there is still no unified adsorption theory, the adsorption isotherm of liquid phase adsorption has always used the gas phase adsorption isotherm. The Langmuir (Langmuir) formula and Freundich (Freundich) formula represent type I adsorption isotherms, and the BET (Brunauer–Emett–Teller) formula represents type II adsorption isotherms, which are now described as follows.

(1) Langmuir Formula

The Langmuir formula is a single molecule adsorption formula derived from a kinetic point of view through some assumptions:

$$q = \frac{\mathrm{ab}C}{(1 + \mathrm{a}C)} \tag{8.2}$$

where q is adsorption amount;

C is equilibrium concentration of adsorbate;

a, b are constants.

For convenience of calculation, the above formula can be changed to a reciprocal formula, namely

$$\frac{1}{q} = \frac{1}{\mathrm{ab}} \cdot \frac{1}{C} + \frac{1}{\mathrm{b}} \tag{8.3}$$

From the above equation, it can be seen that 1/q and 1/C have a linear relationship, and this relationship can be used to find the values of a and b.

(2) Freundlich empirical formula

$$q = \mathrm{K}C^{\frac{1}{\mathrm{n}}} \tag{8.4}$$

In the formula, q and C have the same meaning as in the previous formula;

K and n are constants.

The above formula can be rewritten as a logarithmic formula:

$$\lg q = \lg \mathrm{K} + \frac{\lg C}{\mathrm{n}} \tag{8.5}$$

By plotting the points of C and the corresponding q on a double logarithmic coordinate paper, an approximate straight line is obtained. The intercept of this line is K, and the slope is $\frac{1}{n}$. The smaller $\frac{1}{n}$ is, the better the adsorption performance. It is generally believed that when $\frac{1}{n} = 0.1$–0.5, it is easy to adsorb; when $\frac{1}{n} > 2$, it is difficult to adsorb. When $\frac{1}{n}$ is large, that is, the higher the equilibrium concentration of the adsorbate, the greater the adsorption amount, and the more fully the adsorption capacity is exerted. In this case, it is best to use continuous adsorption operations; when $\frac{1}{n}$ is small, intermittent adsorption operations are often used.

(3) BET formula

The BET formula represents an adsorption model in which multiple layers of solute molecules are adsorbed on the adsorbent, and the adsorption of each layer complies with the Langmuir single-molecule adsorption formula. The formula is as follows:

$$q = \frac{\mathrm{B}Cq_0}{(C_S - C)\left[1 + (\mathrm{B} - 1)\frac{C}{C_s}\right]} \tag{8.6}$$

where q_0 is the saturated adsorption amount of the single-molecule adsorption layer;

C_s is the saturated concentration of the adsorbate, g/L;

B is a constant.

For convenience of calculation, the above formula can be changed to a reciprocal formula, that is,

$$\frac{C}{(C - C_s)q} = \frac{1}{\mathrm{B}q_0} + \frac{\mathrm{B} - 1}{\mathrm{B}q_0} \cdot \frac{C}{C_s} \tag{8.7}$$

From the above formula, it can be seen that $\frac{C}{(C-C_s)q}$ and $\frac{C}{C_s}$ have a linear relationship, and this relationship can be used to find the values of q_0 and B.

The amount of adsorption is an important data for selecting adsorbents and designing adsorption equipment. The size of the adsorption amount determines the length of the regeneration cycle of the adsorbent. The larger the adsorption amount, the longer the regeneration cycle. Although the adsorption amount obtained from the adsorption isotherm is not exactly the same as the actual adsorption amount, the method of obtaining the adsorption amount from the adsorption isotherm is simple and easy to implement, provides comparable data for the selection of adsorbents, and has certain reference significance.

8.1.4 Adsorption Rate

(1) Definition of adsorption rate

The adsorption effect of the adsorbent on the adsorbate is generally measured by the adsorption capacity and adsorption rate. The so-called adsorption rate refers to the mass of the substance adsorbed by the unit mass of the adsorbent in unit time. Solid-phase adsorption from wastewater can be divided into four processes: ① The diffusion of the adsorbate from the liquid phase to the surface liquid film of the solid phase, known as external diffusion. ② The diffusion of the adsorbate through the solid phase surface liquid film to the outer surface of the solid phase, is known as film diffusion. The liquid film is the retention boundary layer of the solid phase surface, and its thickness is related to the stirring strength or flow rate. ③ The diffusion of the adsorbate inside the particles, pollutants enter the internal pores of the adsorbent from the outer surface of the adsorbent, and then diffuse to the inner surface of the solid, composed of diffusion in the pore solution and two-dimensional diffusion on the pore surface, simply known as internal diffusion. ④ The process of the adsorbate being adsorbed on the inner surface of the adsorbent solid phase, known as the surface adsorption process. Internal diffusion is often the rate-limiting step of the entire adsorption process, determining the adsorption rate. In addition, it is generally believed that the rate of physical adsorption is fast, the rate of chemical adsorption is slow, and the rate of chemical adsorption is closely related to the functional groups on the surface of the adsorbent and the adsorption mechanism.

A series of experiments have shown that the external diffusion rate of particles is directly proportional to the solution concentration. The higher the solution concentration, the faster the adsorption rate. For a certain mass of adsorbent, the external diffusion rate is also directly proportional to the size of the specific surface area of the adsorbent, so the smaller the particle diameter, the larger the diffusion rate, and the easier it is to adsorb. In addition, the external diffusion rate is also related to the degree or speed of agitation. Increasing the relative speed between the solution and the particles will thin the liquid film and increase the external diffusion rate. Internal diffusion in particles is more complex, and the diffusion rate is related to factors such as the size and structure of the adsorbent pores, the size and structure of the adsorbate particles. Therefore, for the difference of adsorbents like activated carbon, the state of internal voids is often the main factor affecting adsorption performance. The effect of particle size on internal diffusion is greater than that of external diffusion. It can be seen that the size of adsorbent particles or molecules and ions has a great impact on internal and external diffusion, and the smaller the particles, the faster the adsorption rate. Therefore, from the perspective of improving the adsorption rate, the smaller the particle diameter, the better. However, in actual applications, multiple ions may all undergo adsorption, resulting in competitive adsorption.

(2) Adsorption kinetics and model fitting

By measuring the adsorption amount of the adsorbate on the adsorbent at different adsorption times, a kinetic curve can be obtained, and the adsorption equilibrium time

Table 8.3 Common adsorption kinetic equations

Equation name	Formula	Parameter meaning
Pseudo-first-order kinetic equation	$q_t = q_e[1 - \exp(-k_1 t)]$	q_e is the equilibrium adsorption amount (mg/g); k_1 is the first-order equation adsorption rate constant
Pseudo-second-order kinetic equation	$\frac{t}{q_t} = \frac{1}{k_2 q_e^2} + \frac{t}{q_e} = \frac{1}{v_0} + \frac{t}{q_e}$	q_e is the equilibrium adsorption capacity (mg/g); k_2 is the quadratic equation adsorption rate constant;v_0 is the initial adsorption rate
Elovich equation	$q_t = a + b\ln t$	a and b are Elovich adsorption kinetics rate constants
Particle internal diffusion model	$q_t = k_p t^{0.5} + C = \frac{6q_e}{R}\sqrt{\frac{D}{\pi}}t^{0.5} + C$	k_p is the particle internal diffusion rate constant; R is the particle radius; D is the diffusion coefficient; q_e is the equilibrium adsorption capacity (mg/g);C is a constant
Freundich modified formula	$q_t = kC_e t^{\frac{1}{m}}$	k and m are constants ($m < 1$); C_e is the equilibrium concentration (mg/L)
Parabolic diffusion equation	$q_t = Rt^{\frac{1}{2}} + C_1$	R is the total diffusion coefficient; C_1 is a constant

can be obtained. Adsorption kinetics is an important adsorption characteristic, not only providing the equilibrium time required for other adsorption experiments, but also providing a reference for the bed contact time required for dynamic adsorption experiments. In addition, studying kinetics can also provide a basis for analyzing the adsorption process and adsorption mechanism. Kinetic data can be fitted with various kinetic equations, common equations are shown in Table 8.3.

8.1.5 *Factors Affecting Adsorption*

(1) The effect of temperature on adsorption capacity

According to the Langmuir assumption, adsorption is a dynamic equilibrium. Changes in temperature will cause changes in the adsorption rate, the equilibrium will shift, and the adsorption capacity will change accordingly. The general adsorption process is exothermic, so an increase in temperature reduces the adsorption capacity and weakens the adsorption ability. However, in actual working systems, the impact of temperature should be considered comprehensively according to different situations.

(2) pH value

For non-ionic adsorbates, their adsorption capacity has little to do with pH value; for cationic adsorbates, their adsorption capacity increases with the increase of pH value; for anionic adsorbates, their adsorption capacity decreases with the increase of pH value.

(3) Adsorbent

Since the phenomenon of adsorption occurs on the surface of the adsorbent, under certain circumstances, the larger the specific surface area of the adsorbent, the stronger the adsorption capacity. In addition, the material composition, concentration, and particle shape of the adsorbent can also affect the adsorption effect.

(4) Adsorbate

The concentration, molecular polarity, and molecular weight of the adsorbate all have a certain impact on the adsorption performance. With a fixed dose of adsorbent, increasing the concentration of the adsorbate initially increases the adsorption capacity as the concentration of the adsorbate increases, forming a straight line, then slowly increases, and will not change after reaching a certain adsorption capacity. The more adsorbates that can reduce the surface free energy of the liquid, the easier it is to be adsorbed. The size and unsaturation of the adsorbate molecules also affect the adsorption effect. For example, compared with zeolite, activated carbon is easy to adsorb larger diameter saturated compounds. Synthetic zeolites easily adsorb small diameter unsaturated (carbon–carbon double bonds, carbon–carbon triple bonds, etc.) compounds. It should be pointed out that although the adsorption capacity of activated carbon for homologous organic compounds increases with the increase of the molecular weight of organic compounds, too large molecular weight will affect the diffusion speed. Therefore, when the relative molecular weight of organic matter exceeds 1000, pretreatment is required to decompose it into small molecular weight substances before treatment with activated carbon. The selectivity of other adsorbents for adsorbates is shown in Table 8.4.

Table 8.4 Selectivity of adsorbents

	Non-polar ←→ Polar Saturated ←→ Unsaturated	
Large	Carbon Adsorbent	Silica Dioxide Aluminum Oxide
↑ Molecule	Activated Carbon	Aluminum Oxide Gel Activated Clay
↓ Small	Molecular Sieve	Synthetic Zeolite

8.1.6 Static Adsorption and Dynamic Adsorption

8.1.6.1 Static Adsorption

The adsorption method usually includes three steps: adsorption, separation, and regeneration. According to different operation methods, the adsorption method can be divided into static adsorption and dynamic adsorption. A suitable amount of solid-phase adsorbent is fully stirred and mixed with a suitable amount of wastewater, and the system composed of the two will undergo adsorption. The adsorption amount is measured at regular intervals until the adsorption amount no longer changes and reaches adsorption equilibrium, which is the process of static adsorption. The related theories of static adsorption are briefly described as follows:

(1) Langmuir formula
(2) Freundlich formula
(3) BET (Brunauer–Emmett–Teller) formula.

The related theoretical model parameters can be found in this section's Langmuir monolayer adsorption theory and adsorption isotherm.

8.1.6.2 Dynamic Adsorption

Static adsorption research is a research method for studying the properties of adsorption materials. It has a mature characterization system and can adjust the activation and modification methods according to the situation of each parameter to optimize the adsorption performance. However, in the actual application in the field of water treatment, most situations do not meet the experimental conditions of static adsorption, such as deep treatment in drinking water plants. A considerable part of the application scenarios faces the mode of continuous inflow and outflow of water, and the adsorption performance of static adsorption cannot be equated or converted to characterize its actual performance in water treatment. Therefore, there is a research mode of dynamic adsorption, which examines its actual application background by characterizing various parameters of the adsorption material under the condition of water flow. Wastewater is passed through an adsorption column filled with adsorbent at a certain flow rate, and the adsorbent continuously adsorbs impurities in the flowing wastewater, which is dynamic adsorption. Dynamic adsorption is carried out under flow conditions, which is equivalent to continuous multiple adsorptions, that is, when wastewater continuously passes through the adsorbent filling layer, adsorption removes pollutants in it. It is common in fixed bed, expanded bed, and moving bed devices. Various adsorption devices can run separately, in parallel, or in series. The most widely used is the fixed bed adsorption system, which can be divided into upward flow and downward flow according to the direction of water flow.

The related theories of dynamic adsorption are briefly described as follows:

① Breakthrough curve

The breakthrough curve is drawn based on the data from the adsorption column experiment. The graph is plotted according to the ratio of the solute concentration of the effluent or the ratio of the effluent to the influent solute concentration C_t/C_0 to the filtration time or the volume of filtered water V, as shown in Fig. 8.2. When the effluent concentration reaches the maximum allowable value C_b, it is called breakthrough. The point C_x where the effluent concentration equals 95% is called exhaustion. The S-shaped curve from breakthrough to exhaustion is called the breakthrough curve. Above this curve, it shows the changes in the saturation zone, adsorption zone, and non-adsorption zone of the adsorption column at different filtration times. At the beginning of the dynamic adsorption experiment, most of the pollutants are removed in the adsorption zone at the top of the column, and the concentration of pollutants in the effluent is very low. After filtering a certain amount of water, the upper layer of adsorbent tends to saturate. When the adsorption zone is close to the bottom of the column, the concentration of pollutants in the effluent gradually increases. Once the adsorption column tends to saturate, the concentration of pollutants in the effluent rapidly rises and eventually equals the effluent concentration. When the effluent concentration reaches the maximum allowable effluent concentration C_b, it is called the leakage concentration of the adsorption column. The total amount of water produced is V_b and the corresponding operating time t_b is called the adsorption cycle.

The shape of the breakthrough curve is related to the influent water quality, water volume, and the volume of the adsorbent. When the flow rate and the volume of the adsorption bed are fixed, that is, the contact time is fixed, if the types of pollutants and adsorbents are different, the slope of the curve and the breakthrough time will also change, showing an S-shaped curve, sometimes very steep, sometimes very long. If the water quality requirements of the effluent from the adsorption bed are different,

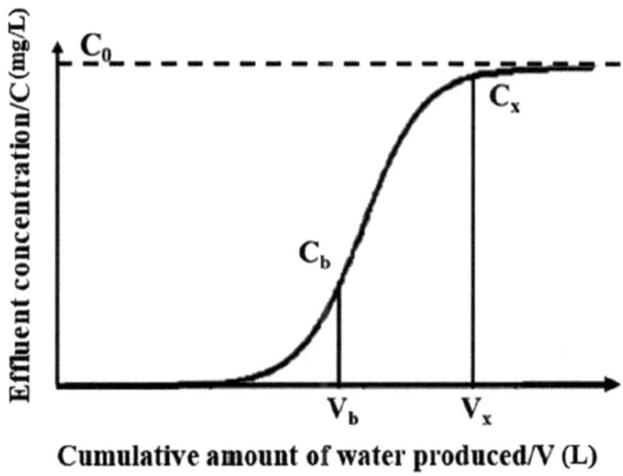

Fig. 8.2 Breakthrough curve of the adsorption column

the time to reach breakthrough will also change. When the thickness of the adsorbent accumulation is fixed, the higher the water quality requirements of the effluent, the earlier the breakthrough, the shorter the operation time of the adsorption column, and the corresponding regeneration frequency increases. Using the breakthrough curve, it can be known that the removal efficiency of the wastewater solution is higher in the initial stage of entering the adsorption column, and the adsorption efficiency gradually decreases with the increase of time until it reaches adsorption equilibrium.

② Thomas model

The Thomas model, Yoon-Nelson model, Adams-Bohart model, Clark model, and Wolborska model are commonly used models for studying the kinetics of dynamic adsorption fixed beds. Especially for dynamic adsorption, the Thomas, Yoon-Nelson and other models are often used in conjunction with the breakthrough curve to study its adsorption kinetics.

In 1944, Thomas proposed the famous Thomas model, which can estimate the equilibrium adsorption amount and adsorption rate constant of the adsorbate. Its expression is as follows:

$$\frac{C_t}{C_0} = \frac{1}{1 + \exp\left(\frac{K_{Th} q_0 x}{v} - \mathrm{K}_{Th} \mathrm{C}_0 \mathrm{t}\right)} \tag{8.8}$$

Its linear form is:

$$\mathrm{Ln}\left(\frac{C_0}{C_t} - 1\right) = \frac{K_{Th} q_0 x}{v} - K_{Th} C_0 t \tag{8.9}$$

where C_t is the concentration at the time of sampling, mg/L;

C_0 is the initial concentration, mg/L;

K_{Th} is the rate constant, mL/(min·mg);

q_0 is the equilibrium adsorption amount, mg/g;

x is the amount of adsorbent in the adsorption column, g;

v is the flow rate, mL/min;

t is the operation time of the adsorption column, min.

③ Adsorption column experimental device

Adsorption columns are generally used to study the effects of flow rate, adsorbent concentration, and adsorbate concentration on dynamic adsorption. Figure 8.3 shows a common dynamic adsorption experimental setup. For example, using activated carbon to adsorb heavy metal Cr(VI) from wastewater, at room temperature, the water sample is pumped at a certain flow rate with a constant flow pump, allowing the water sample to pass through the adsorption column. By examining parameters such as flow rate, total flow, total duration, length of the adsorption column (including adsorbate),

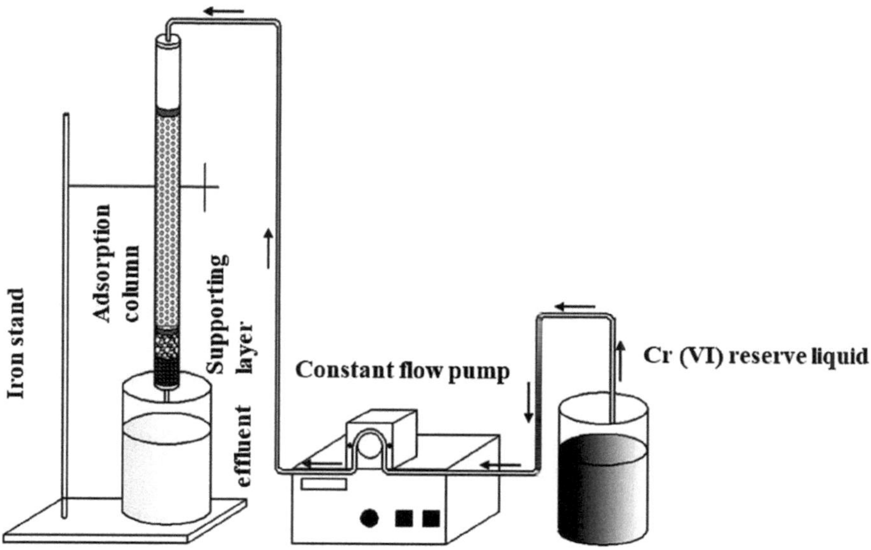

Fig. 8.3 Dynamic adsorption experimental setup for removing Cr(VI) from wastewater using activated carbon

inner diameter of the adsorption column, inlet concentration, and content of filtered substances, the optimal flow rate, optimal adsorbent, and adsorbate concentration are obtained. This allows for the analysis of the dynamic adsorption performance of the adsorbent material on the adsorbate.

8.2 Adsorbents and Activated Carbon

8.2.1 Adsorbents

Based on pore characteristics, commonly used adsorbent materials can be divided into three categories: porous, non-porous, and nano-adsorbent materials. This chapter introduces their structural characteristics, adsorption characteristics, application characteristics in water treatment, and research progress.

8.2.1.1 Porous Adsorbent Materials

(1) Activated Carbon

Activated carbon is a carbon-based adsorbent material. It is made from materials with high organic content such as fruit shells and wood, through a series of processes such as carbonization and activation. Because of its large specific surface area, rich pores, strong adsorption capacity, and low cost, it is an ideal adsorbent material. Its main element is carbon, but it also contains a small number of non-metallic elements such as oxygen, sulfur, and nitrogen. The overall structure is like graphite microcrystals. It can physically adsorb adsorbates through its abundant micropores, and it can also adsorb a certain number of polar substances through its surface oxygen-containing functional groups. Although activated carbon is originally a non-polar adsorbent, its surface functional groups can be modified (such as oxidation, reduction, loading, etc.) to control the content and properties of its surface functional groups, enhancing the adsorption of specific polar substances.

In water treatment processes, activated carbon is widely used in the advanced treatment of drinking water to remove trace pollutants such as disinfection by-products and algal toxins. It can also be used to remove heavy metals, color, and ammonia nitrogen from various types of wastewater. The structure, performance, preparation process, and characterization methods of activated carbon will be detailed in the subsection of Activated Carbon.

(2) Activated Alumina

Activated alumina is a type of alumina with a porous structure, high dispersion, and a large specific surface area. There are known to be nine types of alumina crystals, among which γ-Al_2O_3 is the most important type of activated alumina crystal. Because it has high activity and excellent adsorption performance, activated alumina often refers to γ-Al_2O_3, which is a cheap and efficient adsorbent material.

γ alumina can be prepared by dehydrating gibbsite at a relatively low temperature of 450 ~ 550 °C. The structure of γ-Al_2O_3 consists of Al^{3+} located in the voids randomly dispersed between octahedra and tetrahedra, and the voids are formed by the face-centered cubic close packing of O^{2-}. γ-Al_2O_3 has a larger pore volume and dispersion, which makes its internal surface area as high as 10^2 m^2/g, and it has better activity and adsorption. However, due to the differences in the preparation methods of the precursors, the morphology, specific surface area, and pore volume of activated alumina also differ, and the adsorption capacity also varies. γ-Al_2O_3 is a transitional alumina with low density, defective spinel structure, and excellent reactivity. It is mainly used in petrochemical, organic chemical, rubber, fertilizer, and environmental protection industries.

In the water treatment process, to improve the efficiency of water treatment and reduce operating costs, adsorbents need to have strong adsorption capacity, fast reaction speed, low cost, easy regeneration, easy recovery, and good reuse performance. Activated alumina products have a larger specific surface area (50 ~ 300 m^2/g), which can produce active sites by surface hydroxylation in water, and the microporous surface has strong adsorption capacity, which has strong adsorption capacity for most inorganic ions and some organic substances, and is easy to regenerate with acid and alkali solutions, and the cost is low. Activated alumina is widely used in the water treatment process, mainly used to remove inorganic ions in water, such as arsenic, selenium, phosphorus, and fluoride ions. Industrial activated alumina generally refers to a mixture of various types of transitional alumina and amorphous alumina, generally obtained by dehydration of aluminum hydroxide. Jiang Haiyu

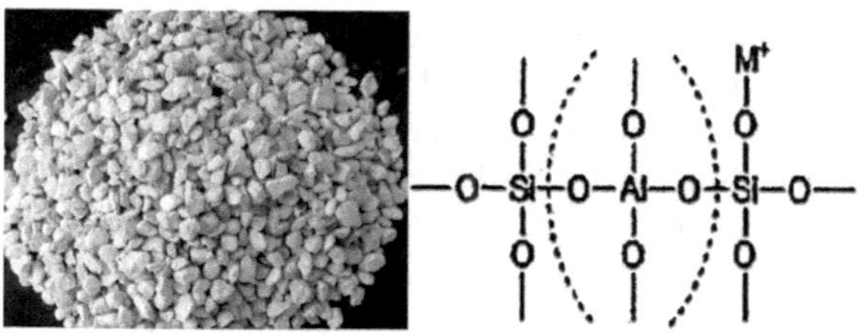

Fig. 8.4 Shape and planar structure of zeolite

et al. used activated alumina particles with a diameter of 2 ~ 4 mm as adsorbents to study the adsorption of arsenic in water by activated alumina. When the arsenic concentration in the water body is 10 mg/L, the arsenic removal rate is higher than 90%, up to 99.2%. Jiemin Cheng et al. used industrial activated alumina to adsorb fluoride ions, with a particle size of 0.38 mm, in granular form, and the adsorption capacity for fluoride ions was 2.74 mg/g. After lanthanum modification, the performance increased to 244%, up to 6.70 mg/g.

(3) Zeolite

Zeolite is a hydrated silicate mineral, characterized by a three-dimensional cage-like framework of SiO_4 and AlO_4 tetrahedra, each oxygen atom is shared by two silicon (aluminum) oxygen tetrahedra on both sides, and the tetrahedra extend infinitely in three-dimensional space in the form of shared vertices. Figure 8.4 shows the shape and planar structure of zeolite.

The advantages of zeolite include: strong ion exchange capacity, the exchangeable cations in the cavity will not cause secondary pollution; relatively large specific surface area; higher thermal stability; stable physicochemical properties, wear-resistant, acid and alkali resistant, high temperature resistant, corrosion resistant; has a certain mechanical strength; has adsorption selectivity; easy to regenerate while maintaining initial performance; low cost. Natural zeolite can also be activated and modified to further enhance its adsorption and ion exchange capacity. The main activation modification methods include high-temperature roasting, acidification, salt/alkalization, surface modification, and skeleton modification. Zeolite is usually used as an adsorbent in the water treatment process to remove ammonia nitrogen, inorganic fluoride ions, polar organic substances, heavy metals, and oxygen-containing anions.

Rao Li et al. used natural zeolite as an adsorbent to treat nitrogen-containing wastewater after blow-off treatment in an iron oxide red factory. The experimental results show that the adsorption capacity of natural zeolite for ammonia nitrogen is about 4 kg/t under certain working conditions, the treatment water volume is as high as 16 tons, the ammonia nitrogen removal rate is about 67%, and the natural zeolite has high regenerability, and its adsorption performance remains stable after regeneration.

(4) Bentonite

Bentonite is a clay mainly composed of montmorillonite, its main mineral components are SiO_2, Al_2O_3, and a small amount of MgO, CaO, K_2O, Na_2O, Fe_2O_3, etc. Montmorillonite is a 2:1 layered hydrated layered aluminosilicate, with one Al-O octahedron sandwiched between two Si–O tetrahedra. Figure 8.5 shows the crystal cell structure of bentonite. Bentonite shows good adsorption performance for metal ions in water bodies, and can adsorb and remove Co^{2+}, Zn^{2+} in wastewater, Pb^{2+}, Cd^{2+}, $Ni^{2+}$and other coexisting metal ions such as Cu^{2+}, Mn^{2+}, Cr^{3} in polluted natural water bodies. However, due to its hydrophilic structure, bentonite has a lower affinity for hydrophobic organic pollutants, so organic modifiers can be used to prepare modified bentonite through surfactant exchange method. Vojkan et al. studied the adsorption and removal of Pb^{2+} from water by carboxymethyl cellulose sodium modified clay, and analyzed the effect of the initial concentration of adsorption acid on lead removal using a solution containing 50–150 mg/L Pb^{2+}. The amount of adsorbent used to remove Pb^{2+} was measured, and when the initial Pb^{2+} concentration was 100 mg/L, the lead removal efficiency was 86.19%.

(5) Diatomite

Diatomite is an amorphous protein mineral formed by the long-term action of diatom remains in the natural environment, with a relatively small relative density, only 0.4 ~ 0.9 g/cm^3. Its chemical composition is mainly silicon dioxide, with a small number of other oxides, such as Al_2O_3 and Fe_2O_3. The color of diatomite is mainly gray, white or grayish white, and it is usually darker due to the presence of iron oxide and organic matter. Diatomite is lightweight, porous, chemically stable, has a large number of nanoscales micropore structures and a large specific surface area, which gives it a strong adsorption capacity, and can effectively remove pollutants in water, such as dyes and heavy metal ions, after calcination. Due to its unique physical and chemical properties, diatomite can bind heavy metal ions in water through physical adsorption, but physical adsorption is a reversible process, and its practical application is not ideal. In order to improve the adsorption capacity of diatomite, metal oxides are usually introduced on the surface of diatomite to change its surface structural characteristics, increase its specific surface area and surface negative charge, thereby achieving the purpose of removing heavy metal ions.

In water treatment technology, diatomite can be used as a filter aid to filter solid suspended matter, colloidal substances, algae, humic acid, and remove color, etc.;

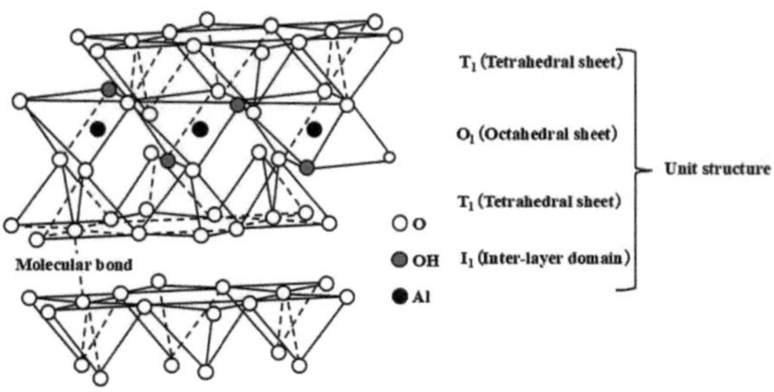

Fig. 8.5 Crystal cell structure of bentonite

in terms of removing heavy metal ions, it has a strong adsorption capacity for Pb^{2+}, Cu^{2+}, Cd^{2+} and Ag^{+} ions; it can also be used to treat various types of wastewaters such as urban domestic sewage, oily wastewater, printing and dyeing or papermaking wastewater. Li Xiaoying et al. modified diatomite with sodium hydroxide and aluminum chloride to remove phosphorus. At a dosage of 12 g/L, the phosphorus removal rate of simulated phosphorus-containing wastewater with a concentration of 20 mg/L can reach 89.45%.

8.2.1.2 Non-porous Adsorbent Materials

(1) Mineral powder

Sepiolite is a fibrous silicate clay mineral rich in magnesium, with a chain structure of hydrated aluminum magnesium silicate minerals. The chain structure contains small units, presenting a 2:1 layered flake structure, with different channels between the unit layers. The pore size of the unit layer of sepiolite ranges from 0.38 to 0.98 nm, which can be expanded to 0.56 ~ 1.10 nm, giving sepiolite excellent physicochemical and processing properties. The schematic diagram of the crystal structure of sepiolite is shown in Fig. 8.6. The main chemical components of sepiolite are silicon and magnesium, and its chemical formula is $Mg_8(H_2O)_4[Si_6O_{16}]_2(OH)_48H_2O$. The SiO_2 content is generally 54 ~ 60%, and the MgO content mainly fluctuates within 21 ~ 25%. It contains a small number of exchangeable cations, such as Mg^{2+} can be replaced by Fe^{2+} or Fe^{3+}, Mn^{2+}, etc. Sepiolite has strong adsorption capacity, good mechanical and thermal stability, and molecular sieve function. It can adsorb a large amount of water or polar substances, including weakly polar substances. At the same time, it can serve as a carrier for flocculants, promoting the precipitation and dehydration of flocculants.

8.2.1.3 Nano Adsorbent Materials

(1) Carbon Nanotubes

Carbon nanotubes (abbreviated as CNTs) were first discovered by physicist S. Iijima of NEC Laboratory (Tsukuba, Japan) in the early 1990s. Carbon nanotubes are hollow tubes made by rolling graphite sheets into a single layer to several tens of layers. Depending on the number of layers, they can be divided into single-walled (SWCNTs) and multi-walled (MWCNTs) categories. The diameter of SWCNTs is generally 1 ~ 6 nm, which makes them highly uniform and prevents the existence of structural defects. The diameter of MWCNTs is generally a few nanometers to tens of nanometers, with 2 ~ 50 layers, and the interlayer distance is about 0.335 nm, which is equivalent to the interlayer distance of graphite. Generally speaking, carbon nanotubes are bundled and sealed at both ends. The schematic diagrams of the structures of single-walled carbon nanotubes and multi-walled carbon nanotubes are shown in Fig. 8.7.

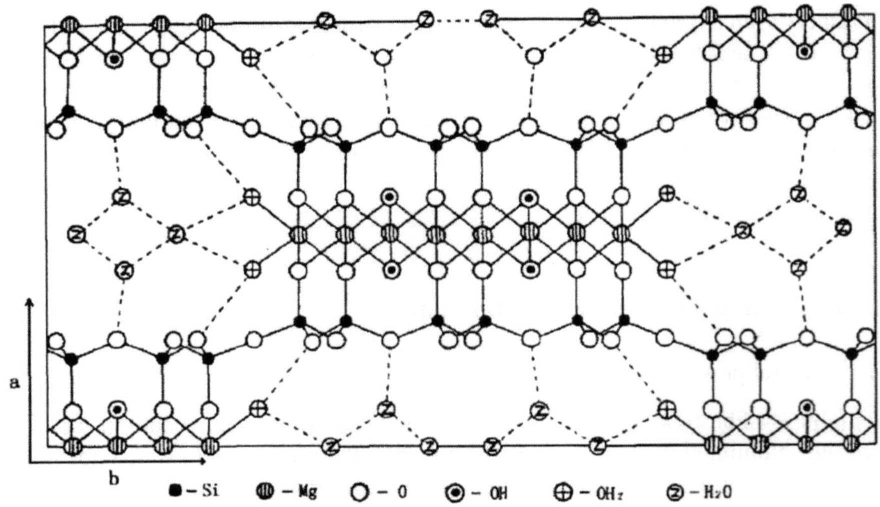

Fig. 8.6 Schematic diagram of the crystal structure of sepiolite

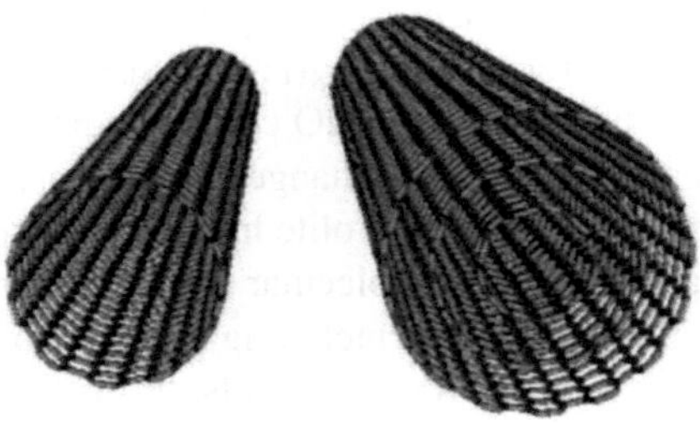

Fig. 8.7 Schematic diagram of the structure of single-walled carbon nanotubes (left) and multi-walled carbon nanotubes (right)

Numerous studies have shown that CNTs have far greater adsorption capacity and affinity for gases, organic pollutants, and heavy metals than other carbonaceous materials (such as activated carbon, C18, etc.). Their superior adsorption capacity is mainly due to two factors: large specific surface area and rich nanopore structure. Theoretical calculations show that the specific surface area of CNTs is 10 ~ 315 m^2/g. The pores of ordinary porous carbon materials are mainly formed by their own structural defects, and the pore size system has a certain randomness, while the appearance of CNTs has led to the formation of completely ordered nanopore structures in the carbon structure, which makes the pore structure of CNTs greatly different from that of ordinary porous carbon; the appearance of CNTs makes it possible to form completely regular nanopore structures in the carbon structure, and they have a high degree of uniformity. Hexagonal nanopores can have a larger specific surface energy and adsorption potential energy, which can adsorb a large amount of gas and liquid as well as metal ions.

Due to factors such as its complete, smooth surface and extremely inert nature, CNTs exhibit extremely strong solvent resistance and are extremely easy to aggregate into bundles in solution. Therefore, the surface can be modified by physical or chemical means to improve the dispersion in solvents, enhance the interface bonding force with the matrix, and improve applicability.

In water pollution treatment, it is used for the treatment of water containing heavy metals such as lead, chromium, or non-metallic poisons such as fluorine, as well as organic chemical poisons such as benzene and phenols. Tofighy et al. studied the maximum adsorption capacity of CNT treated with HNO_3 for removing Pb^{2+} was 101.5 mg/g. Kabbashi et al. also studied the method of using CNT as an adsorbent to remove Pb^{2+} ions. When the concentration of Pb2 + is 40 mg/L, the pH value is 5, and the stirring speed is 50 r/min, the maximum removal rate of Pb^{2+} was 96.03%.

(2) Graphene

Graphene is a two-dimensional carbon nanomaterial, in which carbon atoms are arranged in a honeycomb lattice structure with sp^2 hybridization. Ideal graphene is only 0.335 nm thick and has a large theoretical specific surface area of 2630 m^2/g, making it the basic unit for constructing other size carbon materials. Graphene is mainly a benzene hexagonal ring structure, the carbon–carbon double bond is the most basic chemical bond, the structure is relatively stable, and its benzene ring structure is difficult to break with conventional chemical methods. Due to its unique two-dimensional structure and excellent physicochemical properties, it is widely used in electrical, optical, energy, and environmental fields. Graphene oxide (GO) is an oxide of graphene. It retains the advantage of a large specific surface area of graphene, contains rich oxygen-containing functional groups (hydroxyl, carboxyl, and epoxy groups, etc.), and has good dispersibility in water (the solubility of GO in water is 712 g/L, much higher than the solubility of graphene in water, 0.175 g/L), and is widely used. Due to the presence of a large number of polar groups in the GO layer, it can strongly adsorb substances with hydrogen bonding groups such as amino groups, can efficiently adsorb polar organic molecules, and can bind metal ions of different valence states (Cu^{2+}, Pb^{2+}, Cd^{2+}, Cr^{6+}) through electrostatic attraction and coordination bonds. In water treatment, its good hydrophilicity makes it a great challenge to separate after adsorption, and it needs to be modified before application.

8.2.2 Activated Carbon

8.2.2.1 Physical and Chemical Properties

(1) Physical Structure

① Crystal Structure

Activated carbon is a carbon-based adsorbent material with a complex structure. Unlike graphite and diamond, which have carbon atoms arranged in a fixed pattern, and unlike most carbides with complex macromolecular structures, activated carbon is generally considered to be composed of graphite-like carbon microcrystals and amorphous carbon. They are interconnected to form the bulk and porous structure of activated carbon. However, activated carbon contains both graphite microcrystals and amorphous carbon, and the multiphase substance composed of graphite microcrystals and amorphous carbon determines the unique structure of activated carbon. The basic structure and differences between graphite and activated carbon are shown in Fig. 8.8.

In the graphite structure, carbon atoms are bonded by sp^2 hybridization, and the remaining p orbitals overlap parallel to form a large π bond, forming the planar network structure of graphite. This planar structure is arranged in parallel at fixed intervals (the force between the planes is van der Waals force), forming a regular three-dimensional structure. In this structure, the C–C bond length in the plane is 0.142 nm, and the distance is 0.335 nm. Its structure is shown in Fig. 8.8a, so graphite has characteristics such as electrical conductivity, thermal conductivity, and lubricity.

Activated carbon is essentially non-crystalline carbon, composed of graphite-like microcrystals and the carbon-hydrogen compounds that bind them together. The initial raw materials of activated carbon, wood and coal, are carbonized and activated, forming microcrystalline carbon (the basic crystal of activated carbon) between some carbon atoms of the activated carbon. However, its surface network structure does not adopt the regularity of the layered structure similar to graphite, forming a disordered layered structure as shown in Fig. 8.8b. Besides microcrystalline carbon, non-crystalline carbon remains in the precursor of activated carbon after carbonization and activation, indicating that activated carbon is a multiphase material, composed of microcrystal clusters, individual net-like surfaces that do not form parallel layers, and irregular carbon.

According to X-ray diffraction analysis, two types of microcrystalline carbon with a graphite-like structure have now been found in activated carbon. The size of the carbon changes with the carbonization temperature, composed of three parallel graphite layers, its width is about 9 times the width of the carbon hexagon. Riley believes that these two types of structures exist in most carbon materials, including

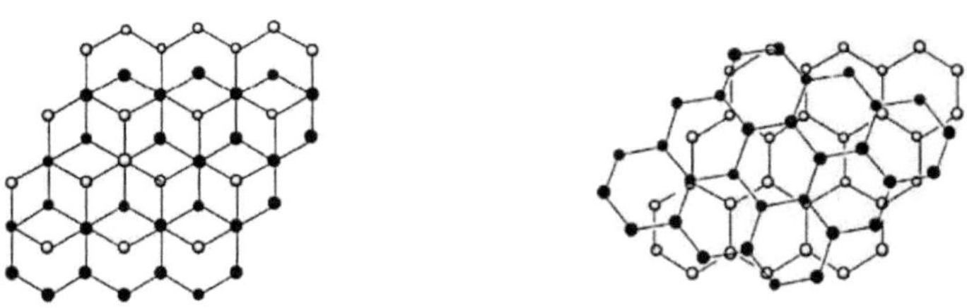

(a) Regularly arranged graphite layers

(b) Graphite layers in the microcrystalline structure of activated carbon: disordered layered structure

Fig. 8.8 Basic structure and differences between graphite and activated carbon

(a) Model of easily graphitizable carbon (b) Model of hard-to-graphitize carbon

Fig. 8.9 Structure of easily graphitizable and hard-to-graphitize carbon

activated carbon, and the final properties of activated carbon depend on which type of structure dominates.

Franklin divided carbonaceous materials other than diamond into two categories: carbonaceous materials that are easy to graphitize and carbonaceous materials that are hard to graphitize (see Fig. 8.9). He also proposed that carbon that is not easy to graphitize is relatively soft, while carbon that is easy to graphitize is relatively hard. Carbon that is easy to graphitize may have a disordered layered structure, while carbon that is not easy to graphitize has a cross-linked lattice structure.

② Pore structure

a. Morphology of the pore structure

In 1960, Dubinin classified the pores of activated carbon into three categories: macropores (pore diameter greater than 50 nm), mesopores (or transition pores, pore diameter 2 ~ 50 nm) and micropores (pore diameter less than 2 nm), a scheme adopted by the International Union of Pure and Applied Chemistry (IUPAC). In activated carbon, these three different sizes of pores interpenetrate each other, forming a tree-like structure. The pore structure of activated carbon is shown in Fig. 8.10.

The inner surface of the macropores classified by Dubinin can cause multilayer adsorption, but for activated carbon, the content of macropores is very low, it mainly serves as a channel for adsorbate molecules to enter the adsorption sites. Therefore, it also has practical importance, it is not always the less the better. Transition pores usually function as channels for adsorbates like macropores, thus determining the

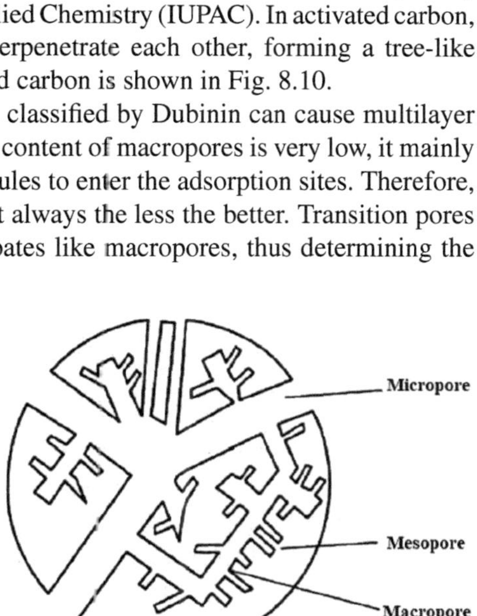

Fig. 8.10 Pore structure of activated carbon

adsorption rate. In addition, it can also serve as an adsorption point for large molecules that cannot enter micropores. The adsorption capacity of activated carbon is determined by micropores, not only most of the adsorption is carried out through micropores, but also the very large specific surface area of activated carbon is contributed by micropores.

b. Calculation of pore volume

In activated carbon, activation increases the volume of pores. The increase in micropores can be considered as an increase in the volume of micropores. If the shape of the pore is fissure-like and composed of parallel planes, then the pore radius is equal to the plane interval. If the specific surface area (S) and pore volume (V) are determined, and the pore shape is assumed to be cylindrical, the pore radius ($\bar{r}$) can be calculated using the following formula:

$$\bar{r} = \frac{2V}{S} \tag{8.10}$$

where r is the radius of the hole, cm;

V is the pore volume, cm^3/g;

S is the specific surface area m^2/g.

c. Determination of pore size distribution

Studying the pore size distribution is the best way to understand the pore structure. The pore size distribution can also represent quite a lot of activated carbon performance. The pore size distribution is usually obtained through methods such as mercury intrusion, electron microscopy, capillary condensation, molecular sieve, X-ray small-angle scattering, etc. The mercury intrusion method is commonly used, the principle of which is that mercury will not wet the fine pore walls of activated carbon, so mercury can be correspondingly pressed into the fine pores, then the following formula holds:

$$Rp = -2\gamma \cos\theta \tag{8.11}$$

where R is the radius of the cylindrical pore, Pa;

p is the pressure applied to the mercury, Pa;

θ is the contact angle of mercury;

γ is the surface tension of mercury, N/m.

Under pressure P, mercury should enter all fine pores with a radius above γ, so the amount of mercury entering due to the increase in pressure can be measured, and from this, the size of each pore can be measured, and the pore size distribution can be determined.

③ Specific surface area

Adsorption occurs on the surface of a solid, and the strength of an object's adsorption capacity depends on the size of its specific surface area. Commonly used methods for measuring the specific surface area include the BET method, flow method, liquid phase adsorption method, wet heat method. In addition, the specific surface area can also be measured by small-angle X-ray scattering, but the BET method is the most used method for measuring the specific surface area of activated carbon, and the specific surface area of activated carbon measured by this method is usually 1000 m^2/g.

(2) Chemical properties

① Elemental composition

The elemental composition of activated carbon can be determined by an elemental analyzer (Table 8.5). Activated carbon contains more than 90% carbon, which to a large extent makes it a hydrophobic adsorbent. The oxygen content is generally a few percent, and it exists in two ways, part of it exists in the ash, and the other part exists on the surface of the carbon in the form of carboxyl groups and other surface functional groups. Due to the presence of oxygen-containing functional groups, activated carbon is not completely hydrophobic, but is hydrophilic to a limited extent. Due to this limited hydrophilicity, activated carbon can be used for water treatment, replacing the air in the pores with water, and adsorbing organic substances dissolved in water.

Plant-derived raw materials usually contain very low levels of nitrogen and sulfur. Most of the proteins and sulfides in the raw materials are thermally decomposed during the carbonization and activation process and evaporate in the form of gas, but a small amount may remain. The residual nitrogen atoms can enhance the catalytic performance of activated carbon under certain conditions. The ash content of activated carbon varies greatly depending on the raw materials and preparation process. Generally, the ash content of wood-based activated carbon is low (generally < 5%). If the ash content of the raw material is high, the raw material must be demineralized before preparing activated carbon.

Trace impurities in activated carbon must be kept at the lowest level, among which arsenic is the most important impurity. If the arsenic concentration in the soil is too high, the arsenic concentration in the raw materials will increase, and the arsenic

Table 8.5 Elemental composition of activated carbon

Activated carbon	Carbon/%	Hydrogen/%	Sulfur/%	Oxygen/%	Ash/%
Steam method A	93.31	0.93	0.00	3.25	2.51
Steam method B	91.12	0.68	0.02	4.48	3.70
Zinc chloride method C	90.88	1.55	0.00	6.27	1.30
Zinc chloride method D	93.88	1.71	0.00	4.37	0.05
Zinc chloride method E	92.20	1.66	1.21[a]	5.61	0.04

Note Nitrogen content is all trace amounts
[a]Experimental, trial-produced sulfurized carbon

concentration in the activated carbon produced from these raw materials will exceed the standard value. Strict control of arsenic content in the production of activated carbon is necessary, and if various waste materials are used in the production of activated carbon, their heavy metal pollution should also be closely monitored.

2. ② Surface Characteristics
a. Surface Functional Groups

The main component of activated carbon is carbon, which is non-polar and has hydrophobic characteristics. However, the surface characteristics of carbon materials can change due to different manufacturing processes and usage environments. This is because the surface of carbon materials can be easily oxidized by oxidants (such as oxygen and water), forming functional groups on the surface. The formation of these functional groups leads to the diversity of interfacial chemical properties in carbon materials.

The functional groups in carbon materials can be determined by organic chemistry and X-ray photoelectron spectroscopy. Generally, the main functional groups in carbon materials are carboxyl, lactonic carboxyl, phenolic hydroxyl, and carbonyl (as showed in Fig. 8.11).

Nitrogen-containing functional groups can also exist on the surface of activated carbon, which generally come from the preparation of nitrogen-containing raw materials or the chemical reaction of activated carbon with artificially introduced nitrogen-containing reagents.

b. Function

The surface chemistry of activated carbon is determined by the chemical functional groups, surface heteroatoms, and compounds present on the surface. The surface chemistry determines the adsorption properties of activated carbon. Since the amount of adsorption varies greatly with different surface functional groups, heteroatoms, and compounds, research on the surface functional groups of activated carbon will be of great help to its practical application.

Activated carbon is a non-polar adsorbent. Generally, due to its hydrophobicity, it only adsorbs various non-polar organic substances in aqueous solutions and does not have the ability to adsorb polar solutes. However, its adsorption performance can be enhanced by introducing and modifying functional groups on its surface. Typically,

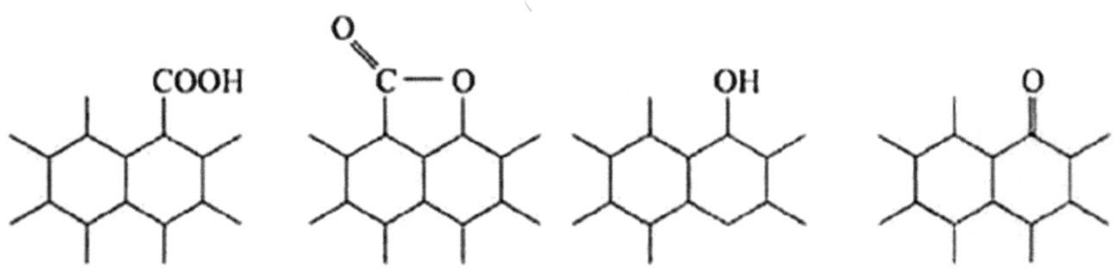

(a) Carboxyl group (b) Lactone group(c) Phenolic hydroxyl group (d) Carbonyl group

Fig. 8.11 Oxygen-containing functional groups on the surface of activated carbon

acidic compounds in the oxygen-containing functional groups on the surface of activated carbon tend to adsorb polar compounds, while basic compounds adsorb substances with weaker polarity and non-polar substances.

The adsorption performance of activated carbon can be changed by modifying the surface functional groups. Increasing the polarity of the surface functional groups will increase the adsorption capacity for polar substances. Correspondingly, increasing the non-polarity of the surface of activated carbon will increase the adsorption capacity for non-polar substances. Therefore, by changing the surface chemistry of activated carbon, more selective and larger amounts of adsorption can occur.

8.2.2.2 Product Classification

Activated carbon can be classified into different types according to manufacturing methods, appearance and shape, usage functions, and pore size. Based on shape, it can be divided into granular activated carbon and powdered activated carbon, and granular activated carbon can be further divided into amorphous and non-amorphous. Depending on the raw materials, activated carbon can be divided into charcoal, petroleum, coal, and resin activated carbon; from different usage functions, it can be divided into liquid adsorption, catalytic performance, gas adsorption activated carbon; from the production methods, there are physical, chemical, and physico-chemical methods. See specific classifications are showed in Tables 8.6, 8.7, 8.8 and 8.9.

Classified by appearance and shape, the common types of activated carbon are shown in Table 8.6.

Based on different raw materials, the classification is shown in Table 8.7.

Based on different manufacturing methods, the classification is shown in Table 8.8.

Based on the function of activated carbon, the classification is shown in Table 8.9.

Table 8.6 Classification of commercially available activated carbon of different

Shape	Characteristics
Powdered	Activated carbon containing a large number of particles with an outer diameter of 0.18 nm (about 80 mesh) or less. This includes powdered activated carbon produced from sawdust, as well as powdered products of granular activated carbon
Granular	Activated carbon containing a large number of particles with an outer diameter of 0.18 nm (about 80 mesh) or more. In terms of shape, they can be divided into crushed, spherical, and hollow microspheres
Crushed	Such as coconut shell activated carbon, coal-based activated carbon. The outer surface of this type of activated carbon is angular due to being crushed
Spherical	Activated carbon prepared from spherical carbonaceous materials or resins
Hollow micro spherical	Mainly made from resin, sometimes with a diameter less than 50 μm, very little powder is produced during use
Fibrous	Activated carbon made from fibrous materials, with fiber diameters of 8 ~ 10 μm. There are filamentous, cloth-like, and felt-like types
Honeycomb	Activated carbon that is extruded into a honeycomb shape
Activated carbon formed objects	Products made by adhering activated carbon powder to substrates such as paper, non-woven fabric, and sponge, as well as formed products made and processed into various shapes from activated carbon alone or in combination with other materials

Table 8.7 Classification of commercially available activated carbon of different raw materials

Type	Raw material
Wooden	Sawdust, charcoal
Shell	Coconut shell, walnut shell, apricot kernel, etc.
Coal	Lignite, peat, bituminous coal, anthracite, etc.
Petroleum	Asphalt, etc.
Regenerated carbon	Waste carbon activated carbon

Table 8.8 Classification of activated carbon by different manufacturing methods

Activation method	Activator
Chemical activation method	Zinc chloride, phosphoric acid, potassium hydroxide, sodium hydroxide, etc.
Strong alkali activation	Method potassium hydroxide, sodium hydroxide, etc.
Gas activation method	Steam, carbon dioxide, air, etc..
Steam activation method	Steam

Table 8.9 Classification of activated carbon by different functions

Activated carbon	Function
High surface area activated carbon	Activated carbon with a high surface area produced by strong alkali activation, with a specific surface area of 2500 m^2/g or more
Molecular sieve activated carbon	Pore activated carbon used for gas separation
Loaded activated carbon	Activated carbon with metal salts or other chemicals adhered to its surface for the purpose of deodorization, catalysis, etc..
Biological activated carbon	Activated carbon that forms a microbial film on its surface and purifies through microbial decomposition. Used in conjunction with ozone treatment for deep water purification

8.2.2.3 Functional Characteristics

(1) Heavy Metal Adsorption

The main methods for treating heavy metal wastewater include electrochemical methods, chemical precipitation, membrane separation, and adsorption. Among them, the adsorption method is one of the most widely used techniques for treating heavy metal wastewater, with the advantages of high treatment efficiency, simple operation, and low investment. A large amount of research and engineering applications show that activated carbon adsorption is very effective in treating heavy metal wastewater. Studies have shown that activated carbon has a good removal effect on single Ni^{2+}, Cr^{3+}, Zn^{2+}, Cd^{2+}, $Pb^{2+}$and Cr^{6+}, as well as their mixed wastewater.

The adsorption of heavy metals by activated carbon can be divided into two types: physical adsorption and chemical adsorption. Figure 8.12 is a diagram of the mechanism of heavy metal cation adsorption by activated carbon. Physical adsorption is based on the pore structure of activated carbon. As shown in Fig. 8.12, heavy metal ions are easily adsorbed by micropores through large pores and mesopores under the action of van der Waals forces. There are three types of chemical adsorption: the first is the occurrence of cation exchange reactions, unprotonated acidic functional groups on the surface of activated carbon such as hydroxyl, hydroxyl hydrogen and heavy metal cations undergo ion exchange adsorption based on valence; the second is surface polar oxygen-containing functional groups such as deprotonated carboxyl, lactone and phenolic hydroxyl due to strong polarity and easy ionization, showing electronegativity, and adsorption by electrostatic attraction with heavy metal cations; the third is the surface neutral functional groups, such as ether, ester, ketone and alcohol hydroxyl, etc., can adsorb some heavy metals through complexation reactions. In addition, basic functional groups (delocalized π electrons) adsorb heavy metal cations through ion exchange with $-C_{\pi}$-H_3O^+ functional groups and electron donor–acceptor complexation with $-C_{\pi}$.

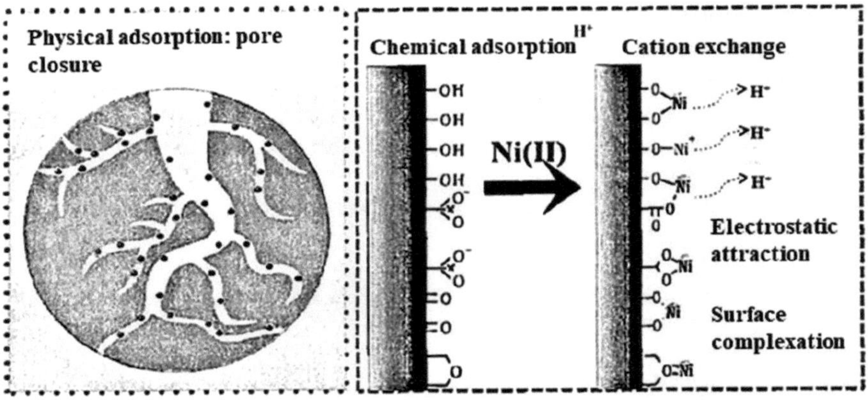

Fig. 8.12 Mechanism of heavy metal cation adsorption by activated carbon

(2) Organic Compound Adsorption

In water treatment processes, the adsorption of organic matter by activated carbon is mainly physical adsorption. In addition to the pore structure and surface properties of activated carbon, the properties of the adsorbate also have a differentiated impact. When activated carbon adsorbs organic matter in water, large molecular substances are more easily adsorbed than small molecular substances, and hydrophobic adsorbates are more easily adsorbed in water than hydrophilic adsorbates. The size of organic molecules and their hydrophobicity are also related to the impact of activated carbon particle size on adsorption capacity. Studies have shown that for small molecular substances, such as phenol (MW is 98 Da), the adsorption capacity of activated carbon is independent of particle size; for saturated large molecular substances, if MW < 3 kDa such as polyethylene glycols, the adsorption capacity is not related to particle size. When MW $= 8$ kDa, such as polyethylene glycol, the adsorption capacity has a weak relationship with particle size. For unsaturated large molecules, such as the high polymer polymer—polystyrene sulfonate, particle size has a significant impact on adsorption capacity. For various natural organic matters (NOM), activated carbon's adsorption capacity for substances with high UV absorption is significantly affected by carbon particle size. Therefore, the adsorption of large molecules with unsaturated chromophores is significantly affected by particle size.

For some small hydrophobic molecules, the adsorption capacity of activated carbon is also affected by carbon particle size. Matsui et al. studied the factors affecting the adsorption performance of activated carbon on two odor organic substances (geosmin and dimethyl isoborneol) in environmental water. The study found that both substances are small organic molecules (MW are 182 and 168 Da respectively), and the adsorption capacity of activated carbon on them is attributed to their hydrophobicity. The former is more hydrophobic than the latter, and the impact on the adsorption capacity of activated carbon for geosmin is greater than that of dimethyl isoborneol. This is because these two substances are low-polarity organic substances, which enter the pores of activated carbon by physical action and can only penetrate a certain distance. When the particle size of activated carbon is much larger than this penetration distance, the internal area of the carbon particles does not play a role in adsorption, so the equilibrium adsorption capacity will increase as the particle size of the carbon decreases, until the particle size is smaller than this penetration distance.

If the adsorbate is an aromatic organic substance, it is easier to be adsorbed compared to non-aromatics; organic molecules containing double bonds are easier to be adsorbed; organic substances containing inorganic elements or inorganic groups are not easily adsorbed; non-polar organic substances are more easily adsorbed. When multiple adsorbates coexist, organic substances will affect each other, possibly causing competition, reducing their adsorption capacity, or inducing each other, increasing their adsorption capacity.

(3) Antibiotic Adsorption

With the large-scale production and application of antibiotics, researchers have been committed to treating wastewater containing antibiotics since the 1940s. Rapid industrial development and stricter discharge standards make traditional primary wastewater treatment difficult to meet current sewage discharge standards. Traditional biochemical and physical treatment processes are still effective for high-concentration antibiotic wastewater, but they do not meet expectations for low-concentration antibiotic water bodies. Therefore, ion exchange, membrane filtration, chemical oxidation, adsorption and other treatment methods have been derived. Activated carbon adsorption is suitable for deep treatment of high and low concentration antibiotic water because the equipment is simple and the investment cost is low. For chloramphenicol, a typical antibiotic, Li Yiran found that the adsorption of chloramphenicol by reed activated carbon is due to the π-π electron donor–acceptor (π-π EDA) interaction with the graphite structure in activated carbon as the π electron donor. The π-π EDA complex is mainly formed by the interaction of the aromatic hydrocarbon structure or graphene structure rich in π electrons on the surface of activated carbon with the benzene ring lacking π electrons in the adsorbate. At the same time, the phenolic hydroxyl group as an electron-donating group can promote the π-π EDA interaction between the graphite structure of activated carbon and the benzene ring of chloramphenicol.

(4) Color Removal

Pigment molecules are also organic compounds. Compared with other organic compounds, they are characterized by a large number of chromophoric and auxochromic groups that express color. Among the many dyes, azo and anthraquinone structures are the most common chromophoric groups. Of these, the former (represented by reactive dyes) accounts for 70% of the total. Based on this, the introduction of various auxochromic groups determines the hue and color strength of the dye. Generally, anthraquinone is more difficult to destroy than azo structures, the latter of which may generate aniline compounds during the degradation process. Due to the complex molecular structure of pigments, most dyes, especially reactive dyes after hydrolysis, have poor biodegradability, with a biochemical treatment decolorization rate of about 10%.

The adsorption of pigment molecules by activated carbon is based on two types: physical and chemical adsorption. Physical adsorption refers to the process where pigment molecules diffuse into the pore structure of activated carbon and adsorb at the active sites. In this case, smaller pigment molecules can reach deeper into the activated carbon and utilize more active sites for fixed adsorption. Therefore, it is important to choose activated carbon with suitable pore size distribution characteristics for different types of pigment molecules. In addition, certain oxygen-containing functional groups inherent in activated carbon also have chemical adsorption capabilities, and strongly polar pigment molecules can be removed through chemical adsorption mechanisms. However, since activated carbon itself is a non-polar adsorbent, its adsorption capacity for polar substances is relatively weak. Therefore, specific activation (mainly zinc chloride and phosphoric acid activation) and modification are needed to improve its decolorization ability.

8.2.2.4 Preparation Process

(1) Activation Method

The activation methods for preparing activated carbon are divided into physical and chemical activation, the difference being whether a chemical reagent is added. Physical activation is an environmentally friendly method, but it requires high temperatures and long heating times during preparation, not only consuming a lot of energy, but also resulting in low yield, poor uniformity, and low adsorption performance. Therefore, the chemical activation method is mainly introduced.

Chemical activation is widely used in the preparation of activated carbon. After the raw material is ground, the activator is mixed with the raw material in a certain proportion. Depending on the activator, selective heating can be carried out in an inert atmosphere to simultaneously complete carbonization and activation. The activators used are mainly zinc chloride, phosphoric acid, alkalis (such as potassium hydroxide, sodium hydroxide), alkali metal carbonates, etc. The roles of these chemical activators in the carbonization and activation process are not yet clear, but it is generally believed that the activator is both a reactant involved in the chemical reaction with the raw material and a catalyst. The roles of these activators in the activation process may vary, but their dehydration effect can greatly reduce the activation temperature of carbonization.

The $ZnCl_2$ method is one of the earliest chemical activation methods for preparing activated carbon. Due to its strong dehydration effect, it greatly reduces the activation temperature of lignin carbonization to 150 ~ 300 °C, changes the pyrolysis process of lignin, reduces the formation of tar, and is conducive to the generation of pores. However, during the preparation process, the volatilization of $ZnCl_2$ can cause serious environmental pollution, and many countries have banned the use of zinc chloride to prepare activated carbon.

The alkali activation method uses alkali substances such as potassium hydroxide and sodium hydroxide. This method was initially mainly aimed at petroleum coke, but it is also effective for other precursors such as coal and nutshells to produce activated carbon. In this method, the alkali is added to the raw material in a certain mixing ratio, crushed and mixed, and then the carbon is heated to 700 ~ 800 °C in an inert gas or closed system to carbonize and activate it, resulting in activated carbon with a large number of cage-like microporous structures and a specific surface area of about 3000 m^2/g.

The H_3PO_4 activation method is a mature process for preparing activated carbon. The activation mechanism is similar to the zinc chloride method, which is conducive to the pyrolysis reaction process and reduces the activation temperature. Phosphoric acid is distributed in the raw material and occupies a fixed position, which can prevent the particles from shrinking under high temperature conditions, inhibit the formation of tar, and when the phosphoric acid is washed, the activated carbon has a developed pore structure. The product prepared by the phosphoric acid activation method has a wide pore size distribution, developed mesopores, and a wide range of applications.

(2) Modification methods

Activated carbon has a rich microporous structure and a huge specific surface area, so it has good adsorption performance for most pollutants. Activated carbon is a broad-spectrum adsorbent, so its selectivity is low, which affects its adsorption performance for certain pollutants. Depending on the type of pollutant to be removed, the structure of activated carbon can be flexibly changed or its surface modified by modification to effectively improve the selective adsorption of specific substances.

① Surface oxidation modification

Under appropriate conditions, the surface of activated carbon can be oxidized with an oxidant to increase the content of oxygen-containing functional groups (such as carboxyl, phenolic hydroxyl, ester, etc.) on its surface, enhance the hydrophilicity i.e., polarity of the activated carbon surface, and improve the adsorption capacity for polar substances. Oxidation modification is a common method for modifying activated carbon. Common oxidants include HNO_3, H_2O_2, H_2SO_4, O_3, $(NH_4)_2S_2O_8$. Depending on the oxidant used, the content and type of oxygen-containing functional groups will change, and the pore structure, specific surface area, volume, and pore size of the modified activated carbon will also change.

Nitric acid is the most commonly used strong oxidant. After oxidation treatment, the adsorption capacity of activated carbon for metal ions is improved, mainly because the acidic functional groups (which can dissociate protons) on the surface can attract metal ions to form complexes through electrostatic attraction. Liu Wenhong et al. used concentrated nitric acid to treat the surface of activated carbon at different temperatures. The results showed that both normal temperature and boiling modification conditions caused more oxygen-containing groups to be produced on the surface of the activated carbon. However, after modification with concentrated nitric acid at room temperature, the specific surface area and pore volume significantly increased, while after modification with boiling concentrated nitric acid, the specific surface area and pore volume significantly decreased. After modification with room temperature concentrated nitric acid, the adsorption reduction reaction of $[Ag(NH_3)_2]^+$ on the surface of activated carbon was improved. When the silver ion concentration was 600 mg/L, the amount of silver adsorbed on the modified activated carbon was more than 5 times that of the original carbon.

② Surface reduction modification

Surface reduction modification mainly uses suitable reducing agents to reduce the functional groups on the surface of activated carbon at appropriate temperatures, increase the content of basic functional groups on the surface of activated carbon, increase the non-polarity of the surface, and improve the adsorption capacity for non-polar substances. Common reducing agents include N_2, H_2 and NaOH, KOH, ammonia water, etc.

The simplest method to prepare alkaline activated carbon is to introduce inert gases, hydrogen and ammonia, during the production process. After introducing NH_3 at 400 ~ 900 °C, alkaline nitrogen functional groups can be generated on the surface of the activated carbon. Under 400 ~ 900 °C, amides, aromatic amines, and protonated amides can be generated; these substances on the surface of the activated carbon will enhance the alkalinity of the activated carbon surface. Gao Shangyu et al. found that hydrogen-modified activated carbon significantly increases the adsorption capacity for phenol and benzenesulfonic acid. During the reduction reaction process, some oxygen-containing functional groups are decomposed into carbon dioxide, carbon monoxide, and water at high temperatures, which significantly reduces the total amount of oxygen-containing functional groups, especially oxygen-containing acidic functional groups. Chen et al. found that high-temperature modification of activated carbon with ammonia can improve its adsorption capacity for perchlorate (ClO_4^-), especially at 650 ~ 700 °C, the adsorption capacity of modified activated carbon is the strongest, reaching 4 times that of the original carbon. This is mainly due to the increase in positive charges on the surface of the activated carbon and the enhancement of the electrostatic attraction between the activated carbon and ClO_4^-. When activated carbon is impregnated with ammonia, nitrogen-rich functional groups can also be obtained. Shaarani and Hameed confirmed that the surface of activated carbon modified with ammonia is more alkaline, the activated carbon surface carries more positive charges, and the adsorption capacity for 2,4-DCP increases from 232.56 to 285.71 mg/g, an increase of 22.86%.

③ Load material modification

The modification of activated carbon by loading is to use the large specific surface area and pore volume of activated carbon to adsorb metal ions and other different atoms into the pores of activated carbon, and use the strong binding force of metal ions and different atoms to adsorbents to improve the adsorption effect of activated carbon on adsorbates. Commonly used metal ions include copper, iron, aluminum, and silver ions. Loaded activated carbon has been proven to have good adsorption capacity for heavy metals, fluoride ions, cyanide, and arsenates. Iron salt solution is the most commonly used impregnating solution for loaded modified activated carbon. Zhu Huijie et al. used trivalent iron salt to modify the loaded nano-iron prepared activated carbon, which has a high removal rate for As(III) at pH = 6 ~ 9; at room temperature, when the initial concentration of As(III) is 2 mg/L and the amount of adsorbent added is 1.0 g/L, the removal rate of As(III) is 99.86%, and the arsenic adsorption amount of loaded activated carbon is 1.997 mg/g. Leyva Ramos et al. used aluminum salt to modify activated carbon and used it for the adsorption of fluoride ions. Compared with the original carbon, the adsorption capacity of the loaded activated carbon for anions increased by 3–5 times. The pH value has a great impact on aluminum adsorption. The best loading pH value is 3.5. When the pH value is less than 3.5 or greater than 4.0, the aluminum salt begins to precipitate and cannot be loaded onto the surface of the activated carbon. Maruyama and Abe found that platinum-loaded activated carbon can improve the adsorption capacity for organic acids during the oxidation–reduction process. Chen et al. reported that humic acid has a specific binding force with copper ions. If copper ion-modified activated carbon is used to treat humic acid wastewater, the adsorption capacity of activated carbon for humic acid can be improved.

8.2.2.5 Characterization Methods

(1) Characterization of activated carbon

① Surface morphology observation

The microscopic surface morphology of activated carbon can reflect the structure and shape of activated carbon, and the degree of activation can be judged by observing the smoothness.

Scanning electron microscope (SEM) is used to observe the surface morphology of activated carbon samples. To improve the conductivity of activated carbon, it must be gold-sprayed before measurement.

② Specific surface area and pore size distribution

Two models are commonly used to calculate the specific surface area: the Langmuir model and the Brunauer–Emmett–Teller (BET) model. The Langmuir model is a single-molecule layer adsorption isotherm, only applicable to the adsorption of microporous materials; for porous materials that not only contain micropores but also mesopores and macropores, the BET model extends the theory of single-layer adsorption to multi-layer adsorption. The specific surface area can be calculated using the formula.

The specific surface area and pore size distribution of activated carbon samples can be determined using a static nitrogen adsorption system. At a temperature of 77 K, liquid nitrogen is used as the adsorbate to complete the nitrogen adsorption/desorption experiment. Before measurement, the activated carbon samples need to be vacuum dried at 200 °C for 2 h to ensure that there are no other impurity gases in the sample tube. The specific surface area selection area is 0.05 ~ 0.3. The specific surface area measurement uses the BET method, the mesopore distribution uses the BJH model, and the micropore distribution uses the t-plot method.

③ Infrared spectroscopy analysis

Fourier transform infrared spectrometry (FT-IR) is used to determine the types of functional groups on the surface of activated carbon. As a classic qualitative analysis, it can determine whether the prepared activated carbon has the required surface functional group modification, and can semi-quantitatively reflect the content of functional groups. Infrared spectroscopy analysis of the sample is performed using an infrared spectrometer. Firstly, about 5 mg of activated carbon is placed in an agate mortar, high-purity potassium bromide is added, and after sufficient grinding, it is pressed into a transparent thin slice with a tablet press and placed in an infrared spectrometer for measurement. Measurement conditions: resolution 2 cm, scanning range is 400 ~ 4000 cm^{-1}, scanned 60 times.

④ Measurement of the content of surface acidic oxygen-containing functional groups

The content of surface acidic oxygen-containing functional groups directly determines the adsorption performance of activated carbon for polar substances, and to a certain extent also determines the hydrophilic/hydrophobic characteristics of activated carbon and its dispersibility in water.

The functional groups on the surface of activated carbon are determined by the Boehm titration method. The Boehm titration method is based on the following assumptions: NaOH can neutralize the carboxyl, lactone, and phenolic hydroxyl groups on the surface of activated carbon, Na_2CO_3 neutralizes carboxyl and lactone groups, $NaHCO_3$ only neutralizes carboxyl groups, so the content of carboxyl groups is the amount of HCl consumed by $NaHCO_3$, the content of lactone groups is the amount of HCl consumed by Na_2CO_3 minus $NaHCO_3$, and the content of phenolic hydroxyl groups is the amount of NaOH minus the amount of HCl consumed by $NaHCO_3$. The content of various functional groups can be obtained by calculation. The specific operation steps are as follows: Firstly, prepare 0.1 M standard alkali solutions of $NaHCO_3$, Na_2CO_3, and NaOH, then weigh 0.5 g of activated carbon sample, put it into a sealed conical flask containing 25 mL of 0.1 mol/L standard alkali solution, shake at 20 °C for 24 h, let it stand for 2 h, then filter, take 10 mL of filtrate, titrate back with 0.1 M HCl solution, record the amount of HCl consumed at the titration endpoint, perform parallel measurements 3 times, and take the average value.

⑤ Measurement of surface isoelectric point

The surface electrical properties of activated carbon in solution can be determined based on the pH value of the aqueous solution and the isoelectric point of the activated carbon surface. If the pH value of the solution is lower than the isoelectric point, it exhibits negative electricity, and vice versa.

The isoelectric point of the activated carbon surface is determined by titration. 0.5 g of activated carbon is added to a series of sealable containers containing 50 mL of a certain concentration of NaCl solution. The initial pH value is adjusted with NaOH or HCl. After shaking on a shaker for 2 h, it is left to stand for 24 h. The supernatant is taken and the equilibrium pH value of the solution is measured. The isoelectric point is the point where the pH value does not change before and after the addition of activated carbon.

⑥ Zeta potential measurement

Zeta potential is an important indicator of the stability of colloidal dispersions, which can reveal the electrical properties of particles in solution and judge the stability of colloidal dispersions based on the absolute value. The larger the absolute value, the more stable the dispersion. In water treatment, the smaller the absolute value of the Zeta potential, the easier it is to aggregate, facilitating various process treatments.

The Zeta potential value of the activated carbon surface can be measured by a microelectrophoresis instrument, using a glass sample pool with a thickness of 0.5 cm. The measurement steps are as follows: firstly, the activated carbon sample is ground to below 200 mesh, so that it can be evenly dispersed in the NaCl solution. During measurement, about 0.5 mL is taken each time in the sample pool, ensuring that no bubbles exist. The instrument uses computer multimedia technology to apply voltage to the particles in the dispersion system, and calculates the surface Zeta potential value through the size of the particle displacement. The direction of movement determines the electrical properties of the surface. Each sample is measured in parallel three times, and the average value is calculated, which is the Zeta potential of the activated carbon at that pH value. By measuring the Zeta potential values at different pH values, the isoelectric point of the activated carbon surface can also be determined.

⑦ X-ray diffraction analysis

X-ray diffraction analysis (XRD) is an effective method for qualitative analysis of crystal structure. X-rays scatter when they encounter regularly arranged atoms or ions in the crystal, and the scattered X-rays are strengthened in certain directions, thus displaying diffraction phenomena specific to the crystal structure. Activated carbon is an amorphous carbon material composed of graphite microcrystals, and the degree of graphitization of different activated carbon varies. Studying the size and structure of graphite microcrystals in activated carbon materials with X-ray diffraction technology can help to understand the structural characteristics of activated carbon materials more deeply. In the preparation process of activated carbon, different preparation conditions such as heating methods, types of activators, raw materials and activation conditions may all affect the structure and related properties of activated carbon. Therefore, it is necessary to use X-ray diffraction technology to study the microcrystalline structure of activated carbon products under different preparation conditions. In the standard graphite X-ray diffraction spectrum, there are nine high-intensity graphite characteristic diffraction peaks, including (002), (100), (101), (004), (102), (103), (110), (112), and (006). The number of characteristic peaks that appear can represent the degree of graphitization. When the degree of graphitization is low, these characteristic diffraction peaks will not appear at the same time. The more diffraction peaks appear in the spectrum, the higher the degree of graphitization of the carbon material and the better the order.

⑧ X-ray photoelectron spectroscopy analysis

X-ray photoelectron spectroscopy (XPS) analysis uses soft X-rays to excite the electron energy spectrum of the sample, which can provide information on the elemental composition of the sample surface, and analyze the valence states of elements through peak fitting results.

(2) Adsorption test characterization

① Preparation of simulated wastewater and drawing of standard curves

The following lists three methods for preparing wastewater for reference. Readers can prepare various types of wastewaters according to characterization needs and measure the content according to the experimental methods stipulated by the relevant national standards.

a. Preparation and plotting of standard curves for wastewater containing antibiotics. Weigh a certain amount of antibiotics, dissolve in 1 L of deionized water, mix evenly, take different volumes of the solution and add to the colorimetric tube, dilute to the scale line, and prepare a series of different concentrations of antibiotics (ciprofloxacin, oxytetracycline) simulated wastewater. Use deionized water as the reference solution, measure the absorbance of a series of concentrations of antibiotic wastewater at their maximum absorption wavelength (the maximum absorption wavelength of ciprofloxacin is 275 nm, and that of oxytetracycline is 275 nm), and obtain the relationship between the absorbance and concentration of ciprofloxacin or oxytetracycline.
b. Preparation and plotting of standard curves for wastewater containing chromate. Weigh a certain amount of dried potassium dichromate, dissolve in 1 L of deionized water, mix evenly, calculate the concentration in terms of chromate element, take different volumes of the solution in the colorimetric tube, and determine the concentration of hexavalent chromium according to the "Water Quality—Determination of Hexavalent Chromium-Diphenyl carbazide Spectrophotometry" (GB 7467-87). The relationship between absorbance and concentration can be obtained.
c. Preparation and plotting of standard curves for wastewater containing nickel. Weigh a certain amount of dried nickel nitrate, dissolve in 1 L of deionized water, mix evenly, calculate the concentration in terms of nickel element, take different volumes of the solution in the colorimetric tube, and determine the concentration of nickel ions according to the "Water Quality—Determination of Nickel—Dimethylglyoxime Spectrophotometry" (GB 11910-1989).

② Adsorption kinetics experiment

The adsorption kinetics experiment can characterize the type of adsorption kinetics of activated carbon for a specific adsorbate. By referring to the content in the first section of this chapter, the type of adsorption of activated carbon for the adsorbate can be judged to analyze the adsorption mechanism and provide a basis for adjusting the activation modification scheme to improve specific adsorption performance.

Prepare a certain concentration of adsorbate solution in a conical flask, add the adsorbent, move to the shaker and shake at a speed of 150 r/min, take samples at a predetermined time, filter through a 0.45 μm filter membrane, and use the above corresponding determination method to determine the concentration of the adsorbate until the concentration no longer changes. The removal rate at time t and the adsorption amount q_t at time t are calculated using formulas (8.12) and (8.13):

$$H(\%) = \frac{C_0 - C_t}{C_0} \times 100 \tag{8.12}$$

$$q_t = \frac{(C_0 - C_t) \times V}{W} \tag{8.13}$$

where H is the removal rate at time t, %;

q_t is the adsorption capacity at time t, mg/g;

C_0 is the initial concentration of the adsorbate;

C_t is the concentration at time t (mg/L);

V is the volume of the adsorbate (L);

W is the mass of activated carbon (g).

③ Adsorption equilibrium experiment

The adsorption equilibrium experiment can characterize the single saturated adsorption amount of activated carbon for a specific adsorbate, which is used to check whether the activated carbon meets specific requirements. It can also be combined with the adsorption kinetics experiment to determine the optimal process conditions for this activated carbon.

The equilibrium experiment is used to explore the effect of different experimental conditions on the adsorption effect. The specific operation method is as follows: take 100 mL of a certain concentration of adsorbate solution in In a 250 mL conical flask, add 0.1 mg of activated carbon, shake at a speed of 150 r/min in a water bath shaker at a preset temperature, ensure that adsorption reaches equilibrium, then filter with a 0.45 μm filter membrane, and measure the concentration of the adsorbate in the filtrate. The removal rate and adsorption amount are calculated using formulas (8.14) and (8.15) respectively:

$$H(\%) = \frac{C_0 - C_e}{C_0} \times 100 \tag{8.14}$$

$$q_e = \frac{(C_0 - C_t) \times V}{W} \tag{8.15}$$

where H is the removal rate at time t, %;

q_e is the adsorption capacity at time t, mg/g;

C_0 is the initial concentration of the adsorbate, mg/L;

C_e is the equilibrium concentration, mg/L;

V is the volume of the solution, L;

W is the mass of activated carbon, g.

8.3 Enhanced Coagulation and Adsorbents

8.3.1 Enhanced Coagulation

Coagulation is one of the most important technical units in water treatment processes, including drug mixing, coagulation, flocculation, etc. As a pretreatment process for solid–liquid separation/removal of pollutants (such as sedimentation, filtration, etc.), coagulation technology is generally used at the front end of water treatment processes. The effect of coagulation treatment directly affects and even largely determines the treatment effect, operating efficiency, and comprehensive cost of the entire purification process. Therefore, the ultimate goal of coagulation treatment is to obtain the best purification ability with the lowest dosage (cost) for different water qualities and to achieve the best treated water quality.

At present, the water quality pollution of surface water and groundwater is still serious; on the other hand, the national requirements for water supply and drainage water quality are still increasing. This requires the coagulation process to play a greater role, and its target pollutants have expanded from traditional particles to sols, polymers, biomacromolecules, and even soluble organic substances. The connotations of traditional turbidity and color sensory indicators are gradually extended to trace pollutants adsorbed by particles and soluble pollutants with optical effects. In addition, the implementation of various combined processes (coagulation-ultrafiltration, micro-flocculation-filtration) has put forward higher technical requirements for the pre-coagulation treatment process, and the standards for the turbidity, color, and even organic matter of the coagulation process effluent are constantly improving. In summary, the increasingly broad functional requirements have exceeded the historical category of coagulation technology mainly used to remove suspended solids and colloidal particles in water, and coagulation needs to undertake more and more fine and efficient water quality purification functions.

Based on the above background, concepts such as enhanced coagulation and optimized coagulation have emerged. In the 1990s, the American Water Works Association first proposed the concept of "enhanced coagulation". Enhanced coagulation technology is an improvement based on coagulation process, mainly based on coagulation chemistry and coagulation kinetics to improve the coagulation process and thus enhance the coagulation effect. Coagulation chemistry starts from the issue of surface charge of colloids, the potential barrier between colloids is reduced by increasing the dosage of coagulant, improving the chemical reaction conditions of coagulant, using new coagulants, etc., and further destroys the stability of colloids. And coagulation kinetics is about the collision problem between particles: greater kinetic energy in the flocculation process is provided by improving hydraulic conditions, improving coagulation equipment conditions, etc., the collision times of floc particles are increased, and the kinetic stability of colloids is reduced. Corresponding to the above mechanisms, there are many methods of enhanced coagulation. In summary, coagulation enhancement includes the development of new efficient coagulants, enhancing the application potential of conventional coagulants, and adding adjuvants as the three main methods.

8.3.1.1 Development of New Efficient Coagulants

Limited by the limitations of traditional coagulants on the types of pollutants such as organic matter and soluble small molecules, the development of new coagulants is the fundamental way to achieve enhanced coagulation. New coagulants maintain stable and efficient turbidity removal and decolorization performance while enhancing other water purification functions, which is of great significance to solving water pollution problems. Professor Wang Dongsheng from the Ecological Center of the Chinese Academy of Sciences once proposed a highly efficient nano-coagulant, focusing on the improvement of effective components such as Al_b (Al_{13}). The team of Gao Baoyu from Shandong University proposed the concept of organic–inorganic composite coagulant, striving to enhance the removal performance of inorganic coagulant for pollutants. Under the support of the Kenyan Water Department and the Chinese Ministry of Science and Technology, the research group of Li Fengting from Tongji University developed new drugs for the raw water of reservoirs under high turbidity and high impact load in the rainy season in Africa. Wang Hongtao, Li Jie et al. successfully developed a series of efficient PAC, PFSS and other new iron/aluminum salt coagulants, and achieved stable large-scale production, ensuring the stable water supply and quality improvement of core cities such as Nairobi. The enhanced removal effect of raw water turbidity under the condition of water quality fluctuation by efficient aluminum salt coagulant is shown in Fig. 8.13.

8.3.1.2 Enhancing the Application Potential of Conventional Coagulants

According to the water quality characteristics of the sample and the types of target pollutants, the selection of coagulant types, increase in dosage, and optimization of operating parameters can further enhance the use efficiency of existing coagulants and improve treatment effects.

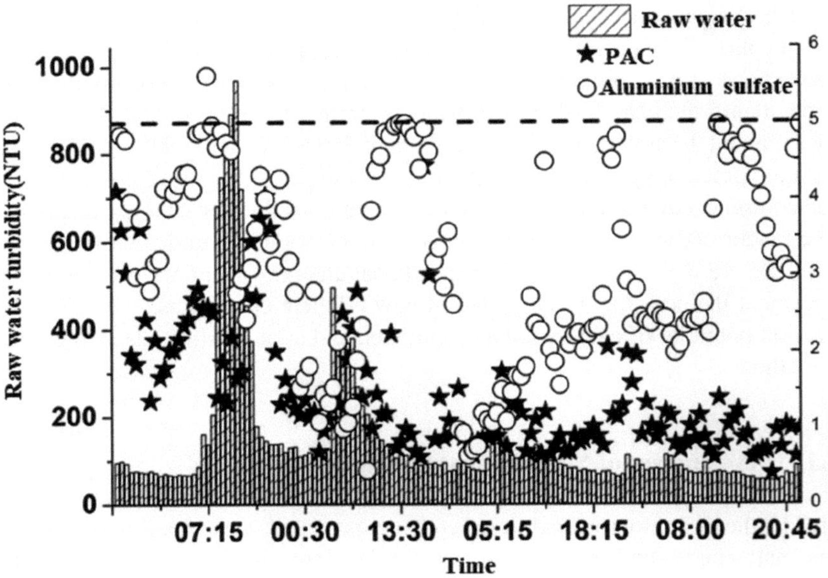

Fig. 8.13 Enhanced removal effect of raw water turbidity under the condition of water quality fluctuation by efficient aluminum salt coagulant

(1) Selection of coagulant types

Different types of coagulants have different hydrolysis properties and are not entirely suitable for the same water quality conditions. Therefore, it is necessary to conduct experiments based on the water quality of the raw water to select the most effective coagulant. According to the type of raw water, water quality characteristics and natural conditions of the site, the coagulant product with the best effect can be selected in many subdivided conditions such as iron salt/aluminum salt, low molecular weight/polymer coagulant, salt base, synthesis process, etc. For example, the Yangshupu Water Plant in Shanghai can achieve good treatment effects by using aluminum sulfate in the summer and autumn seasons, while polyaluminum chloride is usually used in winter. Some scholars believe that inorganic coagulants have a better enhancement effect on the removal of natural organic matter (NOM) than organic coagulants, because the insoluble complexes formed by inorganic metal ions can provide more surface adsorption sites, enhancing the adsorption removal ability of organic matter. There are also reports that iron salt coagulation is better than aluminum salt in removing natural organic matter.

(2) Change the dosage of the coagulant

Increasing the dosage of the coagulant can, on the one hand, allow more colloidal particles in the water to participate in adsorption bridging, greatly increasing the chance of particles being netted and swept, which helps to compress the hydration shell of organic matter. On the other hand, cations will be produced during the hydrolysis process, which will neutralize the anions of organic matter, eliminating the influence of organic matter on inorganic colloids, thereby destabilizing inorganic colloids and achieving the purpose of removing colloids. Liu Cheng et al. conducted experiments on the raw water of the Huangpu River using two kinds of coagulants, examining the effect of enhanced coagulation. The results showed that increasing the dosage of the coagulant can greatly reduce the turbidity and increase the removal rate of organic matter in the water. Xu Yongpeng et al. conducted experiments on raw water samples with different turbidity and organic matter content, and compared the effect of coagulant dosage on the removal of organic matter. The experimental results show that as the dosage of the coagulant increases, the removal rates of turbidity and organic matter also increase correspondingly. When the removal rate of turbidity reaches a maximum of 80%, the removal rates of COD_{Mn} and UV_{254} continue to increase with the increase in the dosage of the coagulant.

However, it should be noted that adding too much coagulant will restabilize the colloid, reduce the coagulation effect, and produce a large amount of sludge during the sedimentation process, causing secondary pollution. This not only affects the effluent effect, but also adds difficulty to the operation and management of the sedimentation tank. In addition, excessive use of aluminum salt can cause excessive aluminum residue, so the amount of coagulant needs to be controlled within the optimal range, ensuring the effluent effect while saving water production costs.

(3) Change the pH value

The pH value (alkalinity) of raw water is one of the core factors affecting the coagulation effect. Its mechanism mainly includes affecting the hydrolysis reaction of the coagulant (chemical characteristics) and the surface potential of colloidal particles (physical and chemical characteristics), thereby having a comprehensive impact on the coagulation effect. Dong Bingzhi et al. found that the pH value is a decisive factor affecting the coagulation effect of Huangpu River water, and the removal effect of organic matter is best within the pH range of 3–6. Jian et al. found that the removal effect of organic matter is best when the pH value is 6.

(4) Add coagulant aids

Single coagulation is difficult to meet higher water quality requirements, so several processes led by coagulation are used for coupling for a complex water treatment problem. That is, by adding other agents to change the coagulation conditions, specifically including coagulant strengthening, oxidant strengthening, and adsorbent strengthening. The methods of use include compound use and combined use of agents. Compound use is to produce in accordance with the optimized ratio, the process is more complex, but it is convenient to use, the effect is stable, and it is beneficial for use in projects with weaker technical management. The combined use of agents refers to the combined use of different types of agents during the coagulation stage. The operation management is more complex, involving the optimization of the order (mode) of dosing, but the cost is relatively low. It should be noted that the compound agents and the combined use of agents are all reacted in a mixed contact manner, the agents are directly added, and the overall operation mode of the coagulation process is still maintained, which is different from the superposition and separate use of the process.

(5) Coagulant Strengthening

Coagulant strengthening mainly includes activated silicic acid, sodium alginate, bone glue, etc. The use of coagulants in combination with coagulants is beneficial to the destruction of the stability of colloidal particles and improves or enhances the coagulation effect. By adding coagulants, the removal efficiency of target pollutants can be effectively improved, and the optimal dosage of coagulants can be reduced. Common coagulants include polyacrylamide, activated silicic acid, sodium alginate, bone glue, etc. Studies have found that adding coagulants to raw water first, and then adding the formed alum particles are larger, evenly hooked, and sink quickly, which saves the amount of coagulant while improving the coagulation sedimentation effect. Research from Liu Zhijie et al. on the water of Kunming Lake in the Summer Palace shows that: chitosan flocculant can be used as a coagulant for treating surface water, and the removal rate of turbidity and organic matter is higher than that of single use, and can reduce the concentration of residual aluminum in the water.

(6) Oxidant Strengthening

Oxidant strengthening mainly includes chlorine dioxide, ozone, ferrate, and permanganate strengthening. It is mainly used to improve the removal efficiency of difficult-to-degrade substances and further improve the subsequent coagulation effect.

When Huo Lijing et al. used chlorine dioxide pre-oxidation combined with polyaluminum chloride to strengthen coagulation treatment of wastewater from coal-fired power plants, they could achieve a TOC removal rate of 50% and a COD_{Cr} removal rate of 73%. Ozone (O_3) as an oxidant replacing chlorine, has the ability to kill drug-resistant pathogens, and is a strong oxidant widely used in water and wastewater treatment. It can either directly oxidize or generate free radicals for indirect reactions. Jiang Tingting's research shows that the effect of ozone-enhanced coagulation is significantly better than conventional coagulation for the deep treatment of urban sewage.

Permanganate (iron) salt can also improve the effect of coagulation effluent by combining pre-oxidation with coagulation. Yang Yanling et al. found that when treating low-temperature and low-turbidity water in the Yellow River in winter, potassium permanganate and chloramine combined pre-oxidation can significantly improve the effect of coagulation sedimentation and filtration. At the optimal mass concentration, the removal rates of turbidity in the post-sedimentation water and post-filtration water reached 94% and 98% respectively. Zhang Jin et al. found through experiments on a eutrophic lake that pretreatment has a higher strengthening effect on algal-containing water, can significantly reduce the number of algae in the effluent, and has a more significant removal effect on odors and organic pollutants caused by algae. The strengthening effect is significantly higher than simple coagulation.

(7) Adsorbent Enhancement

The adsorbent mainly includes powdered activated carbon, diatomaceous earth, clay, etc. This chapter mainly introduces adsorbent enhancement.

Adsorbent enhancement is an effective method to treat pollutants that are not easily removed by coagulation by adding adsorbents. Compared with the method of increasing the advanced treatment unit and biological pretreatment, the adsorbent enhanced coagulation has low cost, will not produce harmful products such as disinfection by-products and algal toxins, and will not occupy land, which is very suitable for the transformation of the original system. The use of powdered adsorbents can also increase the collision rate between particles and improve the coagulation effect. Using adsorbent-enhanced coagulation can significantly save the amount of coagulant added while ensuring the effect of water treatment and overall reducing the amount of net sludge.

8.3.2 Adsorption Enhanced Coagulation

After conventional coagulation treatment, most of the dissolved organic matter is not removed. This part of the organic matter can be removed by adsorption. Activated carbon, diatomaceous earth, attapulgite, coal ash and other materials can all be used as adsorbents in water treatment.

8.3.2.1 Commonly Used Adsorbents for Adsorption-Enhanced Coagulation

(1) Diatomite

Diatomite is a siliceous sedimentary rock mainly composed of diatoms and other single-celled aquatic planktonic organisms. It is a clay-like non-metallic mineral material. Its main chemical component is SiO_2, usually accounting for more than 80%. Diatomite has stable chemical properties, is wear-resistant, acid-resistant, has a porous structure and a large specific surface area, and has good adsorption properties. Diatomite is abundant in reserves and cheap and easy to obtain. In enhanced coagulation, the application of diatomaceous earth not only adsorbs pollutants in the water body, but also improves the settling performance of the flocs.

In recent years, diatomite has been frequently reported as an adsorption-enhancing coagulant, especially showing good performance in strengthening coagulation and algae removal. Jiang et al. have studied diatomite-enhanced coagulation technology to improve the removal of algae and other pollutants. The dosage and ratio of diatomite and aluminum salt are 40 mg/L and 1:1, respectively, and the algae removal rate reaches $98.8 \pm 0.65\%$. It is speculated that the mechanism of diatomite-enhanced coagulation is mainly adsorption bridging and sweeping flocculation. Yang Qi studied the performance of diatomite-enhanced coagulation in algae removal and treatment of polluted raw water. The results show that the addition of diatomite enhances the formation of coagulation flocs, and the performance of diatomite-enhanced aluminum sulfate in removing copper and aluminum microcystis is better than PAC. Under the condition of diatomite/aluminum sulfate addition ratio of 20 mg/L: under the condition of 20 mg/L, the algae removal rate of enhanced coagulation exceeds 95%; in addition, diatomite/aluminum sulfate enhanced coagulation can effectively remove TOC (89%) algal suspension proteins and humic acids and other algal organic matter (AOM). It is speculated that diatomite-enhanced coagulation is achieved by enhancing the aggregation of aluminum salt hydrolysis products on the surface of algal cells, disrupting the stable double electric layer structure of algal cells, and adsorption bridging and net trapping into flocs. Diatomite-enhanced coagulation can also be applied to the treatment of oily wastewater and urban river water. Zhao et al. evaluated the feasibility of treating oily wastewater with polyaluminum chloride (PAC), polyferric sulfate (PFS), and polyacrylamide (PAM) combined with diatomite, and examined it from both cost and performance aspects, preferring the PAC/diatomite enhanced coagulation scheme, and studied its mechanism of enhanced coagulation treatment of oily wastewater, with diatomite adsorbing through the formation of chemical bonds with organic molecules. PAC's enhanced coagulation effect on diatomite is better than PFS/PAM, showing excellent performance in COD/turbidity removal and floc settling characteristics. Wang Saisai et al. studied the application of diatomite and PAC-enhanced coagulation in the treatment of urban river water, and examined the effect of diatomite on PAC coagulation removal of turbidity, organic matter, phosphorus in water, and its improvement of settling performance. The results show that diatomite can effectively enhance the coagulation efficiency of PAC, with the removal rates of turbidity, DOC, and TP reaching 12.8%, 15.5%, and 10.7%, respectively. Diatomite has a very good decolorization effect and can be mixed and stirred with colored wastewater to be treated as a pre-adsorbent. After 5 ~ 10 min, aluminum salt coagulant is added for coagulation. This method is not only very effective for dyeing wastewater, but also very effective for the treatment of various colored raw water. Its only defect is that the amount of sludge will increase. In addition to good adsorption performance, due to the special pore structure of diatomite, it can effectively load microorganisms in the water body, which can play a buffering and enhancing role in microbial treatment, whether for municipal sewage or industrial sewage, this enhancement is very prominent.

(2) Coal ash

Coal Ash refers to the tiny ash particles emitted during the combustion process of fuel (mainly coal). Its particle size is generally between 1 and 100 μm. Coal ash plays a synergistic effect in the flocculation process. As a flocculant aid, coal ash provides flocculation nuclei, providing favorable conditions for the rapid growth of flocs. Moreover, coal ash has a larger density. When coal ash particles are rolled into alum flowers, it can increase the specific gravity of the flocs and accelerate the settling speed of the flocs. Using coal ash to treat wastewater fits the pollution control idea of "waste treatment with waste", effectively solving the environmental pollution problem of coal ash, and has certain economic benefits. The coal ash enhanced coagulation method is widely used in the treatment of industrial wastewater such as mine water, oil field polymer flooding wastewater, and preserved egg wastewater. Liu Fangchi et al. reported a method of treating mine water containing suspended solids by loading flocculation with coal ash as the loading material. Compared with traditional flocculation, the dosage of coal ash loaded flocculation reagents (PAC, PAM) is less, the removal efficiency is high, and the addition of coal ash can adjust the pH value of mine water and promote the flocculation process. Li Bailin et al. prepared a new type of efficient and low-cost coagulant, a new type of aluminum-iron composite coagulant made from coal ash and bauxite, and studied its phosphorus removal performance. When the dosage is 0.40 g/L, the removal rate of TP is 99.58%. Huang Xing verified the feasibility of coal ash enhanced coagulation treatment of oil field polymer flooding wastewater. The results show that the oil removal effect of coal ash enhanced PFS experiment is nearly doubled compared with the use of PFS alone, from 53.6 to 96.2%, and the SS removal rate has increased from 37.5 to 51.2%. Wu Shaolin et al. used PAC + PAM as flocculant and coal ash as adsorbent to strengthen the coagulation process to treat lead-free preserved egg wastewater, and achieved relatively ideal wastewater treatment effects. The removal rate of residual COD in wastewater reached 81.25%, and the removal rates of SS, color, heavy metal ions Cu^{2+} and Zn^{2+} also reached 92.1%, 98.3%, 92.1% and 99.2%, respectively, meeting the standards of wastewater treatment plants. Coal ash, especially the coal ash in the Ordos region, has an alumina content of over 40%, some even over 45%, and can be used as a raw material for the production of aluminum sulfate and polyaluminum chloride. After separating heavy metals and iron salts, it can produce white iron-free aluminum sulfate and polyaluminum chloride.

(3) Attapulgite

Attapulgite, also known as palygorskite, belongs to the sepiolite family in mineralogy. It is a clay mineral with magnesium silicate as the main component and contains elements such as aluminum and iron. The theoretical chemical formula of attapulgite is $Mg_5Si_8O_20(OH)_2(OH_2)_4{\cdot}4H_2O$, and its chemical composition mass fraction is SiO_2: 56.96%, (Mg,Al,Fe)O: 23.83%, H_2O:19.21%. Attapulgite has unique dispersion, high-temperature resistance, alkali resistance and other good colloidal properties and high adsorption decolorization ability. It has a unique layered, chain-like structure, has developed internal pore channels and a large specific surface area, and has good adsorption performance. It is widely used in water treatment and environmental protection industries. Such as algae removal, turbidity removal, humic acid removal, heavy metal removal, and removal of nitrogen and phosphorus in lakes.

The use of attapulgite in combination or compound with coagulants such as polyaluminum chloride and polyacrylamide enhances coagulation. Ling Huishi et al. combined modified attapulgite with the flocculant PAC to explore the effect of enhanced coagulation in removing algae from eutrophic river water. The study showed that adding attapulgite during the PAC coagulation process can significantly increase the density of the flocs, improve the settling performance of the flocs, and compare with using PAC alone, the use of attapulgite in combination has a suppressive effect on the floating of algae in the flocs, and the effect of removing algae and turbidity is significant. Tan Wei studied the effect of modified attapulgite as an adsorbent and coagulant in treating hexavalent chromium solution. The results showed that adding modified attapulgite to the water solution as a coagulant is better than adding PAC alone, and can significantly improve the removal rate of hexavalent chromium as well as color and turbidity. Attapulgite fits the Langmuir isotherm model for Cr^{6+}. Han Zhen et al. used purified-acid heat-modified attapulgite combined with polyaluminum chloride (PAC) to coagulate and remove humic acid from water. When PAC is added at 5 mg/L, the removal rate of humic acid is 3% higher than that of single addition of 12 mg/L PAC, reaching 92.03%. Wang Aiqin et al. used attapulgite and acrylic acid as raw materials, added initiators and crosslinking agents, and made a network-type attapulgite composite adsorbent through graft copolymerization, crosslinking, and drying in a water solution, which can be used for the dynamic removal of nitrogen and phosphorus in lakes. The adsorption product can be used for desertification control and has the function of slow-release nitrogen and phosphorus.

(4) Activated Carbon

The main applications of activated carbon in the environment are in the forms of powdered activated carbon and granular activated carbon. Granular activated carbon is widely used but expensive, while powdered activated carbon is convenient and widely used. Powdered activated carbon has a large specific surface area, fast adsorption speed, high adsorption capacity, and good removal effect on pollutants, especially a high purification rate of dissolved organic matter in water. However, during use, suitable treatment conditions should be selected according to the type of pollutant to be removed, and the separation effect of powdered activated carbon in the precipitation and filtration stages should also be considered. Therefore, when using powdered activated carbon for adsorption pretreatment, both its adsorption removal effect on water pollutants and the feasibility of subsequent treatment should be considered.

① Heavy Metals

Activated carbon mainly achieves adsorption by complexing reactions between surface carboxyl and other acidic functional groups and sulfur-containing functional groups and metal ions. Rosińska evaluated the feasibility of using powdered activated carbon CWZ-30 and polyaluminum chloride to enhance coagulation to remove surface water pollutants. Using 30 mg/L of powdered activated carbon CWZ-30 to enhance coagulation, the removal rates of heavy metals Cu, Cd, Ni, and Pb were increased by 79%, 53%, 26%, and 95% respectively. Li et al. studied the treatment effect of the combined process of coagulation/flocculation and powdered activated carbon adsorption on stable landfill leachate. Under optimal process conditions, the removal rates of COD, Pb, Fe, and toxicity in landfill leachate by the coagulation/flocculation-adsorption combined process were 86%, 97.6%, 99.7%, and 78% respectively. Liu Guimei et al. used enhanced coagulation-adsorption coupling treatment to treat high-concentration mercury-containing produced water from gas fields. Using biomass activated carbon as an adsorbent, the removal rate of mercury was 98.8%, which was significantly higher than the removal rate of adding coagulant alone (80%). The residual mercury concentration was less than 0.05. mg/L, which meets the requirements of the national secondary emission standard. Chen Wei et al. studied the enhancement of lead removal by the combined process of potassium permanganate oxidation and powdered activated carbon adsorption. Under the optimal dosage of 12 mg/L of polyaluminum chloride, the optimal ratio of potassium permanganate oxidation and powdered activated carbon was determined. The removal rate of Pb^{2+} reached 99.77%, and the Pb^{2+} content in the water after sedimentation reached 0.0049 mg/L or less until it was undetectable, fully meeting the national drinking water quality standard in China "Sanitary Standard for Drinking Water" (GB5749-2022) for lead content. Linstedt et al. confirmed that after lime coagulation sedimentation-sand filtration-granular activated carbon adsorption treatment, the removal rates of silver, cadmium, chromium, and selenium were 91.7%, 99.9%, 96.0%, and 99.7% respectively. It is sometimes difficult to meet the standard for the removal of heavy metals. One very important reason is the formation of tiny flocs of heavy metal precipitates suspended in the liquid, especially when controlling the residual concentration at a relatively low level, such as controlling Cr^{3+} below 0.05 mg/L. If a small amount of powdered activated carbon is added at the same time as the heavy metal trapping agent, it promotes coagulation and sedimentation, which doubles the effect of removing heavy metals. If various polyamine-modified thiosulfates are chosen, their adsorption effect on heavy metals will be stronger.

② Organic matter

Activated carbon has a strong adsorption capacity for hard-to-degrade natural organic matter and artificially synthesized organic matter in water. The enhanced coagulation method of activated carbon for the removal of organic matter in water bodies, especially disinfection by-products, has been reported repeatedly. Kristiana et al. applied powdered activated carbon as an adsorbent to enhance coagulation for the removal of natural organic matter (NOM) and control the formation of disinfection by-products (DBPs) in water treatment plants. Adding activated carbon helps to increase the overall removal rate of NOM by 70%, thereby significantly reducing the formation of DBPs (80 ~ 95%). Uyak et al. studied the removal of DBP precursors in lake water through enhanced coagulation and powdered activated carbon adsorption. The results showed that coagulation and powdered activated carbon adsorption are complementary in removing dissolved organic carbon (DOC) and DBP precursors in lake water. The maximum removal rate of DOC by ferric chloride enhanced coagulation is 45%, and this type of natural organic matter (NOM) prefers large organic molecules with negatively charged functional groups. Supplementing PAC adsorption to enhance coagulation can increase the removal rate of DOC to 76%, and activated carbon adsorption mainly removes low molecular weight and uncharged NOM substances. The results show that powdered activated carbon adsorption enhanced coagulation can better meet the requirement of trihalomethanes (THM) 150 μg/L than pure enhanced coagulation. Li Zhenhua et al. studied the deep treatment of urban sewage by coagulation-carbonized sludge adsorption, and the results showed that the combined application of ferric chloride coagulation-carbonized sludge adsorption in the deep treatment of urban sewage has a significantly better treatment effect than using carbonized sludge adsorption alone or using ferric chloride flocculant coagulation alone. The pollution indicators of the effluent can reach the first-class A standard in China ("Urban Sewage Treatment Plant Pollution Discharge Standard", GB18918-2002). Zhan Xiao et al. used powdered activated carbon and modified wheat straw adsorbents in combination with coagulants to increase the removal rate of organic matter, enhance the settling performance of flocs, and achieve better treatment effects. The effluent contains more free residual chlorine to maintain a continuous disinfection effect. He Jiandong conducted an experiment on the removal of polychlorinated biphenyls (PCBs) from polluted water bodies by polyferric sulfate (PFS) and powdered activated carbon enhanced coagulation, and found that PFS and activated carbon enhanced coagulation have good removal effects on PCBs and turbidity. Nie Xiaobao used the combined technology of enhanced coagulation and activated carbon to treat micro-polluted source water, significantly improving the removal rate of organic matter, reducing the turbidity of the effluent, and effectively controlling the residual aluminum content in the effluent.

③ Ammonia Nitrogen

Activated carbon enhanced coagulation can also be applied to the removal of ammonia nitrogen. Bashir used the coagulation-flocculation-pretreatment-adsorption method to treat landfill leachate, and the research found that $FeCl_3$ and powdered activated carbon adsorption enhanced coagulation has a higher removal efficiency for color, COD, and NH_3-N. Liao et al. evaluated the combined performance of powdered activated carbon (PAC) and powdered zeolite (PZ), and the results showed that powdered activated carbon and powdered zeolite showed a synergistic effect, improving the removal effect of ammonia nitrogen. Song Xiangsong et al. reported the study of using activated carbon combined with polyaluminum chloride to remove total nitrogen in raw water. The results showed that the enhanced coagulation process can achieve a removal rate of 88.64% for ammonia nitrogen in raw water, and the turbidity of the raw water also has a significant decrease, from the original 3.2 to 0.18 NTU. In many occasions in China, especially in the emergency treatment of black and odorous water bodies, many construction units often add an appropriate amount of ammonia nitrogen remover to the water body for quick removal of ammonia nitrogen, most of which are chlorine-containing oxidants, such as calcium hypochlorite and superior chlorine net. If an excessive amount is added, a large number of fish deaths often occur.

For some emergency incidents, a variety of composite reagents can be used, which can be combined with iron salts or aluminum salts, or both, supplemented with particulate matter such as diatomaceous earth, polyacrylamide, etc. After adding the solid, it can be quickly stirred, so that the water quality indicators and transparency can be greatly improved within a few hours.

8.3.2.2 Typical Case of Adsorption Strengthened Coagulation Treatment of Industrial Wastewater

(1) Low Temperature and Low Turbidity Water

The water quality in the cold regions of northern China shows low temperature and low turbidity characteristics in winter. The coagulation effect of low temperature and low turbidity water is poor, and the treatment is difficult. Therefore, the treatment of low temperature and low turbidity water has always been an important research topic for experts and scholars. Li Xing et al. studied the combined use of powdered activated carbon and potassium permanganate, combined with PAC to strengthen the coagulation treatment of reservoir low temperature and low turbidity water, UV_{254} and COD_{Mn} removal rate increased by 12 ~ 18% compared with using PAC alone. Potassium permanganate has a certain strengthening coagulation effect, but the effect is general; the addition of powdered activated carbon can enhance the coagulation removal effect of organic matter, and the removal rate of organic matter increases with the increase of dosage. Liu Shuaixia et al. used activated carbon adsorption pretreatment, combined with polyferric sulfate, polyaluminum chloride iron to strengthen coagulation technology to treat winter low temperature, low turbidity, high color Yellow River water with good results. The average removal rate of color was 57.5%, the average removal rate of ammonia nitrogen was 32%, the average removal rate of COD was 32%, and the average removal rate of turbidity was 91.1%. The treatment of low-temperature and low-turbidity raw water was a big problem more than ten years ago, but with the emergence of organic and inorganic polymers, when organic and inorganic polymers are combined with conventional aluminum salts, polyaluminum chloride and aluminum sulfate, it becomes very easy to treat low-temperature and low-turbidity water. In 2010, Professor Li Fengting's team cooperated with a water treatment reagent company in Suzhou to use composite polyaluminum chloride to treat Yangtze River water, and achieved good results. This technology has been used to this day. In winter, the reagent consumption of thousands of tons of water in the lower reaches of the Yangtze River can be as low as 7 kg, and the treatment effect is very good.

(2) Fluoride-containing Wastewater

Calcium salt is the cheapest defluorination agent for high concentration fluoride ions (such as 20 ~ 1000 mg/L). However, due to the limitation of K_{sp}, when the fluoride ion is reduced to a certain level, such as reduced to about 20 mg/L, the calcium salt has no obvious removal effect. For low concentration fluoride ions, iron-aluminum composite coagulant is a good defluorination agent, especially for wastewater generated by the electronics, battery and photovoltaic industries, all have good removal effects. Just by using the method of coagulation and precipitation, the concentration of fluoride ions can be controlled below 1 mg/L, without the need for reverse osmosis, ion exchange and other means, which can save a lot of investment and the treatment efficiency is very high. The team of Professor Li Fengting has promoted this technology in several industrial wastewater projects in the coal chemical, electronics, photovoltaic, and other industries, which can greatly reduce project investment and achieve stable treatment effects. Fluorite calcium fluoride is classified as a strategic mineral resource by the state, so in many defluorination projects, the future solution to recover fluorine elements from precipitated aluminum fluoride and calcium fluoride will be a hot spot for resource utilization.

(3) Municipal Sewage

Wang Donghong et al. introduced coagulants during the mixing stage in the coagulation reaction tank, and adsorbents were added during the flocculation stage to assist coagulation, which can realize the coagulation and adsorption occurring in the same reaction unit. The combined removal of organic pollutants, TP, and fluorescent pollutants in secondary effluent by 150 mg/L PAC-loaded FeOOH powder adsorbent and 50 mg/L PAC is effective, and it can achieve effluent that meets the standards for landscape reuse water and ultra-low concentration phosphorus discharge (below 0.05 mg/L).

(4) Advanced treatment of coking wastewater

Coking wastewater was once the most typical high-concentration and difficult-to-degrade organic wastewater. In response to the difficult problem of cyanide, COD_{Cr}, benzo[a]pyrene, and other characteristic pollutants in coking wastewater that are difficult to meet the discharge standards, Professor Li Fengting's team adopted a method of modified activated carbon and iron salt compound, through enhanced adsorption synergistic action, to treat coking wastewater. The removal rates of cyanide and COD reached 99% and 90% respectively, and the color can be controlled below 15 times. The enhanced removal effect of activated carbon enhanced coagulation on COD and cyanide in coking wastewater is shown in Fig. 8.14 (CP-PFS is an activated carbon fine powder-polyferric sulfate composite agent). This technology has been promoted in more than 30 coking companies nationwide. Interested readers can refer to the paper published by Dr. Li Jie.

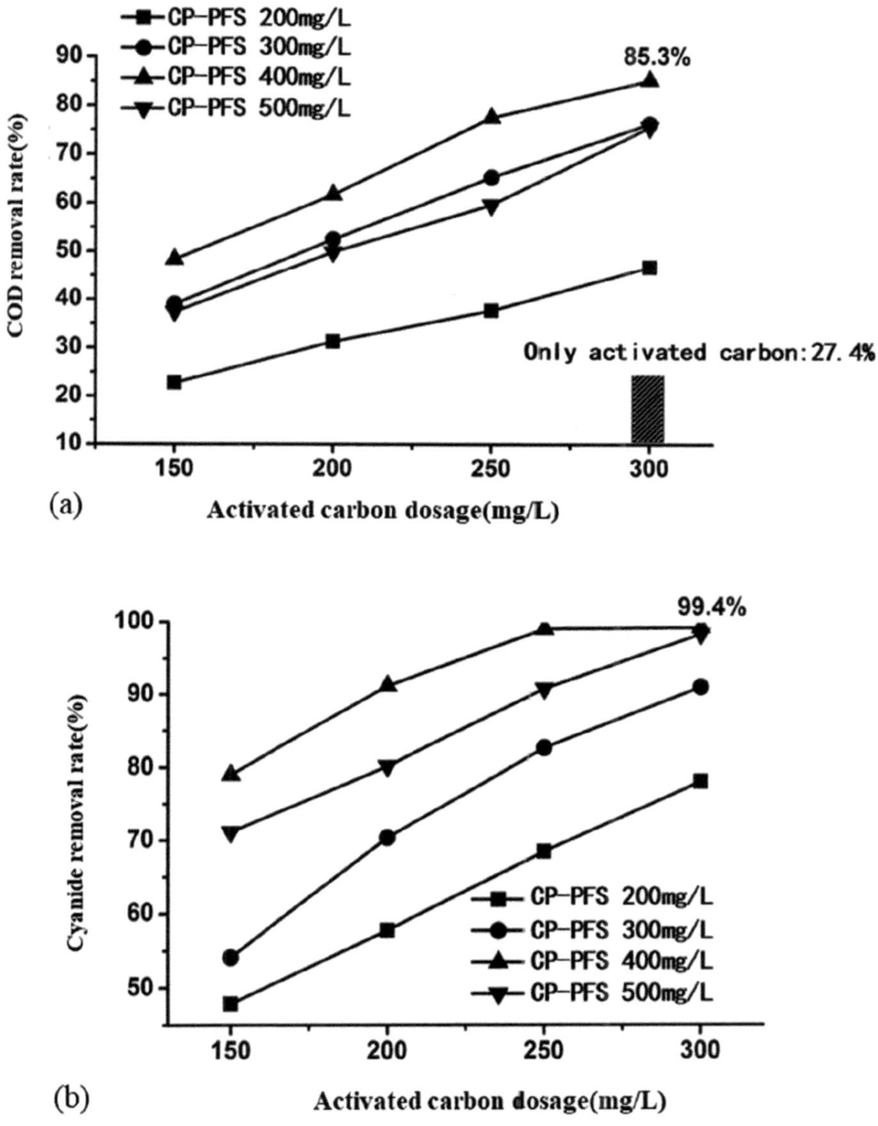

Fig. 8.14 Enhanced removal effect of activated carbon enhanced coagulation on COD (**a**) and cyanide (**b**) in coking wastewater

Chapter 9
Selection and Configuration Methods of Coagulants or Flocculants

9.1 Purpose and Content of Chemical Treatment

As the quality of raw water continues to decline and people's requirements for drinking water quality continue to increase, the cost of obtaining safe and high-quality drinking water is getting higher and higher. The water treatment processes many years ago can no longer meet the requirements of modern people. Especially in the presence of various pesticides, polyhalogenated compounds, etc. in the water body, it is necessary to choose advanced treatment processes to remove the above pollutants. Compared with mechanical treatment, chemical treatment has great advantages, such as low treatment cost, high efficiency, less equipment investment, and small scale, especially when physical treatment requires a very large space, chemical treatment must be chosen.

The chemicals or water treatment agents used in chemical treatment can remove suspended solids, colloids, and some soluble substances in the water, such as removing calcium and magnesium hardness, turbidity, color and odor caused by organic matter, and microorganisms and viruses that may cause human health problems.

When choosing a water treatment agent, the following factors should be considered first: raw water quality, process equipment, and treatment objectives. These three variables can be further classified according to Table 9.1.

The water treatment agent should be selected after fully considering the above factors. The type of raw water and pollutants has a great impact on the type of chemicals used in solid–liquid separation, and two factors must be considered:

(1) Types of pollutants in the water
(2) Turbidity value (can only determine the amount of coagulant needed).

The quality of raw water and wastewater generally changes with the season or production process, so the change in water quality over time must be considered when choosing water treatment agents.

L. Fengting, *Water Treatment Chemicals: Coagulants and Flocculants*,
https://doi.org/10.1007/978-981-96-8221-8_9

Table 9.1 Factors to consider when choosing a water treatment agent

Raw water quality	Process equipment	Treatment objectives
Alkalinity	Sedimentation tank	For drinking water Partial softening Complete softening
pH value	Direct filtration	
Turbidity	Sedimentation + filtration	
Color	Contact clarification	For industrial water General use Ion exchange
Temperature	Flotation	
Hardness	Mixing intensity	
Odor	Sludge disposal	

9.1.1 Alkalinity

The alkalinity of natural water bodies ranges from 10 to 500 mg/L, and domestic wastewater is often alkaline. Detergents and soaps are often used in daily life, introducing alkaline substances into the water. Acid rain has a significant impact on the alkalinity of water bodies, and many surface waters in the southern regions are acidic. The alkalinity of water bodies mostly exists in the form of carbonates and bicarbonates, while hardness exists in the form of calcium carbonate and magnesium carbonate. Since alkalinity is not included in the pollution indicators of drinking water, there is no limit to alkalinity in the drinking water standards of various countries. However, alkalinity is of great significance to fish and aquatic organisms because it reflects the ability of the water body to buffer drastic changes in pH value. Adding coagulant to the water often consumes a certain amount of alkalinity. Figure 9.1 is a curve for titrating alkalinity with sulfuric acid.

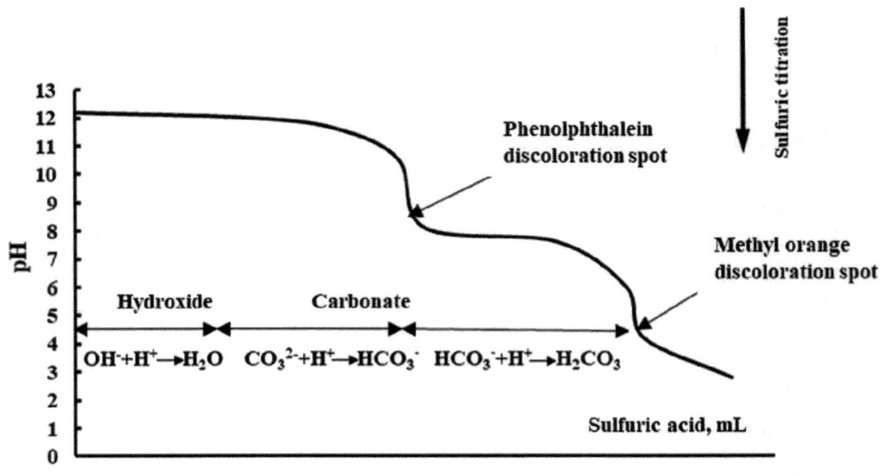

Fig. 9.1 Alkalinity titration curve

Common coagulants include aluminum sulfate, iron sulfate, ferric chloride, ferrous sulfate, and sodium aluminate. In the coagulation process, the first four coagulants will reduce the alkalinity and pH of the solution. To ensure complete coagulation and avoid the presence of residual aluminum ions, it is necessary to ensure that the water body to be treated has sufficient alkalinity. The alkalinity of the water body after general coagulation should be between 25 and 50 mg/L (calculated as calcium carbonate). Sodium aluminate will increase these two indicators. The reactions of coagulants in water bodies are as follows:

$$Al_2(SO_4)_3 + 3Ca(HCO_3)_2 \rightleftharpoons 2Al(OH)_3 + 3CaSO_4 + 6CO_2 \tag{9.1}$$

$$Fe_2(SO_4)_3 + 3Ca(HCO_3)_2 \rightleftharpoons 2Fe(OH)_3 + 3CaSO_4 + 6CO_2 \tag{9.2}$$

$$2FeCl_3 + 3Ca(HCO_3)_2 \rightleftharpoons 2Fe(OH)_3 + 3CaCl_2 + 6CO_2 \tag{9.3}$$

$$FeSO_4 + Ca(HCO_3)_2 \rightleftharpoons Fe(OH)_2 + CaSO_4 + 2CO_2 \tag{9.4}$$

$$Na_2Al_2O_4 + 4Ca(HCO_3)_2 \rightleftharpoons 2Al(OH)_3 + Na_2CO_3 + 4CaCO_3 + 3CO_2 + H_2O \tag{9.5}$$

$$Na_2Al_2O_4 + CO_2 + 3H_2O \rightleftharpoons 2Al(OH)_3 + Na_2CO_3 \tag{9.6}$$

$$Na_2Al_2O_4 + MgCO_3 \rightleftharpoons MgAl_2O_4 + Na_2CO_3 \tag{9.7}$$

The calculation of alkalinity consumed in the reaction process is as follows:

$$\underset{594\,\text{mg}}{Al_2(SO_4)_3 \cdot 14H_2O} + \underset{366\,\text{mg}}{6HCO_3^-} \rightleftharpoons 2Al(OH)_3 + 6CO_2 + 14H_2O + 3SO_4^{2-} \tag{9.8}$$

If 200 mg/L of aluminum sulfate is added to the water to complete the coagulation process,

$$200\,\text{mg } Al_2(SO_4)_3 \cdot 14\,H_2O \text{ consumes } HCO_3^-: (366/594) \times 200 = 123\,\text{mg}$$

Therefore, the consumed alkalinity (calculated as $CaCO_3$): 123 × (50/61) = 101 mg/L.

When choosing metal salt coagulants such as polyaluminium chloride (PAC), aluminium sulphate (PAS), or polysulfate iron, alkalinity is very important. All these substances require some alkalinity to promote hydrolysis reactions, enabling the coagulants to function. If the alkalinity of the water is low, below 50 mg/L, strong acidic metal salts are generally not used. In such cases, there are two options: one

Table 9.2 When ferric chloride and aluminum sulfate are used as coagulants, the alkalinity changes caused by different dosages (After treatment, the alkalinity of the water is 40 mg/L, pH = 6.0)

Alkalinity (mg/L)	1% Ferric chloride solution usage (mL)	1% Aluminium sulphate solution usage (mL)
60	6.4–9.6	7.6–17.6
80	10.6–16.0	12.6–29.2
100	15.0–22.4	17.6–41.0
120	19.2–28.8	22.6–52.6
140	23.6–35.2	27.6–64.4
160	27.8–41.6	32.6–76.0
180	32.0–48.0 (It is recommended not to exceed 40 mL)	37.6–92.6 (It is recommended not to exceed 60 mL)
> 180	–	–

is to increase the alkalinity, for example by adding lime solution, and the other is to choose a coagulant with a high salt base, such as polyaluminium chloride (salt base 60–90%). If the water to be treated has very low alkalinity, it is necessary to add alkalinity. There are cases abroad where acidic or alkaline aluminum salts are used in combination, adding sodium aluminate while adding aluminium sulphate. If the alkalinity is higher than 50 mg/L, there will be enough alkalinity to cause the metal salt to hydrolyze, forming coagulation. If the water body has very low alkalinity, using aluminium sulphate and other coagulants that consume a lot of alkalinities often results in incomplete hydrolysis of aluminium ions, causing the concentration of residual aluminium ions in the effluent to increase, failing to meet the drinking water effluent standard. Several water companies in China have already discovered this problem. In addition, when coagulants were actually used, people were not aware of the effect of the salt base of inorganic polymer coagulants on the treatment effect. Therefore, in order to achieve ideal coagulation results, the amount of coagulant added must be determined. Generally, the higher the alkalinity of the solution, the more coagulant consumed. Table 9.2 shows the changes in alkalinity caused by different amounts of ferric chloride and aluminium sulphate as coagulants.

9.1.2 pH Value

The choice of coagulant greatly depends on the pH value of the water quality. If the pH value of the raw water for drinking is higher than 8.5, and the content of dissolved organic carbon (DOC) is relatively high (i.e., the usual color), then a strong acidic coagulant, such as polysulfate iron, ferric chloride, etc., must be chosen to reduce the pH value to 7.0. For highly alkaline printing and dyeing wastewater, a coagulant composed of ferrous sulfate and ferric sulfate can be used, which can achieve the dual effects of reducing the pH value and coagulation. If the pH value is acidic, to

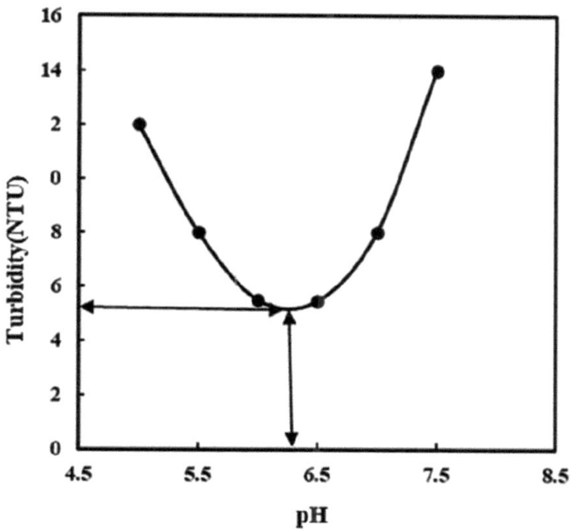

Fig. 9.2 Relationship between residual turbidity and pH value (optimal pH value is 6.2)

ensure the occurrence of coagulation reactions, the coagulation effect of iron salts is better than that of aluminum salts. Even for raw water with the same pH value, there are significant differences due to factors such as alkalinity, ion intensity, etc., so the choice of coagulant is a complex process. According to Fig. 9.2, it can be seen that the residual turbidity is closely related to the pH value, and the residual turbidity reaches the lowest value at a pH value of 6.2.

Alkalinity/pH correlation: Alkalinity represents the ability to neutralize acids and reflects the buffering capacity of a water body. Alkalinity and pH are related; high alkalinity leads to high pH. Metal coagulants are acidic and consume additional alkalinity. For water with low alkalinity, the coagulant may consume all the alkalinity, lowering the pH of the water and reducing the coagulation effect. Water with high alkalinity requires more coagulant to reduce the alkalinity to an appropriate range. Iron sulfate and ferric chloride are more acidic than polyaluminum, thus requiring more additional alkalinity consumption. For PAC, alkalinity consumption is related to salt basicity. High salt basicity PAC will consume less alkalinity than medium and low salt basicity PAC.

pH is the most important factor affecting the coagulation process, influencing:

(1) The surface charge of colloids;
(2) The charge of organic functional groups in the water;
(3) The charge carried by soluble coagulants;
(4) The surface charge carried by flocculated particles;
(5) The solubility of coagulants, etc.

9.1.3 Turbidity

All surface water contains particles of different origins, sizes, and types, which must be effectively removed before water supply. Infectious cysts and parasitic oocysts (Giardia and Cryptosporidium) generally range in size from 3 to 12 μm. These particles can cause health problems, hence the renewed interest in improving particle separation processes. Depending on the raw water quality, typical particle removal includes the following processes:

(1) Rapid or slow filtration;
(2) Coagulation/flocculation/deep bed filtration or coagulation/flocculation/floc separation;
(3) Sedimentation or flotation and rapid filtration.

More than thirty years ago, it was discovered that pre-ozonation before particle removal can significantly improve particle removal efficiency, reduce coagulant dosage, and increase flow rate, as in deep bed filtration. A small amount of ozone, such as 0.5–2 mg/L, should be added before or at the same time as the coagulant (iron salt, aluminum salt coagulant, or cationic polymer). In practice, terms such as "micro flocculation" or "ozone-induced particle destabilization" have been used to describe this process.

Compared to particle removal without ozone oxidation, the improvement effect after ozone oxidation varies greatly. It has been reported that the turbidity of water after filtration can be reduced by 20–90%, and the number of particles will also decrease. The presence of dissolved organic carbon (DOC) in the water is often very necessary, at least 1 mg/L. The effect of ozone pre-oxidation is closely related to the presence of alkaline earth metal ions, especially calcium ions.

Removing algae is usually difficult, and ozone pre-oxidation promotes this. Ozone pre-oxidation can be combined with air flotation, becoming an effective method for separating water algae after coagulation. Low doses of ozone cannot destroy algal cells. In the process of removing particles, the effect of ozone pre-oxidation can be detected whether coagulant is added or not.

Reducing substances, such as Fe^{2+}, Mn^{2+} or NO_2^-, can be rapidly oxidized and can also form precipitates (forming $Fe(OH)_3$, $MnO(OH)_2$), promoting the coagulation process. If the amount of ozone agent for pre-ozonation is determined based on the number of oxidizable substances, then this amount of agent is not sufficient to form residual ozone in the water. However, in cleaner water, it can meet the C-T value required for effective disinfection of the water. This indicates that within a certain reaction time, the concentration of dissolved ozone can maintain a sufficiently high level. The ozone pre-oxidation reactor must deal with particulate matter in the raw water and sometimes should be used in conjunction with a coagulant mixing reaction pool.

Turbidity is a characteristic parameter related to the concentration of suspended particles in water and can conveniently and quite accurately determine the water quality condition. The turbidity index can be used to measure individual treatment processes and the entire water treatment system. Conventional turbidity can be removed through traditional coagulation and flocculation processes. The dosage of coagulant has a great impact on the coagulation effect. As can be seen from Fig. 9.3, the relationship between the dosage of coagulant and the residual turbidity, when the dosage is 12.1 mg/L, the residual turbidity drops to the lowest. Figure 9.4 shows the relationship between the dosage of aluminum sulfate and residual turbidity at different pH values. In raw water, due to the ionization of colloidal particles and the ionization after the particles adsorb organic matter, most particles and colloids carry negative charges. In low turbidity water (< 10 NTU), it is generally used to destabilize as the main, flocculation as the auxiliary, so it often uses inorganic coagulants in combination with organic polymer flocculants. Organic polymer electrolytes, such as high molecular weight polyacrylamide, generally cannot be used because the flocculant cannot function without destabilization. After adding inorganic coagulant, it should be able to quickly form flocs, and under the action of the flocculant, further form large flocs. At present, inorganic–organic composite coagulants have been developed in China, which are very effective for treating low-temperature and low-turbidity water. In addition, the use of cationic organic coagulants is also very effective for treating low-temperature and low-turbidity water, such as polydiallyldimethylammonium chloride, but the treatment cost is higher than that of inorganic coagulants. Some foreign water plants often use direct filtration technology when using these coagulants. In medium turbidity (< 100 NTU) water, it is best to use general inorganic salts, if other conditions are suitable, most will achieve good results. In high turbidity situations, or when the turbidity of surface water increases rapidly, the best choice is to add PAC and polyacrylamide or polyamine mixture. Sludge bed height, sludge volume, dewatering efficiency, and pH reduction are all reasons to consider mixing PAC with a large amount of aluminum sulfate or ferrous sulfate.

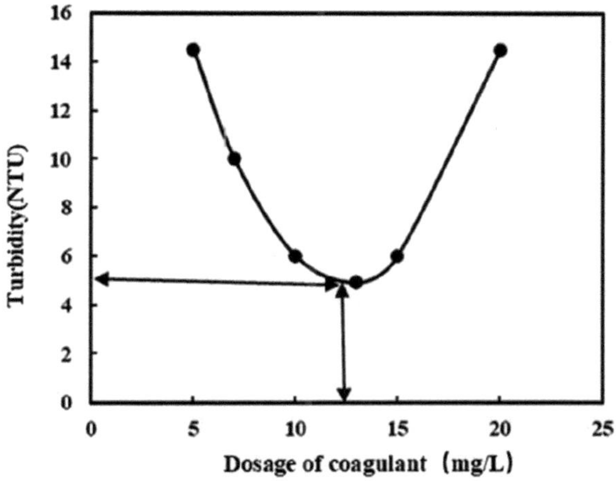

Fig. 9.3 Relationship between coagulant dosage and residual turbidity (optimal dosage is 12.1 mg/L)

Fig. 9.4 Relationship between aluminum sulfate dosage and residual turbidity at different pH values

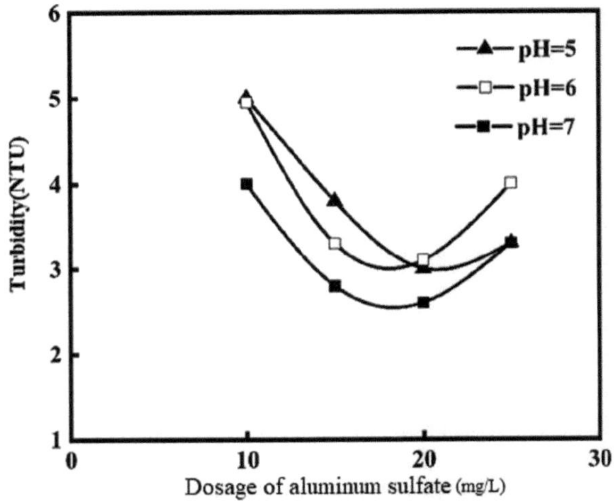

9.1.4 Color

The substances that form color in natural water bodies are mainly humic substances, which are the main components of dissolved organic carbon. Hydrophilic color is always more difficult to remove than hydrophobic mineral turbidity. The complexity of color depends on the pH of the water, the type of color-producing colloid, and the ability of the coagulant to break hydrogen bonds. The selected water treatment agent must make the stability of the color-producing substances in the water the lowest (usually the pH value is between 5.5 and 7.0).

In many applications, a substance itself can hardly achieve good results, especially for inorganic metal salts. In this case, adding a small amount of PDADMAC (Poly dimethyl diallyl ammonium chloride) can achieve good removal effects. When using, PDADMAC can be added separately with inorganic coagulants, or products compounded with PAC and PDADMAC can be chosen. In addition, powdered activated carbon or other inorganic adsorbents can be added, which can also achieve very ideal removal effects.

9.1.5 Temperature

The performance of inorganic metal salts mainly depends on chemical reactions, and temperature has a significant impact on coagulation. The lower the temperature, the slower the reaction; the lower the temperature, the higher the viscosity. Lower temperatures (< 5 °C) will affect aluminum salts and iron salts, and the coagulation effect is poor in low temperatures in winter. Especially the coagulation effect of inorganic low molecular salts is worse, such as aluminum sulfate. In winter, most areas have little rainfall, so low-temperature water is often low-turbidity water, which further increases the difficulty of treatment. In the treatment process, compounded coagulants, or use methods of compounding inorganic low molecular coagulants with organic high molecular flocculants or active silicic acid can be chosen. In addition, there are organic–inorganic composite coagulants and silicon-containing coagulants on the domestic market in China, which have good treatment effects on low-temperature and low-turbidity water. Foreign countries also use the method of heating raw water, but currently no factories in China use heating methods. Almost all coagulants perform well in warm water (10 °C $\leq$ T $\leq$ 25 °C). The Zibo Water Supply Management Bureau in Shandong Province has achieved good results in treating low-temperature and low-turbidity water from the Yellow River by using polyaluminum sulfate and anionic polyacrylamide, and has accumulated rich experience.

9.1.6 *Hardness*

All water has a certain degree of calcium and magnesium hardness. Calcium and magnesium ions do not form flocculation under normal conditions, but they can improve surface neutralization and reduce the amount of flocculant added, improving flocculation performance. In drinking water treatment, methods to increase hardness cannot be used because they will increase the salt content in the water. However, some municipal sewage plants can use this method. Adding lime or magnesium salt to the water is relatively expensive, but if seawater is added for treatment, it is much cheaper, and seawater contains a lot of magnesium. The VEAS sewage treatment plant in Norway adds about 5% of seawater. In Hong Kong, seawater is used to flush toilets, and the amount of flocculant added in subsequent treatment is also the lowest in the world.

Generally, lime is added to form calcium and magnesium precipitation. Lime sludge has a high density and is easy to precipitate. However, it is best to add a small amount of coagulant after adding lime to quickly capture lime particles. Currently, softening technology is not used in drinking water treatment plants in China. But in power plant water treatment, softening technology is used. The added coagulant is only to capture lime particles, not to coagulate raw water turbidity. Lime sludge cannot be returned to the environment, so it needs to be dehydrated. The appropriate flocculant should be chosen according to the water plant equipment.

By controlling the pH value of the treated water, different degrees of softening can be met, generally controlling the pH value at 10.0 ± 0.2. Metal salt coagulants will reduce alkalinity and affect the softening effect. Therefore, it is best to choose a pre-hydrolyzed high salt-based coagulant. High salt-based polyaluminum chloride, especially calcium-containing polyaluminum chloride, is more suitable for lime softening, and the dosage is also relatively small. In addition, polyaluminum sulfate can be used as a coagulant. Because the formed calcium sulfate and aluminum hydroxide have a co-precipitation effect, the coagulation effect will be more ideal.

9.1.7 *Odor and Smell*

Odor and smell can be controlled in many ways, but the most common method is to add powdered activated carbon. Powdered activated carbon can also effectively remove other organic matter and color from the raw water. Most coagulants can settle powdered activated carbon or organic particles, however, because these substances are very fine, the chosen coagulant should ideally form dense sludge, achieving a high settling speed and very low residual floc.

9.2 The Problem of Residual Aluminum in Water

There are varying degrees of residual coagulant ions when using coagulants, and currently there is more research on the issue of residual aluminum ions in drinking water. The toxicity of aluminum is increasingly attracting people's attention. Although there is still much controversy over the relationship between Alzheimer's disease and the accumulation of aluminum in the brain, the World Health Organization, most developed countries, and China have established limit standards for residual aluminum ions. The World Health Organization and China have stipulated that aluminum ions in drinking water cannot exceed 0.2 mg/L. In addition, there are currently no restrictions on residual aluminum and iron ions in wastewater, and their impact on the organisms and ecology of the receiving water bodies is still being studied. Studies have shown that free and polymerized aluminum ions have certain effects on plant roots, aquatic organisms, and fish. Therefore, we must be cautious when using coagulants and minimize the emission of aluminum salts as much as possible. The toxicity of aluminum is increasingly attracting people's attention. Although there is still a lot of controversy about the relationship between Alzheimer's disease and the accumulation of aluminum in the brain, the World Health Organization and developed countries have all restricted the residual aluminum ions in drinking water.

Regarding the impact of residual aluminum ions, foreign countries have also conducted some research. For example, if mice are orally administered 2.5 mg/kg of aluminum, after six months, the animals show mild symptoms of poisoning. If aquatic animals live in water with "aluminum pollution" ranging from 0.07 to 17.8 mg/L, they will also suffer varying degrees of poisoning. For example: when the aluminum content in the water exceeds 0.2–0.5 mg/L, it can cause salmon to die; under weakly alkaline conditions (pH value 8–9), if the concentration of aluminum ions in the water exceeds 0.5 mg/L, it can also cause salmon to die; in addition, newly precipitated aluminum hydroxide colloids can also cause salmon poisoning.

When the dosage is excessive, the change in Zeta potential can well indicate the problem of excessive dosage of the drug, and factories should generally control the Zeta potential around − 5 mV. As shown in Fig. 9.5, the excess of coagulant will cause the Zeta potential to rise to + 5 mV, and it is accompanied by a significant

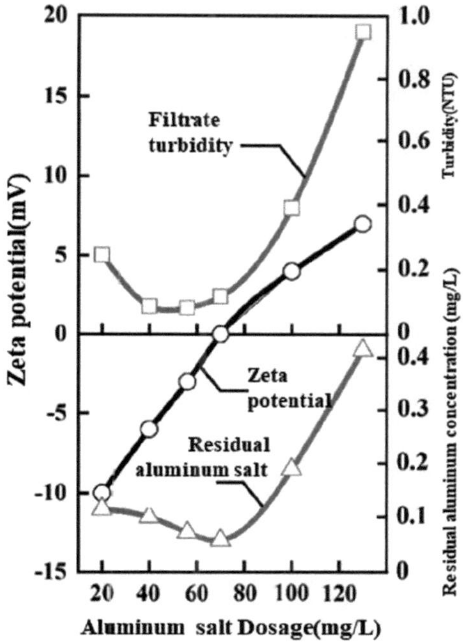

Fig. 9.5 The relationship between dosage and Zeta potential, turbidity, and residual aluminum

increase in the residual aluminum in the final water body. For a factory, because there is a lag of several hours between flocculation and filtration, it is insufficient to control the residual aluminum content by simply monitoring the turbidity of the water, and controlling Zeta is timelier and more effective. As can be seen from Fig. 9.6, when using ferric chloride to treat urban sewage, the best Zeta potential is − 11 mV, at which point the COD removal rate of the sewage is the highest, but if an additional 5 mg/L of ferric chloride is added, the coagulation will rapidly deteriorate. Tables 9.3, 9.4, and 9.5 compare the residual aluminum content and sludge volume of alum and PAC treated wastewater, respectively.

Table 9.6 shows the amount of residual aluminum in the raw water and when polyaluminum chloride is used as a coagulant. If filtration isused directly and simply, the aluminum ion content can be reduced by about 40%. If polyaluminum chloride is used, the aluminum ion content in the water body can be reduced by 96%. If further filtration technology is used, the aluminum ion content can be reduced by 97.5%.

As shown in Table 9.7, the use of iron salts can also effectively reduce the content of aluminum ions in raw water.

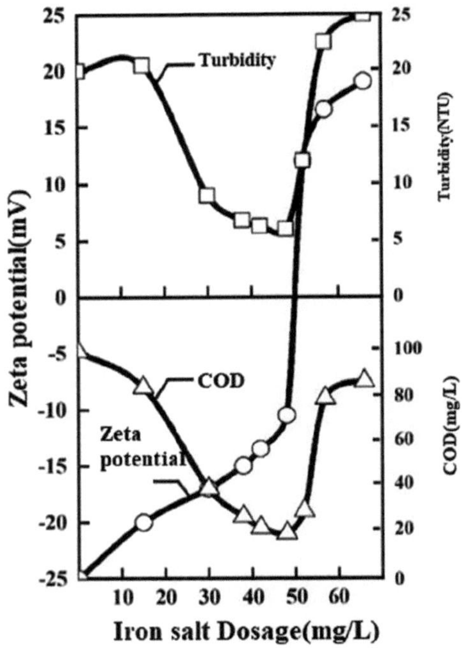

Fig. 9.6 The corresponding curve of ferric chloride dosage and wastewater coagulation effect

Table 9.3 Comparison of residual aluminum after alum and PAC treatment

S. No.	Alum/PAC (dosage)	Residual aluminum content after treatment (mg/L)		pH value after treatment	
		Alum	PAC	Alum	PAC
1	0	0.05	0.05	8.0	8.0
2	200	0.10	0.05	6.5	7.3
3	300	>0.80	0.05	5.0	7.1
4	400	>0.80	0.05	4.8	7.0

Note Raw water turbidity reduced from 150 to 5 NTU

9.3 Beaker Coagulation Experiment

The beaker coagulation test (Jar test or Beaker test) is the most widely used and effective method to evaluate coagulation treatment. Through the beaker test, we can understand the coagulant and coagulation effect, determine the optimal dosage of coagulant or flocculant, as well as kinetic factors, such as stirring intensity, reaction time, etc. The beaker experiment can also provide a basis for preliminary design. Before conducting the beaker test, the experiment must be planned and preliminarily determined:

Table 9.4 Comparison of sludge volume produced by alum and PAC

S. No.	Time (min)	Sludge volume (mL)	
		Alum (120 mg/L)	PAC (80 mg/L)
1	0	0	0
2	30	35	31
3	60	31	27
4	90	28	24
5	120	26	21
6	150	25	20
7	180	23	18
8	210	20	18

Note Raw water turbidity: 1250 NTU, pH value: 8.1, Volume: 1L, raw water from Solapur Water Plant

Table 9.5 Comparison of sludge volume produced by alum and PAC

S. No.	Time (min)	Sludge volume (mL)	
		Alum (80 mg/L)	PAC (50 mg/L)
1	0	0	0
2	5	35	30
3	10	27	21
4	15	24	15
5	20	20	16
6	30	19	15
7	45	17	14
8	60	14	12
9	90	14	12
10	120	13	11
11	150	13	10.5
12	180	12.5	10
13	240	11	9.5
14	300	10.5	9.5
15	1260	9	9.0
16	1440	8.5	8.5

Note Raw water turbidity: 2010 NTU, pH value: 8.0, Volume: 1L, raw water from Narmada River

(1) Selection of instruments;
(2) Types and suppliers of coagulants;
(3) Differences between preliminary experimental schemes and actual schemes;
(4) Measurement indicators and methods of treated water;
(5) Types of data collection and calculation methods.

Table 9.6 Residual aluminum content during water treatment process with aluminum salt as coagulant

Total aluminium content of raw water (surface water) (mg/L)	The total aluminium content of the treated water		
	Direct filtration without medication	After medication, take supernatant	After filtration
1.20	0.68	0.045	0.030
0.93	0.71	0.054	0.033

Table 9.7 Residual aluminum content during water treatment with iron salts as coagulants

Raw water (surface water)	Raw water (μg/L)	Treated water (μg/L)
Total aluminum (mg/L)	245	45
Total dissolved aluminum (mg/L)	140	40
Unstable aluminum (mg/L)	67	32
Low molecular aluminum (mg/L)	95	35

9.3.1 Selection of Beaker Experiment Instruments

In a beaker coagulation test process, the most important instrument is the coagulation stirrer, including the beaker and stirring equipment.

(1) Types of beakers

There are 2 types of available beakers, the most common two are 1L round beaker and 1L square beaker. At present, almost all are 1L round beakers in China, but the round effect is the worst and the least representative of coagulation. If possible, this type of beaker should be avoided as much as possible, and 1L or 2L square beakers with sampling holes are preferred.

The defects of the 1L round beaker are:

① When the beaker test is enlarged to actual engineering, the error is very large
② During the stirring process, the water will rotate with the impeller in the beaker, reducing the effective rate of mixing
③ It is difficult to get good sedimentation data in a 1L beaker, and the amount of test water available is very small, so multiple experiments must be conducted to obtain enough water.

Figure 9.7a is a schematic diagram of a 1L Hudson beaker, and Fig. 9.7b is a 2L Gator beaker, which has a sampling hole 10 cm below the liquid surface. The Gator beaker overcomes the shortcomings of the Hudson beaker and has the following advantages:

① Large volume, reducing the amplification error;
② The square wall reduces the rotation of the water;

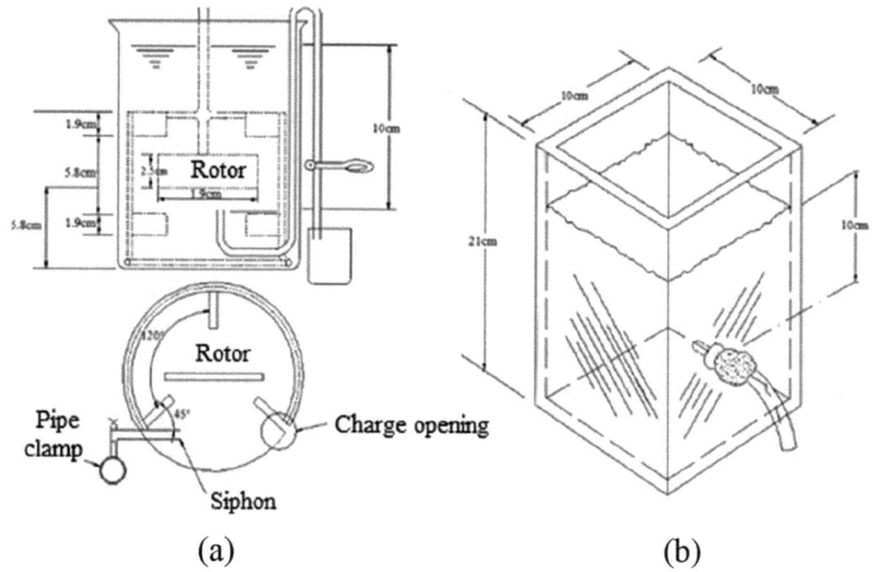

Fig. 9.7 **a** Hudson beaker and **b** Gator beaker

③ There is a large amount of supernatant available for analysis.

(2) Stirrer Drive Systems

There are three basic types of stirrer drive systems available.

① One method is to provide stirring energy to a six or eight-unit stirrer through a motor and transmission device. The motor speed can be adjusted, but the stirring speed transmitted to the stirrer is the same. Therefore, in a beaker experiment, all impellers operate at the same speed, which can vary between 0–800 r/min.

② The second way is to adjust it by a separate motor driven stirrer placed above the beaker, which can make the stirring speed different for each beaker. In addition, each beaker's motor drive can be computer-controlled, allowing for multi-program, multi-speed operation. For example, programming can be done to mix raw water in the beaker at 300 r/min for 1 min; add coagulant and stir again at 250 r/min for 1 min; add flocculant and stir again at 20 r/min for 1 min; stir again at 50 r/min for 10 min, and finally, the coagulation automatically stops and alarms to remind that coagulation is over. In this way, each beaker can run different programs, and multiple dosing, mixing, and coagulant addition can be carried out simultaneously, and six different coagulants can be added at the same time. Therefore, this type of stirrer is very convenient for examining different kinetic conditions and the effects of different coagulants. In addition, this stirrer can also display the G value and G-T value of the coagulation process. Currently, several companies have launched this product on the market, and the product's stability is very good, and it can provide round and square beakers.

③ The third method is to use magnetic stirring. In this way, each beaker can also operate independently under computer control, has the advantages of the second method, and the magnetic stirrer does not occupy the space above the beaker, which is more convenient for observation, sampling, etc., but the intensity of magnetic stirring is often not enough.

(3) Impeller Styles

There are several impeller styles to choose from. The most common are flat blade impellers, frame type, magnetic stirring rods, and turbines. Each stirring method has its own characteristics, the most important of which is the mixing energy or velocity gradient G. Generally, the flat blade impeller has the best velocity gradient. In a 2L square beaker, at a rotation speed of 200 r/min, a G value of 350 s^{-1} can be obtained. For most treatment plants, this is sufficient. The strength of the magnetic stirring rod is smaller, at a rotation speed of 200 r/min, a G value of 200 s^{-1} can be achieved. The efficiency of the turbine is the lowest, in a 2L Gator beaker, at a rotation speed of 200 r/min, the velocity gradient is 70 s^{-1}. For most water treatment plants, the impeller is not suitable.

(4) Temperature Adjustment

When simulating the coagulation effect at different temperatures, the water temperature must be adjusted. At this time, the water to be treated must be heated and cooled, but the beaker experimental devices currently available on the market do not come with heating and cooling devices. Generally, the water temperature is adjusted before coagulation. As mentioned earlier, water temperature affects the hydrolysis process of the coagulant and the hydrodynamic conditions, so the impact of water temperature cannot be ignored.

(5) Analytical Instruments

According to the purpose of water treatment, different water treatment units can be equipped with corresponding testing instruments. Since drinking water treatment involves many indicators, a lot of instruments are needed. For large-scale drinking water treatment plants, in addition to commonly used chemical analysis instruments, they also need to be equipped with liquid chromatography, gas chromatography, atomic absorption spectroscopy, and other instruments. Some water treatment plants have already implemented online monitoring of COD, TP, SS, pH value, UV_{254}, and other indicators. Most sewage treatment plants have fewer control indicators, so fewer corresponding instruments are needed. Regular chemical analysis instruments are usually sufficient, but online control monitoring of sewage discharge has been implemented in many water plants.

9.3.2 *The Determination of Beaker Coagulation Test*

9.3.2.1 Preliminary Formulation of the Test Plan

In the beaker coagulation test, the choice of coagulant and actual operation is important. If the coagulant is chosen improperly, no matter how accurate the operation steps are, they are meaningless. Therefore, the experimenter can determine several preferred coagulants based on water quality and related literature, and the types of coagulants available on the market, and then carry out experimental screening. This is very important. Often, the experimenter determines the type and optimal dosage of a coagulant through orthogonal experiments, but the chosen coagulant is not the best, so such orthogonal experiments are meaningless. This is especially common in sewage treatment. Readers can determine the type of coagulant based on the introduction of coagulant properties, and then carry out detailed screening.

9.3.2.2 Preparation of Reserve Solution

The coagulants or flocculants currently available on the market are generally in liquid (original solution) or solid form, and they must be diluted or dissolved before use. The preparation of the reserve solution should be based on two considerations. One is that the reserve solution must be stable during the experiment and cannot undergo any property changes. The second is that the reserve solution must be easy to use and measure.

Coagulants and flocculants will more or less experience a decrease in performance after dissolution, especially when the concentration of the reserve solution is very low. Therefore, it is best to use a high-concentration reserve solution or original solution, and add it with a micro-adjustment syringe. However, because the amount of drug added is very low, it must be diluted before adding. If dilution is needed, it is best to prepare a 1% (mass ratio) coagulant and 0.1% (mass ratio) flocculant for easy measurement.

9.3.2.3 Coagulant

For aluminum sulfate, PAC, PFS, ferric chloride, and ferrous sulfate, a simple dilution can prepare the reserve solution. Generally, a 1% solution refers to a mass ratio of 1%. But what can easily be confused here is the form of the 1 g coagulant in the 1% solution. For example, polyaluminum chloride can be liquid or solid, so it must be specified whether it is 1 g solid or liquid. The national standard polyaluminum chloride solid product contains 29–30% alumina, and the liquid product contains 10% alumina. Therefore, the same 1% concentration, the effective ingredients differ by 2 times.

If we prepare a 1% reserve solution, then take 1 g of the sample and dissolve it in 99 mL of water. 1 mL of such a 1% solution contains 1/100 × 1000 mg = 10 mg of coagulant. If 1 mL is added to 1L, it is 10 mg/L or 10 kg/1000 m^3 (or 10 mg/L,), if 2 mL is added, the dosage is 20 mg/L. Here we have not considered the state of the added sample, whether it is solid or liquid, because its density is not the same as water, but this error is completely ignored in the above sampling process, and this kind of error is completely acceptable in actual testing.

(1) Polyaluminium chloride

The polyaluminum chloride currently in use is mostly solid products. You can take 1 g of solid sample, dissolve it in 99 g of water, and obtain a 1% PAC solution.

Therefore, 1 mL of reserve solution contains solid polyaluminum chloride 1 × 1000 mg/100 mL = 10 mg/mL.

If the solid polyaluminum chloride contains 29% aluminum oxide, then 1 mL of reserve solution contains aluminum oxide 10 mg/mL × 29% = 2.9 mg/mL.

If liquid polyaluminum chloride is used, and its aluminum oxide content is 10%, then the preparation of 1% reserve solution is as follows:

Take 1 g of liquid polyaluminum chloride, dissolve it in 99 mL of water, then 1 mL of reserve solution contains liquid polyaluminum chloride 1 × 1000 mg/100 mL = 10 mg/mL. If the liquid polyaluminum chloride contains 10% aluminum oxide, then 1 mL of reserve solution contains aluminum oxide 10 mg/mL × 10% = 1.0 mg/mL.

Since there is no unified standard for the addition of coagulants in currently published papers and books, some use solid polyaluminum chloride, some use liquid polyaluminum chloride, and some use aluminum oxide, so it is necessary to specify the state of the polyaluminum chloride sample during use.

(2) Polysulfate Iron

To prepare a 1% polysulfate iron solution, you can directly weigh 1 g of liquid polysulfate iron and dissolve it in 99 mL of water. If you know the density of the solution, you can also measure a certain volume of liquid and then dilute it.

If the density of polysulfate iron is 1.490 g/mL, the Fe^{3+} content is 11.2%.

Therefore, 1 g = 1/1.490 g/mL = 0.671 mL, the mass of 0.671 mL is 1 g.

So, take 0.671 mL of liquid polysulfate iron, add it to 99 mL of water, and you will get a reserve solution of 10 g/mL liquid polysulfate iron. The reserve solution contains Fe^{3+} content of 1.12 g/mL.

Polysulfate iron and other iron salts have a relatively large tendency to hydrolyze. Generally, a 1% solution can be stable for 10 h, and the stability will change with the raw materials. In actual use, polysulfate iron can be prepared into a 5–10% solution to extend the stable time.

(3) Organic Polymer

Because the viscosity of the polymer solution is very high and the usage is very low, a solution of 0.1–0.5% is generally prepared. Weigh 0.1 g, dissolve it in 99.9 mL of water, then 1 mL of reserve solution contains 1 mg of polymer. If 0.5 mL of 0.1%

solution is added to 1L of test volume, then the added dosage is 0.5 mg/L. If 1 mL of 0.1% solution is added to 1L of test volume, then the added dosage is 1.0 mg/L.

For emulsion-type polymers, dilution and dissolution are very easy. For dry powder polymers, to avoid the "fish eye" phenomenon, you can first wet the dry powder with a small amount of ethanol, and then make up the volume to 100 mL. Anionic polyacrylamide, 0.1% of the reserve solution can be stable for 24 h, and degradation occurs as time extends; the stability of cationic polyacrylamide is relatively poor, 0.1% of the reserve solution can be stable for 10 h. In addition, the stability of the reserve solution is related to temperature, low temperature stability is good, high temperature stability is poor. For cationic polyacrylamide, a slightly higher concentration of reserve solution can be prepared, such as a 0.5% solution.

9.3.2.4 Determination of the Experimental Plan

After selecting the type of coagulant, one should consider the impact of hydrodynamic factors on the coagulation effect, which includes mixing strength, mixing time, and settling time.

(1) Mixing Strength

The main parameter characterizing the mixing strength is the velocity gradient (G, represented in s^{-1}). In a beaker coagulation experiment, the velocity gradient corresponds to the stirring speed. The higher the stirring speed, the higher the velocity gradient. In the aforementioned coagulation stirrer, the corresponding G value can be displayed at a determined stirring speed. Once the velocity gradient is determined, the important step is to simulate the same mixing strength at 300 s^{-1}, while slow mixing is at 15–150 s^{-1}, generally at 40 s^{-1}. When purchasing a beaker coagulation testing instrument, a table related to the beaker impeller speed, impeller design, and velocity gradient can usually be found. If the G value under operating conditions is known, the impeller speed can be determined.

(2) Mixing time (Flocculation tank residence time)

Calculating the retention time is quite simple, the following formula can be used to calculate the retention time:

$$Td = V/Q \tag{9.9}$$

where Td is the theoretical retention time, min;

V is the reaction tank volume, m^3;

Q is the flow rate, m^3/min.

The general rapid mixing time is 20–60 s, while the flocculation process is designed for 30 min. Due to the performance differences of coagulants, the actual mixing time and reaction time must be determined through experiments.

(3) Settling Time

When conducting a beaker coagulation experiment, the settling time is one of the largest sources of error. The main error occurs because the settling time is often too long. If the settling time is long, good results can be obtained, but they are often not very valuable. Given time, all coagulated and flocculated particles will settle to the bottom of the stationary beaker. Therefore, the settling time is an operational parameter that must be accurately measured.

The best way to simulate the settling time is to determine the time it takes for a particle to settle 10 cm in the clarifier. This can be confirmed through calculation and actual testing. Once the time for a flocculated particle to settle 10 cm is determined, the settling time of the beaker is also determined. You can choose a 2L Gator beaker with a water hole 10 cm away from the water surface.

To calculate the settling time, the surface load of the clarifier or sedimentation tank must be determined. If the surface load is known, the settling rate can be determined, and then the time required to settle 10 cm can be directly calculated. Table 9.8 summarizes the relationship between the surface load rate and settling time.

Taking water samples from the clarifier and measuring the time for 80% of the particles to settle to 10 cm can confirm the calculation of the settling time. Generally speaking, the estimated settling time here is very close to the calculated sampling time and can serve as a guide in the beaker coagulation experiment.

(4) Selection of Filtration Medium

In almost all cases, the water after sedimentation must pass through a filter pool before entering the distribution system. Therefore, there should be a filtration step in a beaker test procedure. If the beaker test is running well, taking a water sample from 10 cm below the surface after sedimentation and filtering it will yield valuable results. Recently, due to concerns about deadly microbial spore substances, people have been paying more attention to the concept of "filterability". After coagulation, the flocs in the water can be further separated, reducing turbidity.

(5) Compatibility study of preliminary experimental scheme

Once the experimental parameters are known, the mixing strength and mixing time of rapid mixing and flocculation stirring, and the sedimentation time have been calculated, the treatment process can be preliminarily determined. In addition, correlation

Table 9.8 Relationship between Surface Load Rate and Settling Time

Surface load rate m/h	Settling rate cm/min	Sampling time for 10 cm mins
0.3	0.5	20
0.6	1.0	10
1.2	2.0	5
2.4	4.0	2.5
6.0	10.0	1

studies need to be conducted to compare the differences between beaker tests and actual scale applications in water plants.

Based on the results of the water plant sedimentation experiment and the beaker test, data is analyzed. Generally, turbidity, color, pH value, UV_{254}, etc. need to be analyzed. A simple bar chart can be drawn to compare the results of the actual water plant and the beaker test.

(6) Selection of chemical treatment methods

Following the work guidelines discussed earlier, we can avoid many troubles in the experiment. In addition, we must achieve the following points:

① Fully understand the operation process of the water plant;
② Preliminary screening of coagulants has been carried out, and the coagulants have good performance;
③ The analysis parameters have been fully determined, and the analysis instruments are ready;
④ A beaker test plan has been formulated, which can yield valuable results;
⑤ The filtration plan has been determined;
⑥ Test relevance.

Now there are only three questions: Which combination of coagulant/flocculant has the best effect? What is the dosage of coagulant/flocculant? What is the total cost of treatment?

(7) Selection of coagulants

The first step is to choose a coagulant and estimate the appropriate dosage.

The selection and dosage of coagulants are obtained through observation and measurement. The goal is to generate "flocs", agglomerating hydrophobic mineral turbidity and hydrophilic organic color. Depending on the choice and dosage of coagulant, the DOC content is often a decisive factor. Initially, for every 1.0 mg/L DOC, 5 mg/L Al_2O_3 or 2 mg/L Fe^{3+} can be added.

Suggested procedure:

a. Set the impeller speed to 10 r/min and add oxidant if necessary. Add coagulant at the required dosage and add a pH adjuster if necessary. Set the impeller to the speed required for rapid mixing and the subsequent predetermined mixing time and observe the performance of the coagulant in removing color during the rapid mixing stage. Set the impeller to the speed required for flocculation, and the subsequent predetermined mixing time, observe the formation of flocs, and record the formation time.
b. If flocs form quickly but do not grow after initial formation, make a record. Observe the color of the water between the flocs, record; observe whether there is any floating matter in the beaker. Adjust the impeller speed to zero and remove it from the beaker. Start the sedimentation stage.

c. Classify the flocs according to their size; classify them according to their settling rate; observe whether any flocs float to the surface and record; record whether there is a temperature rise and whether it causes the release of gas; observe whether any flocs are suspended in the center of the beaker and record.
d. After a predetermined time, take a sample 10 cm from the liquid surface. Perform all necessary analysis tests on the sample. Reobserve the suspended matter in the beaker afterwards.
e. Slowly insert the impeller, increase the speed until the sludge bed rises, and record the speed of each chemical system.
f. Turn off the impeller and observe the settling rate of the sludge bed.
g. If analyzing a reaction determined by contact time, such as the formation of trichloromethane (THMs), it may be necessary to first place the sample so that it goes through a complete retention time, which may take several hours. In this case, several 500 mL beakers are needed.
h. Select several of the best coagulants and add them in the optimal dosage.

(8) Selection of flocculants

As shown in Fig. 9.8, the Zeta potential curve corresponding to the cationic coagulant can be used to evaluate the electrical neutrality of the cationic polymer.

During the coagulation process, flocculants may or may not be needed. Once the coagulant and its optimal dosage are determined, the above steps can determine whether a flocculant can significantly increase the coagulation effect or reduce processing costs. Recently developed low-temperature low-turbidity water coagulants in China, such as polyaluminum chloride (for example, Tongji University's 820 series of polyaluminum chloride), have very good flocculation effects and can eliminate the use of flocculants. In addition, during the use of flocculants, it should

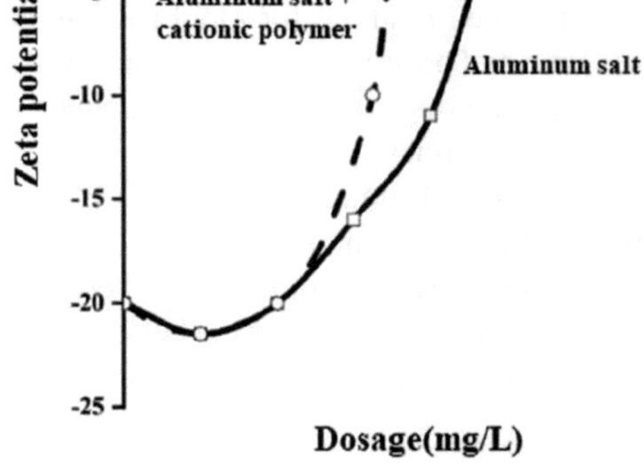

Fig. 9.8 Zeta potential corresponding to cationic coagulant at different concentrations

be avoided to store them for a long time after dissolution, which leads to the deterioration of the solution, avoid too long dissolution time or excessive stirring, and avoid the appearance of "fish eye" phenomenon.

9.3.3 Selection of the Final Treatment Process

If the chemical treatment process is determined, a series of final tests should be conducted to determine the advantages of this flocculant. A set of eight tests should be conducted as follows:

(1) Plant's settling water;
(2) Settling water in the beaker without flocculant;
(3) Settling water in the beaker with flocculant;
(4) Settling water in the beaker after the existing chemical treatment process;
(5) Plant's filtered water;
(6) Filtered water in the beaker without flocculant;
(7) Filtered water in the beaker with flocculant;
(8) Filtered water in the beaker after the existing chemical treatment process.

At this stage, a chemical treatment procedure should be formulated.

If done correctly, a beaker coagulation test is useful. The key is to take into account the target and the plan developed to achieve the treatment objectives. A beaker coagulation test can provide the following information:

(1) Which coagulant should be used in the plant;
(2) Which coagulants are not suitable;
(3) Ranking of coagulant efficiency;
(4) To achieve the processing goal, the dosage is sequenced;
(5) Economic ranking of coagulants.

9.3.4 Evaluation of Coagulants

(1) Good sedimentation;
(2) pH adaptability;
(3) Temperature adaptability;
(4) Low chromaticity;
(5) High removal rate of organic matter, COD, UV_{254};
(6) Sludge is compact and easy to dewater;
(7) High performance ratio.

9.4 Preparation, Measurement, and Addition of Medicinal Solution

In water treatment, the addition of coagulants to raw water can disrupt the stability of colloidal particles in the water. Under certain hydraulic conditions, through the collision and aggregation between colloidal particles and other microparticles, flocs that are easy to separate from the water are formed. The coagulation effect is determined by both the performance of the coagulant and the fluid dynamics of the structure. In engineering practice, coagulation is often divided into two stages: coagulation and flocculation, followed by separation. Coagulation equipment includes mixing equipment, reaction equipment, and separation equipment. The two stages of coagulation and flocculation are completed in the mixing equipment and reaction equipment respectively. These three types of equipment can be independent or integrated (such as a clarifier), and before the mixing equipment, a medicinal solution storage device and a metering device are usually required. Separation equipment includes sedimentation, air flotation, filtration, and other equipment.

9.4.1 Preparation, Measurement, and Addition of Medicinal Solution

There are usually two methods to add coagulants to the water to be treated, namely dry addition and wet addition. Dry addition refers to the process of crushing solid coagulants into powder and adding them to the water in a fixed amount. The process is usually as follows: drug delivery → crushing → lifting → metering → drug mixing. Wet addition refers to dissolving the coagulant first, preparing it into a solution of a certain concentration, and then adding it to the water to be treated in a fixed amount. The process is shown in Fig. 9.9.

From an operational perspective, both additional methods have their advantages and disadvantages. The dry addition method has a small footprint, no corrosion problems with the equipment, and better drug quality (less pollution and deterioration). However, the dry addition method has unavoidable disadvantages: when

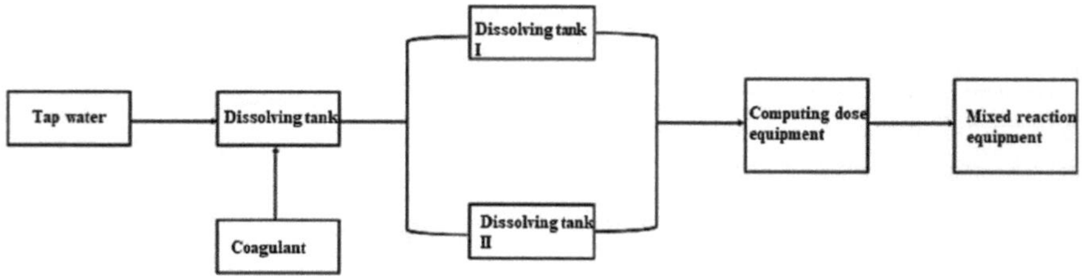

Fig. 9.9 Preparation and addition process of medicinal solution

a large amount of medication is used, crushing equipment is needed, and when a small amount of medication is used, it is not easy to adjust. The contact conditions between the drug and water are poor, and the working conditions are worse. It is not suitable for hygroscopic and slow-dissolving coagulants. For micro-polluted water, the removal efficiency of liquid coagulants is currently relatively low, with a UV_{254} removal rate of only 10%, but if solid composite coagulants (sometimes with adsorbents) are added, better results are often achieved. For example, during the outbreak of blue-green algae in Taihu Lake, the deodorization and organic matter removal effects of solid adsorption coagulants were much better than conventional Al and Fe. The wet addition method is suitable for various coagulants, the additional amount is easy to control and adjust, it is easy to mix fully with water, and the operation is convenient. Its disadvantages are that it occupies more equipment area and larger land area, and the equipment is also prone to corrosion. At present, the wet addition method is commonly used, and the dry addition method is used less. This section focuses on the wet addition method.

9.4.2 Preparation of Medicinal Solution

The preparation of coagulant solution mainly considers the concentration of coagulant and the amount of addition. The concentration of inorganic coagulant solution is usually prepared to be 5–20%, and the concentration of organic polymer solution is usually prepared to be 0.5–1.0%. The daily addition amount of coagulant can be calculated by the following formula:

$$W = 24QC\ (\text{kg/d}) \tag{9.10}$$

where, W is the amount of coagulant needed per day, kg/d;

Q is the amount of raw water, m^3/h;

C is the amount of coagulant added, kg/m^3

The solution preparation equipment includes a dissolving tank and a solution tank. When choosing, the main considerations should be the equipment volume and the mixing method.

The volume of the solution tank can be calculated by the following formula:

$$V = \frac{24 \times 100aQ}{1000 \times 1000\,\text{bn}} = \frac{aQ}{417\,\text{bn}} \tag{9.11}$$

where, V is the effective volume of the dissolving tank, m^3;

a is the maximum dosage of coagulant, mg/L;

Q is the amount of raw water, m^3/h;

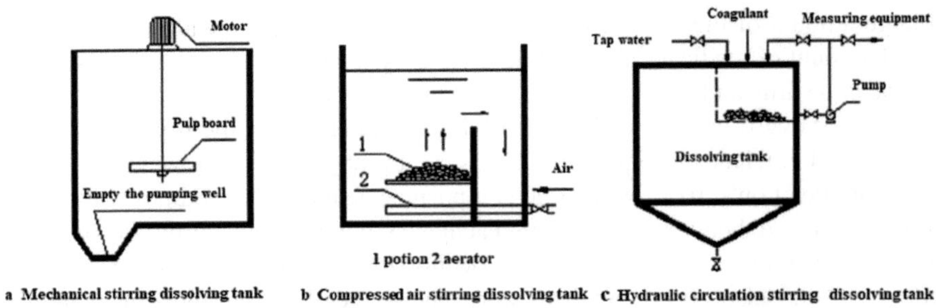

Fig. 9.10 Coagulant dissolution mixing methods

b is the content of the solution, %;

n is the number of times the solution is prepared per day, generally 2–6 times.

The volume of the dissolving tank can be calculated by the following formula:

$$V' = (0.2 - 0.3)\,V \tag{9.12}$$

To accelerate the dissolution of solid coagulants and ensure uniform concentration of the solution, mixing equipment needs to be set up in the solution tank. There are three common mixing methods: mechanical mixing, hydraulic mixing, and compressed air mixing.

As shown in Fig. 9.10a, mechanical mixing is done by an electric motor driving a paddle or turbine, and is generally suitable for situations where a large amount of reagent is used; as shown in Fig. 9.10b, compressed air mixing involves introducing compressed air into the dissolving tank for mixing. The air pressure is generally 1–2 kg/cm^2, and the air consumption is generally 0.2 m^3 (air)/m^3 (solution) min. Compressed air mixing equipment; as shown in Fig. 9.9c.

It must be noted that the solution preparation equipment and pipes should consider corrosion prevention or use corrosion-resistant materials, especially when using trivalent iron salt coagulants.

References

1. Tang Hongxiao, Qian Yi, Wen Xianghua. Characteristics and control technology principle of particulate Matter and refractory organic Matter in water [M]. China Environmental Science Press, 2000.
2. H. Helmholtz, Uber die auf das Innere Magnetischoder dielektrisch polarisierte Korperwirkenden Krzfte [J]. Annalen der Physik und Chemie. 1881, 13: 385–406
3. Gouy, G. Sur la constitution de la charge electrique a la surfaced'un electrolyte [J]. Journal de Physique et le Radium. 1910, 9:457–468
4. Chapman, D. L. A contribution to the theory of electrocapillarity [J]. Philosophical Magazine, 1913, 25:475–481
5. Stern O., Zür theorie der electrolytischen doppelschicht, Z [J]. Electrochemistry, 1924, 132, 508–516
6. Johnson P.N. & Amirtharajah A. Ferric Chloride and Alum as Single and Dual Coagualants Jour [J]. Journal AWWA, 1983, 75:5:232.
7. Juske, Horita, and, et al. Liquid-vapor fractionation of oxygen and hydrogen isotopes of water from the freezing to the critical temperature [J]. Geochimica Et Cosmochimica Acta, 1994, 58(16): 3425.
8. Amirtharajah A. & Mills, K.M. Rapid-Mix Design for Mechanisms of Alum Coagulation Jour [J]. Journal AWWA, 1992, 74:4:210.
9. Li Fengting, Wang Zheng, Wang Sen, et al. Special coagulant for low temperature and turbidity water—polyaluminum ferric chloride. 2004 National Water Treatment Technology Seminar and Proceedings of the 24th Annual Meeting [C]. 2004: 742.
10. Li Fengting, Li Zitong, Li Jie. The 60 year development history and future development trend of China's inorganic aluminum iron salt water treatment agent industry [J]. Inorganic Salt Industry, 2020, 52(10): 25–29.
11. Wang Zhiyong, Xu Xue, Zhang Shidong, et al. A new clean production process for manufacturing aluminium sulfate by decomposing bauxite with industrial sulfuric acid. CN 111137913A[P]. 2020-05-12.
12. Wang Zhiyong, Xu Xue, Zhang Shidong, et al. A new clean production process for producing aluminum sulfate by industrial sulfuric acid decomposition of bauxite [P]. Hebei Province: CN111137913A, 2020-05-12.
13. Liu Yanling, Wei Yanli. The process of preparing aluminum hydroxide micro powder by two-stage seeding method [J], Journal of Hebei University (Natural Science Edition), 2021, 41(4): 378–383.
14. Xia Jupei, Gu Fang. A method for efficient utilization of high alumina coal gangue [P]. Yunnan: CN108975367A, 2018-12-11.
15. Tang Weizhen, Li Fengting, Wang Jialei, et al. Research on Phosphorus Removal by Polysilicate Aluminum Iron Sulfate Coagulation from Wastewater Treatment Plant effluent [J]. Water Treatment Technology, 2011, 37(4):72–75+83.

L. Fengting, *Water Treatment Chemicals: Coagulants and Flocculants*,
https://doi.org/10.1007/978-981-96-8221-8

16. Yu Hongfa, Wu Jinyong, Wu Chengyou, et al. A process method for producing high-purity low iron aluminum sulfate from fly ash and its comprehensive utilization [P]. Inner Mongolia: CN102101689A, 2011-06-22.
17. Hu Qiuming, Xu Renyi, Zhang Jinbin, et al. Industrial production method of pharmaceutical grade ferrous sulfate [P]. Hubei Province: CN110894081A, 2020-03-20.
18. GB/T 10531–2016. Water Treatment Agent Ferrous Sulfate [S]. Beijing: China Standard Press, 2016.
19. Tian Qinghua, Zhang Lei, Li Dong. A method for regenerating high acidity nickel containing ferric chloride et alhing wastewater [P]. Hunan: CN106958021A, 2017-07-18.
20. Liang Ximing. A method for producing ferric chloride using acid washing waste liquid [P]. Shandong Province: CN103073067B, 2015-09-02.
21. HG/T 4538-2022 Water Treatment Agent Ferrous Chloride [S]. Beijing: China Standard Press, 2022.
22. Lu Guangjie, Qu Jiuhui. Research on the Synthesis of High Efficiency Polyaluminum Chloride by Electrodialysis Method [J]. Environmental Science in China, 2000, 20(3): 4.
23. Kang Yan, Li Fengting. Study on the influence of polymer types of polyaluminum chloride on coagulation effect [J]. Chemical Technology Market, 2008, 31(9): 4.
24. Luan Zhaokun, Qu Jiuhui, Tang Hongxiao. Research on the Morphological Stability and Electrokinetic Properties of Polyaluminum [J]. Environmental Chemistry, 1997, 16(6): 9.
25. Li Fengting, He Yan, Pan Hongjie, et al. Study on the influencing factors of coagulation performance of polyaluminum chloride [J]. Industrial Water Treatment, 2008, 28(10): 4.
26. Parker D R, Bertsch P M. Identification and quantification of the "Al13" tridecameric aluminum polycation using ferron [J]. Environmental Science & Technology, 1992, 26(5): 908–914.
27. Xu Zuxin, Chen Xuan, Li Xia. Polyaluminum chloride and its application and prospects in industrial wastewater treatment [J]. Shanghai Environmental Science, 2003, 22(5): 5.
28. Tang Hongxiao Several new understandings of inorganic polymer flocculants [J]. Industrial Water Treatment, 1997, 17(004): 1–5.
29. Tang Hongxiao. Flocculation morphology of hydroxy polyaluminum chloride [J]. Journal of Environmental Science, 1998, 18(1): 1–10.
30. Johansson G, Dorm E, Seleborg M, et al. The Crystal Structures of $[Al_2(OH)_2(H_2O)_8](SO_4)_2.2H_2O$ and $[Al_2(OH)_2(H_2O)_8](SeO_4)_2.2H_2O$ [J]. Acta Chemica Scandinavica, 1962, 16: 403–420.
31. Brosset C, Biedermann G, Sillén L G, et al. Studies on the Hydrolysis of Metal Ions. XI. The Aluminium Ion, Al^{3+} [J]. Acta Chemica Scandinavica, 1954, 8(10): 1917–1926.
32. Van Cauwelaert F H, Hall W K. Studies of the hydrogen held by solids. XX. The chromatographic separation of the isotopic and allotropic hydrogens on alumina and fluorided alumina [J]. Journal of Colloid and Interface Science, 1972, 38(1): 138–151.
33. Vermeulen A C, Geus J W, Stol R J, et al. Hydrolysis-precipitation studies of aluminum (III) solutions. I. Titration of acidified aluminum nitrate solutions [J]. Journal of Colloid and Interface Science, 1975, 51(3): 449–458.
34. Matijević E, Mathai K G, Kerker M. Detection of metal ion hydrolysis by coagulation. V. Zirconium [J]. The Journal of Physical Chemistry, 1962, 66(10): 111–114.
35. Hsu P H, Bates T F. Fixation of Hydroxy-Aluminum Polymers by Vermiculite 1 [J]. Soil Science Society of America Journal, 1964, 28(6).
36. Rausch W V, Bale H D. Small-Angle X-Ray Scattering from Hydrolyzed Aluminum Nitrate Solutions [J]. The Journal of Chemical Physics, 2004, 40(11): 3391–3394.
37. Letterman R D, Iyer D R. Modeling the effects of hydrolyzed aluminum and solution chemistry on flocculation kinetics [J]. Environmental Science & Technology, 1985, 19(8): 673–681.
38. Akitt J W, Greenwood N N, Khandelwal B L, et al. ^{27}Al nuclear magnetic resonance studies of the hydrolysis and polymerisation of the hexa-aquo-aluminium(III) cation [J]. Journal of the Chemical Society, Dalton Transactions, 1972, (5): 604–610.
39. Seichter W, Moegel H J, Brand P, et al. Crystal Structure and Formation of the Aluminum Hydroxide Chloride $[Al_{13}(OH)_24(H_2O)_{24}]\ Cl_{15} \times 13H_2O$ [J]. Berichte der deutschen chemischen Gesellschaft, 1998, (6):795–797

40. Bottero J Y, Cases J M, Fiessinger F, et al. Studies of hydrolyzed aluminum chloride solutions. 1. Nature of aluminum species and composition of aqueous solutions [J]. The Journal of Physical Chemistry, 1981, 84(22) 2933–2939.
41. Li Fengting, Zhu Qian, Liu Chang Production technology and quality control issues of polyaluminum chloride in China [J]. Industrial Water Treatment, 2014, 34(12): 3.
42. Zhang Xuejin Research on the Preparation of High Salinity Polyaluminum Chloride [J]. China Chlor Alkali, 1995, (4): 29–32.
43. Ning Xun'an, Li Kai, Li Runsheng Revision of the National Standard for Water Treatment Agent Polyaluminum Chloride [J]. Water Supply and Drainage, 2005, 31(3): 5.
44. Luo Ziqin, Chen Shiqian, Zhou Jin Preparation and Properties of Polyaluminum Chloride [J]. Inner Mongolia Petrochemical Industry, 2006, (09): 29–30.
45. Lu Xiuguo, Zhai Jian, Jiao Ling, et al. Experimental study on one-step acid dissolution method for preparing polyaluminum chloride from aluminum ash [J]. Water Supply Technology, 2007, (04): 17–19.
46. Liu Xixiang, Wu Qilin, Shi Bingfang, et al. Research on the preparation of polyaluminum chloride from waste aluminum slag in aluminum profile factories [J]. Inorganic Salt Industry, 2014, 46(04): 53–55.
47. Chao Xi, Zhang Ting'an, Zhang Yubin, et al. Research on the Preparation of Polyaluminum Chloride by Secondary Aluminum Ash Acid Leaching [J]. Nonferrous Metals Science and Engineering, 2021, 12(05): 1–9.
48. Zhang Yeling, Zheng Xianqiang, Sun Yanan, et al. A system for preparing polyaluminum chloride from aluminum ash [Z]. 2022
49. Li Fengting, Zhou Qiutao, Wang Ying, et al. A drying equipment and drying method for polyaluminum chloride series salts [Z]. 2018
50. Li Fengting, Tao Xiaoping, Gao Yan, et al. Production technology and research progress of polyaluminum chloride [J]. Inorganic Salt Industry, 2004, 36(6): 4.
51. Zhang Shujie. Discussion on hygiene standards for production enterprises of chemical treatment agents for drinking water [J]. Jiangsu Health Care, 2001, 3(1): 2.
52. Implementation of mandatory national standard for polyaluminum chloride used in drinking water (GB 15892–2020) [J]. Water purification technology, 2021, 40(9): 118–118.
53. Association A. AWWA standard for liquid polyaluminum chloride [J]. Awwa Standards, 1993, ansi/awwa b408-93.
54. Li Fengting. The current status of technological development of coagulant polymeric ferric sulfate in China [J]. Industrial Water Treatment, 2002, (01): 5–8.
55. Li Fengting, Chen Fujun, Du Xirong, et al Discussion on the reaction mechanism of coagulant polymerization of ferric sulfate [J]. Water Supply and Drainage, 1994, (08): 39–40+35.
56. US 4284611 [P/OL]. 1981-08-18
57. US 4536665 [P/OL]. 1985-08-20
58. CA 1123306 [P/OL]. 1979-11-14
59. CA 1203664 [P/OL]. 1986-04-29
60. CA 1203665 [P/OL]. 1986-04-29
61. US 4877597 [P/OL]. 1989-10-24
62. US 3544476 [P/OL]. 1970-12-01
63. US 3909439 [P/OL]. 1975-09-30
64. US 4082685 [P/OL]. 1978-4-4
65. US 5069893 [P/OL]. 1991-12-03
66. US 5149400 [P/OL]. 1992-09-22
67. Li Fengting. Discussion on the current situation and problems of coagulant application in China [J]. Water Purification Technology, 2018, 37(07): 1–3+13.
68. Wang Hongtao, Li Fengting, Zhang Lu, et al. A red mud adsorbent and its preparation method and application, CN109107524B [P/OL].
69. Fengt Li, Yun Lu, Qi Xue. The Preparation of Inorganic Coagulant—Poly Ferric Sulfate [J]. Journal of Chemical Technology & Biotechnology: International Research in Process, Environmental and Clean Technology, 1997, 68, 219–221.

70. EP 1385972 [P/OL]. 2004–2–4
71. US 4929655 [P/OL]. 1990–5–29
72. US 5006590 [P/OL]. 1991-04-09
73. US 5605970 [P/OL]. 1997-2-25
74. US 5171783 [P/OL] 1992-12-15
75. Ruchhoft C C, Mcnamee P D, Butterfield C T. Studies of Sewage Purification: VII. Biochemical Oxidation by Activated Sludge [J]. Public Health Reports (1896–1970), 1938, 53(38):1690–1718.
76. Nakamura J, Miyashro S, Hirose Y. Screening, isolation and some properties of microbial cell flocculants [J]. Agricultural and Biological Chemistry, 1976, 40 (2): 377–383.
77. Takagi H, Kadowki K. Flocculant production by Peacilomyces sp. taxonomic studies and culture conditions for production [J]. Agricultural and Biological Chemistry, 1985, 49(11): 3151–3157.
78. Xiong Xingying. Research on microbial flocculant produced by Aspergillus Niger cultured in starch wastewater and its application [D]. Chengdu: Chengdu University of Technology, 2016. (in Chinese)
79. Jiang Zhi, Yi Xiaochang, Qin Liu, et al. Screening and culture conditions of Saccharomyces Cerevisiae with high flocculation activity [J]. Environmental Science and Technology, 2014, 37(01):53–55. (in Chinese)
80. Wan Yingxin, Xu Jing, Yang Xuejing, et al. Study on flocculation characteristics of Geotrichum candidum, a flocculant producing bacterium [J]. Earth and Environment, 2013(2):6.
81. Wu Xiaochang, Zheng Jingfan, Li Riqiang, et al. Identification and flocculation characteristics of microbial flocculant producing bacteria [J]. Journal of Shanxi University (Natural Science Edition), 2019, 42(02):435–442. (in Chinese)
82. Wang Bo, Wang Hong, Ouyang Xiaofang, et al. Uv mutagenic selection and optimization of flocculating conditions in plateau environment [J]. Research of Environmental Sciences, 2018, 31(09):1603–1611. (in Chinese).
83. Yu X, Wei X, Chi Z, et al. Improved production of an acidic exopolysaccharide, the efficient flocculant, by Lipomyces starkeyi U9 overexpressing UDP-glucose dehydrogenase gene [J]. International Journal of Biological Macromolecules, 2020, 165(Pt B):1656–1663.
84. Xing Jie. Removal efficiency and mechanism of carbamazepine by protein-based microbial flocculant [D]. Harbin: Harbin Institute of Technology, 2014. (in Chinese)
85. Chen Ting. Removal of heavy metal ions from water by polysaccharide microbial flocculant and its mechanism [D]. Harbin: Harbin Institute of Technology, 2017. (in Chinese)
86. Li Ningjie, LAN Qi, Chen Zhongwei, et al. Analysis of flocculation characteristics and mechanism of PCF-1767 flocculant produced by BKMF-1767 of Chrysospora flavespora [J]. Chinese Journal of Microbiology, 2019, 47(02):431–439.
87. Crabtree K , Mccoy E , Boyle W C , et al. Isolation, Identification, and Metabolic Role of the Sudanophilic Granules of Zoogloea ramigera1 [J]. Applied Microbiology, 1965, 13(2):218
88. Friedman B A, Dugan P R, Pfister R M, Remsen C C. Structure of exocellular polymers andtheir relationship to bacterial flocculation. [J]. Journal of bacteriology, 1969, 98(3):1328–1334.
89. Wang Kun. Screening of microbial flocculant producing bacteria and study on their flocculation ability [D]. Jinan: Qilu University of Technology, 2015.
90. Zhang Wei. Isolation, identification and treatment of an efficient microbial flocculant producing bacterium in eutrophic water [D]. Chengdu: Sichuan Normal University, 2015. (in Chinese)
91. Li Ningjie, LAN Qi, Chen Zhongwei, et al. Analysis of flocculation characteristics and mechanism of PCF-1767 flocculant produced by BKMF-1767 of Chrysospora flavespora [J]. Chinese Journal of Microbiology, 2019, 47(02):431–439.
92. Tian Lian-Sheng, Chen Xiu-Qing. Artificial breeding and treatment of bioflocculant producing bacteria [J]. Industrial WATER TREATMENT, 2016, 36(03):47–50. (in Chinese)

93. Liu Jiewei, Ma Junwei, Liu Yanzhong, et al. Culture optimization of M-C11 flocculant produced by Klebsiella and its application in sludge dewatering [J]. Environmental Science, 2014, 35(03):1183–1190.
94. Li Mengqian, Long Jieyun, CAI Chengru, et al. Study on preparation and properties of microbial flocculants of Bacillus polymyxides [J]. Journal of Hydrometallurgy, 2019, 39(05):424–428. (in Chinese)
95. Zhang Xiaofei, Sun Yunlong, Han Hong, et al. Screening and identification of a flocculant producing bacterium Bacillus Morcellularis Z11 [J]. Industrial Microbiology, 2019, 50(06):18–24. (in Chinese)
96. Peng Cuizhen, Zong Xuyan, Xu Yong, et al. Screening of microbial flocculant strains from wine wastewater and optimization of fermentation conditions [J]. China Brewing, 2017, 36(09):92–97. (in Chinese)
97. Guo Junyuan, Zhou Xin-Tian. Study on preparation of microbial flocculant from slaughter wastewater and improvement of sludge dewatering performance [J]. China Environmental Science, 2017, 37(07): 2615–2622. (in Chinese)
98. Lu Cui. Optimization, composition analysis of exopolysaccharide flocculant produced by Bacillus B69 and its application in wastewater treatment [D]. Hangzhou: Zhejiang University, 2014.
99. Luo Laipeng. Characteristics and fermentation optimization of flocculant from Raoultella ornithinolytica 160-1 [D]. Hefei: Anhui Medical University, 2020.
100. Tang Jiayi. Study on fermentation optimization, isolation, physicochemical properties and application of polysaccharide flocculant of Bacillus glia [D]. Guangzhou: South China University of Technology, 2014.
101. Zhang Meiping. Study on bioflocculation and metal ion adsorption properties of Arthrobacter ps-5 extracellular polysaccharide [D]. Dalian: Dalian Polytechnic University, 2012.
102. Feng Pei, Wang Shimei, Zhou Lixiang. Isolation and identification of bioflocculant producing bacteria and its effect on turbidity removal of drinking water [J]. Chinese Journal of Environmental Sciences, 2009, 29(8):1666–1671. (in Chinese)
103. Zhang Longyan. Study on the application of a new microbial flocculant to the treatment of groundwater with high hardness and high salinity [D]. Taiyuan: Shanxi University, 2015. (in Chinese)
104. Qi Hua, Tian Shun, Xie Enliang. Experimental study on the application of microbial flocculant B-16 to water treatment [J]. Water Supply Technology, 2008, 2(3):10–12. (in Chinese).
105. Xia X , Lan S , Li X , et al. Characterization and coagulation-flocculation performance of a composite flocculant in high-turbidity drinking water treatment [J]. Chemosphere, 2018, 206(SEP.):701–708.
106. Li Z , Zhong S , Lei H Y , et al. Production of a novel bioflocculant by Bacillus licheniformis X14 and its application to low temperature drinking water treatment. [J]. Bioresource Technology, 2009, 100(14):3650–3656.
107. Zhang Chao, Luan Xing-she, Chen Wen-Bing, et al. Experimental study on the treatment of domestic wastewater by novel biological flocculants [J]. Industrial WATER TREATMENT, 2013, 33(09):31–33. (in Chinese)
108.] Wang Lan, Tang Jing, Liu Yunjie. Treatment of domestic wastewater with microbial flocculant XM09 combined with PAC [J]. Journal of Tianjin University, 2011, 44(11):984–988. (in Chinese)
109. Zhang Shasha, Xu Jianfeng, Liu Zengshan. Screening of microbial flocculant producing bacteria and analysis of flocculation effect on pig farm wastewater [J]. Heilongjiang Animal Husbandry and Veterinary Science, 2021, (13):10–15+20+143
110. Peng Guixiang, Lu Qiuyan, Kong Huiqing, et al. Screening and fermentation optimization of strains with high bioflocculation activity [J]. Journal of South China Agricultural University, 2014, 35(02):66–72. (in Chinese)
111. Xiong Xingying. Research on microbial flocculant produced by Aspergillus Niger cultured in starch wastewater and its application [D]. Chengdu: Chengdu University of Technology, 2016. (in Chinese)

112. Shi Chunfang, Leng Xiaoyun. Application of microbial flocculants in the treatment of pharmaceutical wastewater [J]. Modern Chemical Industry, 2015, 35(09):85–87+89. (in Chinese)
113. Li Wenpeng, Ren Xiaoli, Xiang Xuemin, et al. Study on the effect of microbial flocculant on the treatment of papermaking wastewater [J]. Industrial WATER TREATMENT, 2013, 33(11):13–16. (in Chinese)
114. Kaur R, Roy D, Yellapu S K, et al. Enhanced Composting Leachate Treatment Using Extracellular Polymeric Substances as Bioflocculant [J]. Journal of Environmental Engineering, 2019, 145(11).
115. Ye Chunsong, Hao Hongduo, Wang Tianping, et al. Effect principle and industrial application test of microbial microbial agent on treating circulating cooling water [J]. Environmental Engineering, 2019, 37(08):42–46. (in Chinese)
116. Fang Yaling. Research on microbial flocculant and its cheap preparation of coal chemical wastewater [D]. Taiyuan: North University of China, 2016.
117. Qin Minjie. Study on the preparation and characteristics of flocculant by Lead-zinc-Chryseobacterium sp. [D]. Guilin: Guilin University of Technology, 2018.
118. Murugesan, Kumarasamy, Yu, et al. Sludge conditioning using biogenic flocculant produced by Acidithiobacillus ferrooxidans for enhancement in dewaterability [J]. Bioresource Technology: Biomass, Bioenergy, Biowastes, Conversion Technologies, Biotransformations, Production Technologies, 2016.
119. Dai Xingbao. Study on the efficacy of mycelium pellets for regulating filamentous bulking sludge [D]. Handan: Hebei University of Engineering, 2021.
120. Guo Junyuan, Zhou Xin-Tian. Study on preparation of microbial flocculant from slaughter wastewater and improvement of sludge dewatering performance [J]. China Environmental Science, 2017, 37(07):2615–2622. (in Chinese)
121. Li Y , Xu Y , Song R , et al. Flocculation characteristics of a bioflocculant produced by the actinomycete on microalgae biomass [J]. BMC Biotechnology, 2018, 18(1).
122. Zeng S Y, Pan X S, Shen L, et al. A novel bioflocculant for the enrichment of Desmodesmus sp.F51 from Pyrophorus palustris [J]. Chinese Journal of Microbiology, 2019, 47(01):35–42.
123. Sirasit S , Arunothai C , Supavadee C , et al. A rapid method for harvesting and immobilization of oleaginous microalgae using pellet-forming filamentous fungi and the application in phytoremediation of secondary effluent [J]. International Journal of Phytoremediation, 2018, 20(10):1017–1024.
124. Li Y, Xu Y, Liu L, et al. Flocculation mechanism of Aspergillus niger on harvesting of Chlorella vulgaris biomass [J]. Algal Research, 2017, 25:402–412.
125. Yang L, Zhang H Y, Cheng S Z, et al. Enhanced Microalgal Harvesting Using Microalgae-Derived Extracellular Polymeric Substance as Flocculation Aid [J]. Acs Sustainable Chemistry & Engineering, 2020, 8(10): 4069–4075.
126. Yang J, Wu D, Li A, et al. The Addition of N-Hexanoyl-Homoserine Lactone to Improve the Microbial Flocculant Production of Agrobacterium tumefaciens Strain F2, an Exopolysaccharide Bioflocculant-Producing Bacterium [J]. Appl Biochem Biotechnol, 2016, 179(5):728–739.
127. Chen Z, Liu P, Li Z, et al. Identification of key genes involved in polysaccharide bioflocculant synthesis in Bacillus licheniformis [J]. Biotechnology and Bioengineering, 2016, 114(3).
128. Ma L, Liang J, Liu Y, et al. Production of a bioflocculant from Enterobacter sp. P3 using brewery wastewater as substrate and its application in fracturing flowback water treatment [J]. Environmental Science and Pollution Research, 2020, 27(15):18242–18253.
129. Guo H, Hong C, Zhang C, et al. Bioflocculants' production from a cellulase-free xylanase-producing Pseudomonas boreopolis G22 by degrading biomass and its application in cost-effective harvest of microalgae [J]. Bioresource Technology, 2018:171.
130. Liu W, Dong Z, Sun D, et al. Bioconversion of kitchen wastes into bioflocculant and its pilot-scale application in treating iron mineral processing wastewater [J]. Bioresource Technology, 2019, 288:121505.

131. Ndikubwimana T, Zeng X, Murwanashyaka T, et al. Harvesting of freshwater microalgae with microbial bioflocculant: a pilot-scale study [J]. Biotechnology for Biofuels, 2016, 9(1):1–11.
132. Liu C, Wang K, Jiang J H, et al. A novel bioflocculant produced by a salt-tolerant, alkaliphilic and biofilm-forming strain Bacillus agaradhaerens C9 and its application in harvesting Chlorella minutissima UTEX2341 [J]. Biochemical Engineering Journal, 2015, 93:166–172.
133. Chen Z, Meng T, Li Z, Liu P, Wang Y, He N, Liang D. Characterization of a beta-glucosidase from Bacillus licheniformis and its effect on bioflocculant degradation [J]. AMB Express, 2017, 7(1):1–7.
134. Zhang C, Wang X, Wang Y, et al. Synergistic effect and mechanisms of compound bioflocculant and $AlCl_3$ salts on enhancing Chlorella regularis harvesting [J]. Applied Microbiology & Biotechnology, 2016, 100(12):5653–5660.
135. Guo J, Chen C. Sludge conditioning using the composite of a bioflocculant and PAC for enhancement in dewaterability [J]. Chemosphere, 2017, 185(oct.):277–283.
136. Wang Yina, Wang east, look, et al. Study on the cultivation of complex bioflocculant producing bacteria in wine wastewater [J]. China Water Supply and Drainage, 2007(01):38–42. (in Chinese)
137. Zhou Chi Jun, Xu Wei, Ren Xiaoying, et al. Study on coagulation and decolorization of cationic inorganic particles system with microbial flocculants [J]. Industrial WATER TREATMENT, 2006(04):50–53.
138. Carvajal-Zarrabal O, Nolasco-Hipolito C, Barradas-Dermitz D M, et al. Treatment of vinasse from tequila production using polyglutamic acid [J]. Journal of Environmental Management, 2012, 95(suppl): S66-S70.
139. Ma Fang, Duan Shuyue, Kong Xiangzhen, et al. Research status and development trend of microbial flocculants [J]. China Water Supply and Drainage, 2012,28(02):14–17. (in Chinese)
140. Yang Jianzhou, Lin Li, Zhang Rongli. Study on the treatment of papermaking white water by high substitution cationic starch [J]. Industrial WATER and Wastewater, 2003, 34 (6): 37–39. (in Chinese)
141. Wang Chen, Chen Jierong, Zong Gang, et al. Preparation of cationic starch flocculant by microwave drying method and its application [J]. Advances in Chemical Industry, 2003, 22 (11): 1217–1221
142. Ma Yongmei. Preparation and application of natural modified cationic flocculants [J]. Sichuan Chemical Industry and Corrosion Control, 2003, 6 (5): 17–21. (in Chinese)
143. Weihai. SAH cationic natural polymer modified flocculant [J]. Chemical Industry Information Everyday, 2003(8):1.
144. Liu Minghua, Zhang Hong. Study on flocculation performance and application of a composite flocculant [J]. Chemical Research and Application, 2003, 15 (4): 475–478. (in Chinese)
145. Tan Fengzhi, Guan Jianfeng, Ma Xichen. Synthesis and application of anionic polymer flocculant based on corn starch [J]. Journal of Dalian Institute of Light Industry, 2001, 20 (4): 254–256. (in Chinese)
146. US 89716497A[P]. 1999-11-23.
147. Guo Ling, Jin Zhihao. Synthesis of starch modified flocculant and its application in wastewater treatment [J]. Journal of Shanxi Normal University (Natural Science Edition), 2003, 17 (4): 58–62. (in Chinese)
148. Yang Bo, Duan Yunbiao, Zhao Yulin. Flocculation and purification of yellow phosphorus wastewater by starch acrylamide grafted copolymer [J]. Chemical Industry Environmental Protection, 2002, 22 (6): 367–369. (in Chinese)
149. Feng Yunsheng, Zhao Xin, Dong Guowen. Preparation and flocculation effect of modified cationic flocculants for potato starch [J]. Advances in Science and Technology, 2002, 10:39–41
150. Zhuang Yunlong, Shi Xiuchun. Starch grafted polyacrylamide for the treatment of papermaking wastewater [J]. Paper and Papermaking, 2003, 3:60–61
151. Zhang Shijun, Li Sijian. Synthesis of grafted copolymer of starch and acrylamide from taro and its treatment of wastewater from wine tank [J]. Environmental Protection, 2000, 5:45–46
152. Chang Wenyue, Han Xue. Synthesis and application of grafted starch polymer flocculant [J]. Environmental Protection Science, 1996, 22 (4): 4–7. (in Chinese)

153. Zhang Shuyuan, Li Zifa. Treatment of wastewater containing chromium [J].1993, 19 (5): 293–296. (in Chinese)
154. US 4,560,416[P]. 1985-12-24
155. US 4,564,004[P]. 1986-1-14
156. US 20030145763[P]. 2003-07-08
157. I. Dogu, A.I. Arol, Separation of Dark-colored Minerals from Feldspar by Selective Flocculation Using Starch [J]. Powder Technology, 2004,139:258–263
158. US 4,678,585[P]. 1987-07-07
159. US 6,726,845[P]. 2004-04-27
160. US 4,269,975[P]. 1981-05-26
161. Zhang Yuxia, Li Yuling, Xue Lingfen. Modification of sesame stalk cellulose and adsorption of aniline from wastewater [J]. Journal of Xinyang Normal University (Natural Science Edition), 2004,17 (1): 38–39. (in Chinese)
162. US 5,990,216[P]. 1999-11-23
163. Gong Shengzhao, Yun Zhimian. Study on optimization of the treatment conditions of heavy metal wastewater by cellulose xanthate [J]. Industrial Water and Wastewater, 2002,33 (3): 30–31. (in Chinese)
164. Cheng Jianguo, Zhang Leqin. Preparation of cellulose xanthate (IPX) from soybean residue and its adsorption properties and application [J]. Hangzhou Chemical Industry, 2001,31 (1), 17–19
165. Dao Zhou, Lina Zhang, Jinping Zhou and et al., Cellulose/Chitin Beads for Adsorption of Heavy Metals in Aqueous Solution [J]. Water Research, 2004(38): 2643–2650
166. Wan Junmin, Hu Zhiwen, Chen Wenxing, et al. Application of loaded β-cyclodextrin cellulose fiber in wastewater treatment [J]. Environmental Pollution and Control, 2004,26 (1): 57–59. (in Chinese)
167. Wang Qing-lei, Guo Juan-Juan, Song Qing-feng, et al. Complexation of humic acid and cellulose to heavy metal ions in wastewater, Journal of Shanxi Normal University (Natural Science), 2003,17 (1): 69–71
168. Jiang is quite big. Chitin [M]. Beijing: Chemical Industry Press, 1st edition, March 2003Hu Jiazhen, Chen Donghui, Chen Liang. Application of chitosan in water treatment [J]. Journal of China Textile University, 2000,26 (5): 124–127. (in Chinese)
169. Ravi Divakaran, V.N. Sivasankara Pillai, Mechanism of Kaolinite and Titanium Dioxide Flocculation Using Chitosan—Assistance by Fulvic Acids? [J].. Water Research, 2004(38):2135–2143
170. US 5,543,056[P]. 1996-08-06
171. AU6865901A [P]. 2002-01-08
172. H. Ganjidoust, K. Tatsumi, T. Yamagishi and R.N. Gholian, Effect of Synthetic and Natural Coagulant on Lignin Removal from Pulp and Paper Wastewater [J]. Water Science and Technology. 1997(35):291–296
173. Jill Ruhsing Pan, Chihpin Huang, Shuchuan Chen, Ying-Chien Chung, Evaluation of a Modified Chitosan Biopolymer for Coagulation of Colloidal Particles [J]. Colloids and Surfaces, 1999(147): 359–364
174. Chihpin Huang, Shuchuan Chen and Jill Ruhsing Pan, Optimal Condition for Modification of Chitosan: a Biopolymer for Coagulation of Colloidal Particles [J]. Water Research, 2000(34): 1057–1062
175. Gruber James V. Synthesis of N - [(3′- Hydroxy - 2′,3′- dicarboxy) - ethyl] chitosan : A New Water - Soluble Amphoteric Chitosan Derivative [J]. Advances in Chitin Science, 1996 ,(1) :230 – 235
176. Zeng Defang, Li Juan, Yuan Jizu. Application of natural polymer flocculants in the treatment of industrial wastewater [J]. Industrial WATER TREATMENT, 2004, 24 (2): 20–22. (in Chinese)
177. Li Yuling, Yang Haixia. Study on purification of calcium bentonite by flocculation [J]. Journal of Xinyang Normal University (Natural Science Edition), 2002,15 (4): 443–444. (in Chinese)
178. Chen Jianchu. Study on preparation and application of chitosan [J]. Starch and Starch Sugars, 1994, 1:24–30.

179. Chen Jianchu. Study on preparation of protein flocculant from chitin and its application [J]. Application Technology Market, 1994,12:12–13.
180. L. Guerrero, F. Omil, R. Méndez and J.M. Lema, Protein Recovery during the Overall Treatment of Wastewaters from Fish-meal Factories [J]. Bioresource Technology, 1998,63(3): 221–229
181. US 5,433,865[P]. 1995-07-18
182. Galo Cardenas , Parra Orlando , Taboada Edelio. Synthesis and applications of chitosan mercaptanes as heavy metal retention agent [J]. International Journal of Biological Macromoleacules ,2001,28(2) :167–174.
183. John J . Lenhart , Linda A. Figueroa , Byuce D. Honeyman , et al . Modeling the adsorption of U(VI) onto animal chitin using coupled mass transfer and surface complexation [J]. Colloids and Surfaces A: physicochernical and Engineering Aspects , 1997 , 120(1 - 3) : 243 - 154.
184. Qu Rongjun, Liu Qingjian. Research on Natural Polymer Adsorbents III Preparation and Characteristics of Modified Chitosan [J]. Research and Development of Natural Products , 1996, 8(2):49–53
185. Cheng Jianguo, Zhang Leqin. Preparation of cellulose xanthate (IPX) from soybean residue and its adsorption properties and application [J]. Hangzhou Chemical Industry, 2001,31 (1), 17–19
186. US 5,543,058[P]. 1996
187. US 4,455,257[P]. 1984-06-19
188. US 4,775,744[P]. 1988-10-04
189. US 4,781,840[P]. 1988-03-28
190. Lei Zhongfang, Lu Yongsen. Experimental study on coagulability of lignin from straw pulp by caustic soda process [J]. Journal of Tongji University, 1995,23 (5): 547–551. (in Chinese)
191. Liu Qianjun, Zhan Huaiyu, Liu Mengru, et al. Preparation and application of lignin-based flocculation decolorizing agent [J]. Paper Science and Technology, 2003,22 (6): 125–127. (in Chinese)
192. Chen Junping. Synthesis and application of alkali lignin anionic polymer flocculant [J]. Journal of Hubei University of Technology, 1994,9 (2): 90–93. (in Chinese)
193. Wu Bingyan, Yu Gang. Preparation of a new decolorizing flocculant lignin quaternary amine salt and Study on its flocculation performance and mechanism [J]. Chemical Environmental Protection, 1997,17:268–272
194. Zhang Zhilan, Lu Yongsen. Study on properties and application of lignin coagulants [J]. Water Treatment Technology,1997,23 (1): 38–44. (in Chinese)
195. Fang Guizhen, Wu Chenliang, Song Zhanqian. Optimization of synthesis conditions for cationic flocculant lignin quaternary amine salt [J]. China Paper, 2003,22 (11): 24–27
196. Huang Minsheng, Julie. Experimental study on the flocculation-adsorption pretreatment of monosodium glutamate wastewater [J]. Water Treatment Technology, 1998,24 (5): 299–302. (in Chinese)
197. US 4,734,216[P]. 1988–03–29
198. US 5,659,002[P]. 1997-08-19
199. Mitchell, David Brian, Minnis et al., Treatment of Aqueous systems using a chemically modified tannin, US5,830,315, 1998
200. US 5,659,002[P]. 1997-08-19
201. Ji Yongliang. Modified Denningmanich polymer [J]. Water Treatment Information Report, 1999,2:33–36
202. Zhang Shuyun. Plant flocculants for removing colloidal substances from water [J]. Water Treatment Information Report, 2003,3:18–19
203. Mahmut Özacar, I. Ayhan Sengil, Evaluation of Tannin Biopolymer as a Coagulant Aid for Coagulation of Colloidal Particles [J]. Colloids and Surfaces A: Physicochemical and Engineering Aspects. 2003(229): 85–96
204. Mahmut Özacar, I. Ayhan Sengil, Effectiveness of Tannins Obtained from as a Coagulant Aid for Dewatering of Sludge [J]. Water Research. 2000,34(4): 1407–1412

205. Mahmut Özacar, I. Ayhan Sengil, Enhancing Phosphate Removal from Wastewater by Using Polyelectrolytes and Clay Injection [J]. Journal of Hazardous Materials. 2003, 100(1/3): 131–146
206. Zhao Lizhi, Zhao Yongpeng, Dai Jialin. Preparation and performance evaluation of cationic tannin flocculants, Chongqing Environmental Science, 1993,15 (6): 33–36
207. US 5,908,921[P]. 1999-06-01
208. M. P. Bell, T. B. Neff. Animal collagen and gelatin [P]. CN 1420892A, 2003
209. Ma Jun, Tang Minghui, Guan Jianning. Halogenated bone glue flocculant and its manufacturing method [P]. CN 1243808A, 2000
210. Qin Lamei, Zhang Zhuang, Niu Fuling, etc., flocculating agent in the process of traditional Chinese medicine extraction technology, applied research [J], Chinese journal of experimental formulas of Chinese medicine, 2000, 6 (3) : 3–5
211. Yasuhiko Tabata, Yoshito Ikada, Protein Release from Gelatin Matrices [J]. Advanced Drug Delivery Reviews, 1998(31): 287–301
212. Li Fengting, Lu Xuefei, Zhang Bingru. Recent development of decolorization technology of printing and dyeing wastewater [Z]. Proceedings of the first International Symposium on chlorine dioxide and Water Treatment Technology. Shanghai, 2001: 22–26
213. Lin Feng. Development and Prospect of Dicyandiamide Formaldehyde Polymer coagulants [J]. Industrial Water Treatment, 2004, (01): 1–4+11.
214. Sun Ximeng. Synthesis, characterization, and performance study of melamine formaldehyde series polymer decolorizing flocculants [D]; Henan University, 2003.
215. Li Zaibo. Research on the synthesis and application of dicyandiamide formaldehyde flocculant decolorizer [D]; Guangdong University of Technology, 2003.
216. Javaid Mughal M, Saeed R, Naeem M, et al. Dye fixation and decolourization of vinyl sulphone reactive dyes by using dicyanidiamide fixer in the presence of ferric chloride [J]. Journal of Saudi Chemical Society, 2013, 17(1): 23–28.
217. Yuan Xiao, Li Fengting. Research and application prospects of decolorization methods for industrial wastewater [J]. Shandong Chemical Industry, 2016, 45(04): 129–132.
218. Yang, K., L.Z. Zhu, and B.S. Xing, Adsorption of polycyclic aromatic hydrocarbons by carbon nanomaterials [J]. Environmental Science & Technology, 2006. 40(6): p. 1855–1861.
219. Langmuir I. The adsorption of gases on plane surfaces of glass, mica and platinum [J]. Journal of American Chemistry Society,1918,40:1361–1403.
220. Freundlich H. Concerning adsorption in solutions [J]. Zeitschrift fur Physikalische Chemie-Stochiometrie und Verwandtschaftslehr.1906.57:385–470.
221. Nestle N F E I, Kimmich R. NMR imaging of heavy metal absorption in alginate, immobilized cells, and kombu algal biosorbents [J]. Biotechnology and Bioengineering.1996,51(5):538–543.
222. Guibal E, Saucedo I, Roussy J. Uptake of uranyl ions bynew sorbing polymers : Discussion of adsorption isotherms and pH effect [J]. Reactive Polymer.1994.23(2–3):147–156.
223. Jossens L, Prausnitz J M, Fritz W. Thermodynamics of multi-solute adsorption from diluteaqueous-solutions [J]. Chemical Engineering Science,1978,33(8):1097–1106.
224. Hill T L. Theory of multimolecular adsorption from a mixture of gases [J]. Journal of Chemical Physics, 1946,14:268–275.
225. Polanyi M. Adsorption of gases by a non-volatile adsorbent [J]. Verhandlungen der Deutschen Physikalischen Gesellschaft.1916.18(2–3):55–80.
226. Leborans G F, Novillo A. Toxicity and bioaccumulation of cadmium in Olisthodiscus luteus (raphidophyceae) [J]. Water Resources, 1996,30(1):57–62.
227. Trivedi H C, Patel V M, Patel R D. Adsorption of cellulose triacetate on calcium silicate [J]. European Polymer Journal,1973,9(6):525–531.
228. Pan B, Xing B S. Adsorption kinetics of 17α-ethinyl estradiol and bisphenol A on carbon nanomaterials. I. Several concerns regarding pseudo-first order and pseudo-second order models [J]. Journal of Soils Sediment, 2010, 10(5):838–844.
229. Garg V. Oliveira L C A, Rios R V R A. et al. Activated Carbon/Iron Oxide Magnetic Composites for the Adsorption of Contaminants in Water [J]. Carbon, 2002, 40(12):2177–2183.

230. Jiang H Y, Study of adsorption of the aquatic arsenate by activated alumina in dynamic conditions [J]. Journal of Qingdao Technological University, 2014, 35(2):53–58
231. Cheng J, Meng X, Jing C, et al. La3+-modified activated alumina for fluoride removal from water [J]. Journal of Hazardous Materials, 2014, 278: 343–349.
232.] Wei Ruixia, Pang Ruizhi, Li Yanxia. Recovery of phenol from coking wastewater by resin adsorption [J]. Industrial WATER TREATMENT, 2008, 28 (12): 65–69. (in Chinese
233. Zhu ZhaoLian, Chen JinLong, Li Aimin, et al. Study on the treatment of o-benzoylamine wastewater by resin adsorption [J]. Environmental Pollution and Control, 2004, 26 (1): 60–62
234. Rao Li, Wang Xiaojun. Pilot study on natural zeolite treatment of ammonia nitrogen wastewater [J]. Water Treatment Technology 2016, 42(04): 104–106+111.
235. Miljkovic V. The removal of lead (II) ions from aqueous solutions by acid-activated clay modified with sodium carboxymethyl cellulose [J]. Applied Ecology and Environmental Research, 2017, 15: 1461–1472.
236. Li Xiaoying. Research on the treatment of phosphorus containing wastewater with modified diatomaceous earth [J]. Liaoning Chemical Industry, 2013, 42(06): 598–600.
237. Tang Xinhong, Song Min, Sun Fei. Study on adsorption of activated carbon fiber to organic wastewater [J]. Industrial WATER and Wastewater, 2016, 47(04):53–57. (in Chinese)
238. Meng Jie, Li LAN, Chen Pan. Research on the application of activated carbon fiber in petrochemical wastewater treatment [J]. China Rural Water Resources and Hydropower, 2015(05): 81–84+90.
239. Tofighy M A , Mohammadi T. Adsorption of divalent heavy metal ions from water using carbon nanotube sheets [J]. Journal of Hazardous Materials, 2011, 185(1):140–147.
240. Kabbashi N A, Atieh M A, Al-Mamun A, et al. Kinetic adsorption of application of carbon nanotubes for Pb(II) removal from aqueous solution [J]. Journal of Environmental Sciences, 2009, 21(4): 539–544.
241. Matsui Y, Nakao S, Sakamoto A, et al. Adsorption capacities of activated carbons for geosmin and 2-methylisoborneol vary with activated carbon particle size: Effects of adsorbent and adsorbate characteristics [J]. Water Research, 2015, 85: 95–102.
242. Li Yiran. Preparation, in-situ modification and adsorption mechanism of cattails activated carbon for antibiotics and heavy metals [D]. Shandong University, 2018.
243. Liu Wenhong, Yuan Huaibo, Lu Jianping. Effect of HNO3 modification on silver adsorption by activated carbon at different temperatures [J]. The Chinese Journal of Nonferrous Metals, 2007, 17(4): 663–667. (in Chinese)
244. Chen, W., F.S. Cannon, and J.R. Rangel-Mendez, Ammonia-tailoring of GAC to enhance perchlorate removal. II: Perchlorate adsorption [J]. Carbon, 2005. 43(3):581–590.
245. Shaarani, F.W. and B.H. Hameed, Ammonia-modified activated carbon for the adsorption of 2,4-dichlorophenol [J]. Chemical Engineering Journal,2011. 169(1–3):180–185.
246. Zhu Huijie, Jia Yongfeng, Wu Xing, et al. Study on removal of As(III) from drinking water by supported nano-iron adsorbent [J]. Environmental Science,2009,30(06):1644–1648. (in Chinese)
247. Leyva Ramos,R., J. Ovalle-Turrubiartes, and M.A. Sanchez-castillo, Adsorption of fluoride from aqueous solution on aluminum-impregnated carbon [J]. Carbon, 1999. 37(4):609–617.
248. Maruyama, J. and I. Abe, Enhancement effect of an adsorbed organic acid on oxygen reduction at various types of activated carbon loaded with platinum [J]. Journal of Power Sources, 2005.148(0):1–8.
249. Chen, J.P. and S. Wu, Simultaneous adsorption of copper ions and humic acid onto an activated carbon [J]. Journal of Colloid and Interface Science, 2004.280(2):334–342.
250. Huo Lijing, Zhang Yuling, Wei Hanxiao, et al. Study on the removal of water quality stabilizer from desulfurization wastewater by CLO2-PAC [J]. Industrial Water Treatment, 2019, 39(11): 54–57.
251. Jiang C, Ding W, Zhu W, et al. Diatomite-enhanced coagulation for algal removal in polluted raw water: performance optimization and pilot-scale study [J]. Environmental Science and Pollution Research, 2021, 28(36): 50204-50216.

252. Yang Qi. Study on diatomite enhanced coagulation treatment of micro-polluted raw water and algae removal performance [D]. Zhejiang University, 2015.
253. Zhao S, Huang G, Fu H, et al. Enhanced coagulation/flocculation by combining diatomite with synthetic polymers for oily wastewater treatment [J]. Separation Science and Technology, 2014, 49(7): 999–1007.
254. Wang Saisai, Zhang Zhenya, Wen Xiaodong. Experimental study on treatment of urban river water by diatomite enhanced coagulation [J]. Industrial Water Treatment,2019,39(06):52–56. (in Chinese)
255. Liu Fangchi, Qi Kai, Li Xiangdong. Experimental study on the treatment of coal mine water by fly ash loading flocculation [J]. Energy and Environmental Protection, 2022, 36(01):44–50.
256. Li Bolin, Liang Ya Nan, Zhang Chengchen, et al. Study on Phosphorus removal and coagulation mechanism of aluminum-iron composite coagulant prepared by flyash-bauxite modification [J]. Journal of Environmental Sciences, 2016, 36(7): 2503–2511.
257. Huang Xing. Treatment of oilfield polymer flooding wastewater by enhanced coagulation with fly ash zeolite [D]. Northeast Petroleum University, 2021. 2021.000295.
258. Wu Shaolin, Li Haizhu, Ma Ming. Treatment of lead-free preserved egg wastewater by a novel composite coagulant and adsorption reagent [J]. Chinese Journal of Environmental Engineering, 2011, 5(02):369–372.
259. Ling Huishi, Wang Zhihong, He Xiaomei, Liu Lifan, Ruan Caiqun, Qiu Yongting. Enhanced coagulation treatment of raw water with high algal content and low turbidity based on modified attapulgite [J]. Chinese Journal of Environmental Engineering, 2014, 8(07): 2765–2771.
260. Tan Wei. Study on enhanced removal of hexavalent chromium from water by modified attapulgite [D]. Southwest Jiaotong University, 2016.
261. Han Zhen. Study on removal of humic acid from feed water by mineral compounding PAC coagulation [D]. Qingdao: Qingdao University of Science and Technology, 2009:41–47. (in Chinese)
262. Wang Aiqin, Zheng Yi 'an, Zhang Junping. Network attapulgite clay composite adsorbent and its preparation method [P]. China: 10166466A, 2010–03–10.
263. Qu Jiuhui. Technical Principle of Drinking water safety Assurance [M], Beijing: Science Press, 2007: 103–104.
264. Rosińska A, Dąbrowska L. Enhancement of coagulation process with powdered activated carbon in PCB and heavy metal ions removal from drinking water [J]. Desalination and Water Treatment, 2016, 57(54): 26336–26344.
265. Li W, Hua T, Zhou Q, et al. Treatment of stabilized landfill leachate by the combined process of coagulation/flocculation and powder activated carbon adsorption [J]. Desalination, 2010, 264(1–2): 56–62.
266. Liu Guimei, Zeng Yubin, Peng Mengjiao, Xie Liang, Yang Pingping. Study on mercury removal from gas field water by enhanced coagulation-adsorption coupling method [J]. Water Treatment Technology, 2019,46(07):16–19. (in Chinese)
267. Chen Wei, Li Xing, Pang Yali, Yang Yan-ling, Li Gui-bai. Removal of lead from water by enhanced coagulation with potassium permanganate and powdered activated carbon [J]. Journal of Civil Engineering, 2010, 43(S2):457 460. (in Chinese)
268. Linstedt K D, Houck C P, O'Connor J T. Trace element removals in advanced wastewater treatment processes [J]. Journal (Water Pollution Control Federation), 1971: 1507–1513.
269. Kristiana I, Joll C, Heitz A. Powdered activated carbon coupled with enhanced coagulation for natural organic matter removal and disinfection by-product control: Application in a Western Australian water treatment plant [J]. Chemosphere, 2011, 83(5): 661–667.
270. Uyak V, Yavuz S, Toroz I, et al. Disinfection by-products precursors removal by enhanced coagulation and PAC adsorption [J]. Desalination, 2007, 216(1–3): 334–344.
271. Li Zhenhua, Yang Kai, Wu Yanhua. Advanced treatment of municipal wastewater by adsorption of coagulation-carbonized sludge [J]. Chinese Journal of Environmental Engineering, 2012,6(7):2356–2360.
272. Zhanxiao. Study on the treatment of water from diversion reservoir by coagulation and its combination with adsorption [D]. Jinan: Shandong University, 2011.

273. He Jiandong. Study on the treatment efficiency of polychlorinated biphenyls in micro-polluted water by enhanced coagulation process [D]. Lanzhou Jiaotong University, 2017.
274. Nie Xiao-Bao, Zhang Peng. Treatment of micro-polluted source water by enhanced coagulation and activated carbon [J]. Environmental Engineering, 2007, 25 (3): 90–91, 74. (in Chinese)
275. Bashir M J K, Xian T M, Shehzad A, et al. Sequential treatment for landfill leachate by applying coagulation-adsorption process [J]. Geosystem Engineering, 2017, 20(1): 9–20.
276. Liao Z, Chen H, Zhu B, et al. Combination of powdered activated carbon and powdered zeolite for enhancing ammonium removal in micro-polluted raw water [J]. Chemosphere,2015, 134: 127–132.
277. Song Xiangsong, WU Zhaoyang, Zhu Tiequn. Study on adsorption characteristics of several inorganic materials for ammonia nitrogen in wastewater [J]. Guangdong Chemical Industry, 2015, 22:109–110.
278. Liu Shuaixia, Wang Rui. Study on water productivity of Yellow River in winter treated by enhanced coagulation technology [J]. Environmental Science and Technology, 2005, 28 (1): 32–33, 116. (in Chinese)
279. Wang Donghong, Chen Zhiqiang, Wen Qinxue. Study on removal of pollutants from secondary effluent by hydroxy-iron oxide series adsorbents [J]. Environmental Science and Management, 2021, 46(07): 105–109. (in Chinese)
280. Liu Zhijie, Yu Gang, Liu Man-Hong. Effect of chitosan flocculant on coagulation of polyaluminum chloride [J]. Environmental Chemistry, 2004, (03): 306–309.
281. Jiang Tingting. Experimental study on secondary effluent from municipal wastewater treatment plant with ozone enhanced coagulation [D]. Qingdao University of Technology, 2010.
282. Yang Yanling, Li Xing. Pretreatment of low temperature and turbidity water by combination of potassium permanganate and chloramine [J]. Journal of Safety and Environment,2007(05):57–59.
283. Zhang Jin, Chen Zhonglin, Fan Jie, et al. Comparison of potassium permanganate and its compound chemicals for enhanced coagulation, algae and odor removal [J]. Journal of Harbin Institute of Technology,2004(06):736–738.
284. Anthony S. Greville, How to Select a Chemical Coagulant and Flocculant, Alberta Water& Wastewater Operators Association, 22th Annual Seminar March 11–14, 1997
285. Li Fengting, Zhao Yan, Zhang Bingru. Quality and Standard of coagulants for drinking water and wastewater treatment, Shanghai Standardization, 2003(11): 32–34.
286. Hu Wanli. Coagulation· Coagulant· Coagulant Equipment [M]. Beijing: Chemical Industry Press, 1st edition, April 2001

MIX
Papier aus verantwortungsvollen Quellen
Paper from responsible sources
FSC® C105338

If you have any concerns about our products,
you can contact us on
ProductSafety@springernature.com

In case Publisher is established outside the EU,
the EU authorized representative is:
Springer Nature Customer Service Center GmbH
Europaplatz 3, 69115 Heidelberg, Germany

Printed by Libri Plureos GmbH
in Hamburg, Germany